TABLES OF

PHYSICAL AND
CHEMICAL CONSTANTS

AND SOME MATHEMATICAL FUNCTIONS

BY

G. W. C. KAYE

M.A., D.Sc., Capt. R.E. (T.)

THE NATIONAL PHYSICAL LABORATORY

AND

T. H. LABY, M.A.

PROFESSOR OF NATURAL PHILOSOPHY, THE UNIVERSITY OF MELBOURNE

THIRD EDITION

LONGMANS, GREEN AND CO.
39 PATERNOSTER ROW, LONDON
FOURTH AVENUE & 80th STREET, NEW YORK
BOMBAY, CALCUTTA, AND MADRAS
1918

THERMOCHEMISTRY. By JULIUS THOMSEN, late Professor of Chemistry in the University of Copenhagen. Translated by KATHARINE A. BURKE, B.Sc. (Lond.). Crown 8vo, 8s. net.

A SYSTEM OF PHYSICAL CHEMISTRY. By WILLIAM C. McC. LEWIS, M.A. (R.U.I.), D.Sc. (Liv.), Brunner Professor of Physical Chemistry in the University of Liverpool: formerly Lecturer in Physical Chemistry, University College, London. With numerous Diagrams. 2 vols. Crown 8vo, 9s. 6d. net each.

THE PHYSICAL PROPERTIES OF COLLOIDAL SOLUTIONS. By E. F. BURTON, B.A. (Cantab.), Ph.D. (Toronto), Associate Professor of Physics, University of Toronto. With 18 Illustrations. 8vo, 6s. 6d. net.

NEW IDEAS ON INORGANIC CHEMISTRY. By Dr. A. WERNER, Professor of Chemistry in the University of Zurich. Translated, with the Author's sanction, by EDGAR PERCY HEDLEY, Ph.D., A.R.C.Sc.I. 8vo, 8s. net.

NEW REDUCTION METHODS IN VOLUMETRIC ANALYSIS. A Monograph. By EDMUND KNECHT, Ph.D., M.Sc.Tech., F.I.C., Professor of Technological Chemistry at the Victoria University of Manchester, and EVA HIBBERT, Demonstrator in Chemistry, Municipal School of Technology, Manchester. Crown 8vo, 3s. 6d. net.

ANALYTICAL MECHANICS, comprising the Kinetics and Statics of Solids and Fluids. By EDWIN H. BARTON, D.Sc. (Lond.), F.R.S.E., A.M.I.E.E., Professor of Experimental Physics, University College, Nottingham. 8vo, 10s. 6d. net.

A HISTORY OF THE CAVENDISH LABORATORY, 1871-1910. With 3 Portraits in Collotype and 8 other Illustrations. 8vo, 7s. 6d. net.

LONGMANS, GREEN AND CO.

LONDON, NEW YORK, BOMBAY, CALCUTTA, AND MADRAS

PREFACE TO FIRST EDITION

THE need for a set of up-to-date English physical and chemical tables of convenient size and moderate price has repeatedly impressed us during our teaching and laboratory experience. We have accordingly attempted in this volume to collect the more reliable and recent determinations of some of the important physical and chemical constants. ·

To increase the utility of the book, we have inserted, in the case of many of the sections, a brief *résumé* containing references to such books and original papers as may profitably be consulted.

Every effort has been made to keep the material up to date ; in many cases a full reference to the original paper is given, while, failing such reference, the year of publication is almost always indicated.

The scope of the volume calls for little comment on our part. We have dipped a little into Astronomy, Engineering, and Geology in so far as they border on Physics and Chemistry. It will be noticed that considerable space has been allotted to Radioactivity and Gaseous Ionization : it is hoped that the collection of data, which we believe to be the first of the kind, will be of assistance to the numerous workers in a field whose phenomenal and somewhat transitional growth is a little dismaying from our present point of view.

Attention has been paid to the setting and accuracy of the mathematical tables ; these are included merely to facilitate calculations arising out of the use of the book, and limitations of space have cut out all but a few of the more essential functions. The convenience of the student of the newer physics has been studied by the inclusion of a table of values of e^{-x} reduced from Newman's original results.

We began this book while at the Cavendish Laboratory, Cambridge, and Dr. G. A. Carse shared in its inception. To Mr. G. F. C. Searle, F.R.S., we feel we owe much for his encouragement and suggestions when the scope of the book was under consideration. We record gratefully the help of a number of friends who have seen the proof-sheets of sections dealing with subjects with which their names are associated. Dr. J. A. Harker, F.R.S., and Mr. R. S. Whipple read the sections on Thermometry ; Mr. F. E. Smith revised the account of Electrical Standards, and Mr. C. C,

PREFACE

Paterson that of Photometry ; Mr. A. Campbell criticized the section on Magnetism ; and Professor Callendar, Principal Griffiths, and Dr. Chree have elucidated various points in Heat and Terrestrial Magnetism.

We owe thanks to Dr. Glazebrook for his permission to utilize the values of a number of constants recently determined at the National Physical Laboratory. Finally, we are greatly indebted to Mr. E. F. F. Kaye, M.Sc, who has given us valuable assistance in preparing the manuscript and revising the proof-sheets.

It was decided to keep the volume within reasonable limits, partly for the reader's convenience, and partly with the hope that the task of subjecting it to frequent revision in the future might not be impossible. We have consequently had to pick and choose our data, and it is scarcely likely that our selection will meet every individual requirement. That some sections are inadequately treated we fully realize, and we shall be very glad to receive suggestions and to be informed of any mistakes which, despite every care, have eluded us.

G. W. C. K.

T. H. L.

September, 1911.

PREFACE TO SECOND EDITION

WE regret that the difficulties of the times have not permitted the complete revision which we had contemplated. We have had to content ourselves with removing those mistakes of which, by the courtesy of many readers, we had become aware, and inserting a number of the more fundamental constants which contemporary research has yielded since 1911. A few tables have been thoroughly revised.

G. W. C. K.

T. H. L.

September, 1916.

PREFACE TO THIRD EDITION

IN the few months that have elapsed since the publication of the last edition, we have not found it possible to do more than bring a few primary constants up to date.

G. W. C. K.

T. H. L.

December, 1917.

CONTENTS

INTERNATIONAL ATOMIC WEIGHTS FOR 1918 (O = 16)

(See F. W. Clarke, "A Recalculation of the Atomic Weights," 1910)

Element.	Symbol.	Atomic Weight.	Element.	Symbol.	Atomic Weight.
Aluminium	Al	27·1	Neodymium	Nd	144·3
Antimony	Sb	120·2	Neon	Ne	20·2
Argon	A	39·88	Nickel	Ni	58·68
Arsenic	As	74·96	Niobium †	Nb	93·1
Barium	Ba	137·37	Niton (Ra. Em.)	Nt	222·0
Beryllium *	Be	9·1	Nitrogen	N	14·01
Bismuth	Bi	208·0	Osmium	Os	190·9
Boron	B	11·0	Oxygen	O	16·00
Bromine	Br	79·92	Palladium	Pd	106·7
Cadmium	Cd	112·40	Phosphorus	P	31·04
Cæsium	Cs	132·81	Platinum	Pt	195·2
Calcium	Ca	40·07	Potassium	K	39·10
Carbon	C	12·005	Praseodymium	Pr	140·9
Cerium	Ce	140·25	Radium	Ra	226·0
Chlorine	Cl	35·46	Rhodium	Rh	102·9
Chromium	Cr	52·0	Rubidium	Rb	85·45
Cobalt	Co	58·97	Ruthenium	Ru	101·7
Copper	Cu	63·57	Samarium	Sa	150·4
Dysprosium	Dy	162·5	Scandium	Sc	44·1
Erbium	Er	167·7	Selenium	Se	79·2
Europium	Eu	152·0	Silicon	Si	28·3
Fluorine	F	19·0	Silver	Ag	107·88
Gadolinium	Gd	157·3	Sodium	Na	23·00
Gallium	Ga	69·9	Strontium	Sr	87·63
Germanium	Ge	72·5	Sulphur	S	32·06
Gold	Au	197·2	Tantalum	Ta	181·5
Helium	He	4·00	Tellurium	Te	127·5
Holmium	Ho	163·5	Terbium	Tb	159·2
Hydrogen	H	1·008	Thallium	Tl	204·0
Indium	In	114·8	Thorium	Th	232·4
Iodine	I	126·92	Thulium	Tm	168·5
Iridium	Ir	193·1	Tin	Sn	118·7
Iron	Fe	55·84	Titanium	Ti	48·1
Krypton	Kr	82·92	Tungsten	W	184·0
Lanthanum	La	139·0	Uranium	U	238·2
Lead	Pb	207·20	Vanadium	V	51·0
Lithium	Li	6·94	Xenon	Xe	130·2
Lutecium	Lu	175·0	Ytterbium	Yb	173·5
Magnesium	Mg	24·32	Yttrium	Y	88·7
Manganese	Mn	54·93	Zinc	Zn	65·37
Mercury	Hg	200·6	Zirconium	Zr	90·6
Molybdenum	Mo	96·0			

* Beryllium or Glucinum (Gl). † Niobium or Columbium (Cb).

The following atomic weights for 1911 (see pp. 109, 127) include only those which have been subsequently changed :—

Ca, 40·09 ; C, 12·00 ; Er, 167·4 ; He, 3·99 ; Fe, 55·85 ; Kr, 82·9 ; Pb, 207·10 ; Lu, 174·0 ; Hg, 200·0 ; Pr, 140·6 ; Ra, 226·4 ; S, 32·07 ; Ta, 181·0 ; Th, 232·0 ; Sn, 119·0 ; U, 238·5 ; V, 51·06 ; Yb, 172·0 ; Y, 89·0.

THE ELEMENTS IN THE ORDER OF ATOMIC WEIGHTS (1918)

Symbol	Atomic Weight.	First isolated by	Date.	Symbol	Atomic Weight	First isolated by	Date.
H	1·008	Cavendish	1766	Ru	101·7	Claus	1845
He	4·00	Ramsay & Cleve *	1895	Rh	102·9	Wollaston	1803
Li	6·94	Arfvedson	1817	Pd	106·7	Wollaston	1803
Be§	9·1	Wöhler and Bussy	1828	Ag	107·88	—	P.
B	11·0	Gay-Lussac&Thénard	1808	Cd	112·40	Stromeyer	1817
C	12·005	—	P.	In	114·8	Reich and Richter	1863
N	14·01	Rutherford	1772	Sn	118·7	—	P.
O	16·00	Priestley and Scheele	1774	Sb	120·2	Basil Valentine	15 centy.
F	19·0	Moissan	1886	I	126·92	Courtois	1811
Ne	20·2	Ramsay and Travers	1898	Te	127·5	v. Reichenstein	1782
Na	23·00	Davy	1807	Xe	130·2	Ramsay and Travers	1898
Mg	24·32	Liebig and Bussy	1830	Cs	132·81	Bunsen and Kirchhoff	1861
Al	27·1	Wöhler	1827	Ba	137·37	Davy	1808
Si	28·3	Berzelius	1823	La	139·0	Mosander	1839
P	31·04	Brand	1674	Ce	140·25	Mosander	1839
S	32·06	—	P.	Pr	140·9	Auer von Welsbach	1885
Cl	35·46	Scheele	1774	Nd	144·3	Auer von Welsbach	1885
K	39·10	Davy	1807	Sa	150·4	L. de Boisbaudran	1879
A	39·88	Rayleigh & Ramsay	1894	Eu	152·0	Demarçay	1901
Ca	40·07	Davy	1808	Gd	157·3	Marignac	1886
Sc	44·1	Nilson and Cleve	1879	Tb	159·2	Mosander	1843
Ti	48·1	Gregor	1789	Dy	162·5	U. & D.	1907
V	51·0	Berzelius	1831	Ho	163·5	L. de Boisbaudran	1886
Cr	52·0	Vauquelin	1797	Er	167·7	Mosander	1843
Mn	54·93	Gahn	1774	Tm	168·5	Cleve	1879
Fe	55·84	—	P.	Yb	173·5	Marignac	1878
Ni	58·68	Cronstedt	1751	Lu	175·0	Urbain	1908
Co	58·97	Brand	1735	Ta	181·5	Eckeberg	1802
Cu	63·57	—	P.	W	184·0	Bros. d'Elhujar	1783
Zn	65·37	Ment. by B. Valentine	15 centy.	Os	190·9	Smithson Tennant	1804
Ga	69·9	L. de Boisbaudran	1875	Ir	193·1	Smithson Tennant	1804
Ge	72·5	Winkler	1886	Pt	195·2	—	16 centy.
As	74·96	Albertus Magnus	13 centy.	Au	197·2	—	P.
Se	79·2	Berzelius	1817	Hg	200·6	Md. by Theophrastus	300 B.C.
Br	79·92	Balard	1826	Tl	204·0	Crookes	1861
Kr	82·92	Ramsay and Travers	1898	Pb	207·20	Mentd. by Pliny	P.
Rb	85·45	Bunsen and Kirchhoff	1861	Bi	208·0	Mtd. by B. Valentine	15 centy.
Sr	87·63	Davy	1808	Nt	222	M. & Mme. Curie	1900
Y	88·7	Wöhler	1828	Ra	226·0	Curies and Bemont	1902
Zr	90·6	Berzelius	1825	Th	232·4	Berzelius	1828
Nb	93·1 ‡	Hatchett	1801	U	238·2	Peligot	1841
Mo	96·0	Hjelm	1790				

P., Prehistoric ; * Lockyer (in sun), 1868 ; U. & D., Urbain & Demenitroux ; § Be or Ge ;
‡ Nb or Cb.

C.G.S. UNITS AND DIMENSIONS

References: Mach, "Science of Mechanics;" Everett, "C.G.S. System of Units;" Maxwell "Theory of Heat."

The metric standards of length and mass are kept at the International Bureau of Weights and Measures in the Pavillon de Breteuil, Sèvres, near Paris. The Bureau is jointly maintained by the principal civilized governments as members of the Metric Convention. The use of metric weights and measures was legalized in the United Kingdom in 1897.

LENGTH

Unit—the **centimetre**, 1/100 of the international metre, which is the distance, at the melting-point of ice, between the centres of two lines engraved upon the polished "neutral web" surface of a platinum-iridium bar of a nearly X-shaped section, called the **International Prototype Metre**.

The alloy of 90 Pt, 10 Ir used (also for the International Kilogramme) has not a large expansion coefficient (see p. 53), is hard and durable, and was artificially aged. Pt-Ir copies of this metre, called **National Prototype Metres**, were made at the same time, and distributed by lot about 1889 to the different governments. The international metre is a copy of the original Borda platinum standard—the mètre des archives. This was intended to be one ten-millionth of the quadrant from the equator to the pole through Paris, and was legalized in 1795 by the French Republic. But as the value of a quadrant came to be more accurately determined, and moreover is changing, the actual bar constructed was made the standard.[*]

The international prototype metre has been measured (1894 and 1907) in terms of the wavelengths of the cadmium rays (see p. 75), and equals 1,553,164·1 wave-lengths of the red ray in dry air at 15° C. (H. Scale) and 760 mm. pressure. (See Michelson's "Light Waves," 1903.)

References: Guillaume, "La Convention du Mètre," and Chree, *Phil. Mag.*, 1901.

MASS

Unit—the **gramme**, 1/1000 of the **International Prototype Kilogramme**, which is the mass of a cylinder of platinum-iridium.

The international kilogramme is a copy of the original Borda platinum kilogramme—the kilogramme des archives—which was intended to have the same mass as that of a cubic decimetre of pure water at the temperature of its maximum density. More exact measurements revealed the incorrectness of the relation (see p. 10), and so the kilogramme was subsequently defined as above.

As with the metre, Pt-Ir copies of the international standard—**National Prototype Kilogrammes**—have been distributed to the different governments.

TIME

Unit—the **second**, which may be defined simply as 1/86,164·09 of a **sidereal day**. For all practical purposes the sidereal day may be regarded as the period of a complete axial rotation (360°) of the earth with respect to the fixed stars.[†]

The second is usually defined as $1/(24 \times 60 \times 60)$ of a **mean solar day**, *i.e.* 1/86,400 of the **average** value of the somewhat variable interval (the apparent solar day) between two successive returns of the sun to the meridian (see p. 15).

Strictly, the sidereal day is the interval between two successive transits of the first point of Aries[‡] across any selected meridian.[§] The true period of rotation of the earth is actually about 1/100 second longer than the sidereal day; the difference arises from the slow and continual change of direction ("precession") of the earth's axis in space.

A **tropical or solar year** is the average interval between two successive returns of the sun to the first point of Aries; it is found to equal 365·2422 mean solar days. Our modern (Julian) calendar assumes that in 4 successive civil years, 3 consist of 365 days, and 1 of 366; the average thus being 365·25 days. The Gregorian correction (that century years are not to count as leap years unless divisible by 400) reduces this value to 365·2425 mean solar days, and thus the **average civil year** is a close approximation to a tropical year.

---

[*] According to the latest estimates, the *mean* meridian quadrant = 10,002,100 metres (see p. 13).

[†] Tidal friction is retarding the rotation of the earth, so that the above (sidereal) definition of the second, while practically justified, is theoretically not quite perfect.

[‡] The first point of Aries is that one of the two nodes of intersection of the ecliptic and the celestial equator where the sun (moving in the ecliptic) crosses the equator from south to north (at about March 21). The ecliptic is the apparent yearly track of the sun in a great circle on the celestial sphere.

[§] Neglecting small irregularities, this is true also for any star.

A **sidereal year** is the time interval in which the sun appears to perform a complete revolution with reference to the fixed stars ; *i.e.* it is the time in which the earth describes one sidereal revolution round the sun. Owing to precession, a sidereal year is longer than a tropical year.

	h.	m.	s.	
Mean solar day =	24	0	0	= 86,400 secs.
Sidereal day =	23	56	4·0906	= 86,164·0906 secs.

Tropical year = 365·2422 mean solar days.
Sidereal year = 365·2564 ,, ,, ,, (epoch 1900).
= 366·2564 sidereal days.

Reference : Newcomb, "Astronomy."

BRITISH IMPERIAL STANDARDS.

(From information supplied by Major MacMahon, F.R.S., Board of Trade, Standards Office.)

According to the Weights and Measures Act, 1878, the **yard** is the distance, at 62° F., between the central transverse lines in two gold plugs in the bronze bar, called the **Imperial Standard Yard**, when supported on bronze rollers in such manner as best to avoid flexure of the bar.

The defining lines are situated at the bottom of each of two holes, so as to be in the median plane of the bar, which is of 1 inch square section and 38 inches long. Its composition is 32 Cu, 5 Sn, 2 Zn. Copper alloys are now known not to be suitable for standards of length, and in 1902 a Pt-Ir ✕-shaped copy of the yard was made.

The **pound** is the **weight** in vacuo of a platinum cylinder called the **imperial standard pound**.

The imperial standard yard and pound are preserved at the Standards Office of the Board of Trade, Old Palace Yard. A number of official copies have been prepared, and are in the custody of the Royal Society, the Mint, Greenwich Observatory, and the Houses of Parliament.

The **gallon** contains 10 lbs. weight of distilled water weighed in air against brass weights at a pressure of 30 inches, and with the water and the air at 62° F.

[NOTE.—No mention is made in the Act of the density of the brass weights, or of the humidity of the air.]

BRITISH AND METRIC EQUIVALENTS

The present legal equivalents are those legalized by the Order in Council of May 19, 1898, and derived at the International Bureau of Weights and Measures, by Benoît in 1895 in the case of the yard and the metre, and by Broch in 1883 for the pound and the kilogramme. (See *Trav. et Mém. du Bur. Intl.*, tomes iv., 1885, and xii., 1902.)

Imperial Standard.		International Prototype.	(Reciprocal.)
1 yard	=	·914399 metre	1·093614
1 pound	=	·45359243 kilogramme	2·2046223

[NOTE.—The yard is defined at 62° F., the metre at 0° C.]

DERIVED C.G.S. UNITS AND STANDARDS
GENERAL AND MECHANICAL UNITS

Area :—*Unit*—the square centimetre.

Volume :—*Unit*—the cubic centimetre (c.c.). The metric unit is the **litre**, now defined as the volume of a kilogramme of pure, air-free water at the temperature of maximum density (see p. 22) and 760 mm. pressure (*Procès Verbaux*, 1901, p. 175). The litre was originally intended to be 1 cubic decimetre or 1000 c.cs. ; the present accepted experimental relation is that 1 kilogramme of water at 4° C. and 760 mm. pressure measures 1000·027 c.cs. (see p. 10).

Density :—*Unit*—grammes per c.c. **Specific gravity** expresses the density of a substance relative to that of water, and is objectionable in requiring two temperatures to be stated.

Velocity :—*Unit*—1 cm. per second.   **Angular Velocity** :—*Units*—1 radian ($57°\cdot296$) per sec. ; 1 revolution per sec.

Acceleration :—Time rate of alteration of velocity.   *Unit*—(1 cm. per sec.) per sec.   **Angular Acceleration** :—*Units*—1 radian per sec.² ; 1 revolution per sec.²

Momentum : —Mass multiplied by velocity.   *Unit*—1 gm. cm. sec.⁻¹.

Moment of Momentum :—Momentum multiplied by distance from axis of reference.   *Unit*—1 cm.² gm. sec.⁻¹.

Moment of Inertia :—Σmd^2, where m is the mass of any particle of a body, and d its distance from the axis of reference.   *Unit*—1 cm.² gm. (see p. 16).

Angular Momentum :—Moment of inertia multiplied by angular velocity round axis of reference.   *Unit*—1 cm.² gm. sec.⁻¹.

Force :—Measured by the acceleration it produces in unit mass.   *Unit*—the **dyne** = cm. gm./sec.²   *Gravitational unit*—the weight of 1 gram = g dynes.

Couple, Torque, Turning Moment :—Force multiplied by distance from point of reference.   *Unit*—1 dyne cm.

Work :—Force multiplied by distance through which point of application of force moves in direction of force.   *Unit*—the **erg** = 1 dyne cm.; 1 **joule** = 10⁷ ergs. [1 calorie = 4·18 joules].   *Gravitational unit*—weight of 1 gm. × 1 cm. = g dyne cms. = g ergs.

Energy :—Measured by the work a body can do by reason of either (1) its motion—**Kinetic Energy** (= $mv^2/2$) or (2) its position—**Potential Energy**. *Unit*—the **erg**.   (See "Work.")   1 **Board of Trade Unit** = 1 kilowatt hour = $3\cdot6 \times 10^6$ watt-secs.

Power :—Work per unit time.   *Unit*—1 erg per sec.   1 **watt** = 10⁷ ergs per sec. = 1 joule per sec. = 1 volt-ampere.   1 kilowatt = 1·34 horse-power.

Pressure, Stress :—Force per unit area.   *Unit*—1 dyne per cm.²   1 **megabar** = 10⁶ dynes per cm.² = 750 *mm. mercury at 0° C., lat. 45°, and sea-level (g = 980 6).   1 **atmosphere** = 760 mm. mercury at 0° C., lat. 45°, and sea-level = 759·4 mm. mercury at 0° C. in London = 1·0132 × 10⁶ dynes per cm.² = 14·7 lbs. per inch² = 0·94 ton per foot².   * Correct to 1 part in 5000.

Elasticity :—Ratio of stress to resulting strain.   *Unit*—1 dyne per cm.², since the dimensions of a strain are zero.

HEAT UNITS

Temperature :—The melting-point of pure ice under 1 atmosphere is defined as 0° C., and the boiling-point of water under 1 atmosphere as 100° C.   This fundamental interval is divided into 100 parts by use of the constant-volume hydrogen thermometer (see p. 44) ; each part is a degree Centigrade.   Dimensions of temperature are not required, as it is defined independently of mass, length, and time.

Heat :—*Dynamical unit*—the **erg**.   *Thermal unit*—**the calorie** = heat required to raise the temperature of 1 gramme of water from $t°$ C. to $(t + 1)°$ C.   The **20° calorie** ($t = 20°$) = 4·180 × 10⁷ ergs.   The **15° calorie** ($t = 15°$) = 4·184 × 10⁷ ergs.   The **mean calorie** (= 1/100 heat required to raise 1 gramme of water from 0° to 100° C.) = 4·184 × 10⁷ ergs. (see pp. 55, 56).   1 *watt-minute* = 14·3 calories.   The large calorie = 1000 calories.

Gas Constant R., in $pv = R\theta/m$, where p is the pressure, v the volume, θ the absolute temperature of a gram-molecule (*i.e.* m grams) of a gas of molecular weight m.   For 1 gram-molecule of an ideal gas of density ρ,

$$R = \frac{pvm}{\theta} = \frac{p}{\theta} \cdot \frac{m}{\rho} = \frac{1\cdot0132 \times 10^6 \times 22412}{273\cdot1} = 83\cdot15 \times 10^6 \text{ ergs per gm. mol.}$$ (Berthelot, see p. 106).   This value is a constant for all ideal gases.   To derive R for 1 gram of a gas, this figure should be divided by the molecular weight (oxygen = 16) of the gas.   R has the dimensions of a specific heat in dynamical units.

ELECTRICAL AND MAGNETIC UNITS

Reference :—J. J. Thomson, "Mathematical Theory of Electricity and Magnetism."   The fundamental basis of the electrostatic system of units is the repulsive force between two quantities of like electricity.   In the electromagnetic system the repulsion between two like magnetic poles is taken as the basis.

The electromagnetic system (or one based on it) is universally employed in electrical engineering ; the electrostatic is used only in certain special cases.

ELECTROSTATIC UNITS

Quantity or **Charge** :—*Unit*—that quantity which placed 1 cm. distance from an equal like quantity repels it with a force of 1 dyne.

Current:—*Unit*—Unit quantity flowing uniformly past a point in unit time.

Potential Difference and Electromotive Force:—*Unit*—that P.D. which exists between two points when the work done in taking unit quantity from one point to the other is 1 erg.

Capacity :—*Unit*—the charge on a conductor which is at unit potential ; or in the case of a condenser, when its plates are at unit P.D.

Dielectric Constant, Inductivity, or **Specific Inductive Capacity** of a medium is the ratio of the capacity of a condenser having the medium as dielectric, to the capacity of the same condenser with a vacuum as dielectric (p. 84).

ELECTROMAGNETIC UNITS

Magnetic Pole Strength or Quantity :—*Unit*—that quantity which, placed 1 cm. distance from an equal like quantity, repels it with a force of 1 dyne.

Magnetic Force or Field Strength :—*Unit*—the force which acts on unit magnetic pole.

Magnetic Moment of magnet = pole strength × length of magnet.

Intensity of Magnetization = magnetic moment per unit volume.

Permeability of a medium is the ratio of the magnetic induction in the medium to that in the magnetizing field (p. 89).

Susceptibility:—*Unit*—intensity of magnetization per unit field (p. 89).

Electric Current :—*Unit*—that current which produces unit magnetic force at the centre of a circle of radius 2π cms.

Quantity = current × time.

Potential and E.M.F. :—*Unit*—that P.D. which exists between two points when the work done in taking unit quantity from one point to the other is 1 erg.

Electrostatic Capacity = quantity/potential difference.

Resistance = potential difference/resulting current.   (Ohm's law is assumed.)

Conductance :—Reciprocal of resistance.

Specific Resistance :—Resistance of prism of unit area and unit length.

Conductivity :—Reciprocal of specific resistance.

Coefficient of Self-induction of a circuit is the E M.F. produced in it by unit time-rate of variation of the current through it.

Coefficient of Mutual Induction of two circuits is the E.M.F. produced in one by unit time-rate of variation of the current in the other.

PRACTICAL ELECTRICAL UNITS

At an International Conference on Electrical Units and Standards held in London, October, 1908, it was resolved that—

1. The magnitudes of the fundamental electrical units shall, as heretofore, be determined on the electromagnetic system of measurement with reference to the centimetre, gramme, and second (c.g.s.). These fundamental units are (1) the **Ohm**, the unit of electrical resistance, which has the value 10^9 c.g.s.; (2) the **Ampere**, the unit of electric current, which has the value 10^{-1} c.g.s.: (3) the **Volt**, the unit of electromotive force, which has the value 10^8 c.g.s.; (4) the **Watt**, the unit of power, which has the value 10^7 c.g.s. [For absolute electrical units, see p. 8.]

2. As a system of units representing the above, and sufficiently near to them to be adopted for the purpose of electrical measurements, and as a basis for legislation, the Conference recommends the adoption of the International Ohm, the International Ampere, and the International Volt.

3. The **Ohm** is the first primary unit. The **International Ohm** is defined as the resistance offered to an unvarying electric current by a column of mercury at 0° C., 14·4521 grammes in mass, of a constant cross-section, and of a length of 106·300 cms.

4. The **Ampere** is the second primary unit. The **International Ampere** is defined as the unvarying electric current which, when passed through a solution of nitrate of silver in water, in accordance with authorized specification, deposits silver at the rate of ·00111800 gramme per second.

5. The **International Volt** is defined as the electrical pressure which, when steadily applied to a conductor whose resistance is one International Ohm, will produce a current of one International Ampere.

6. The **International Watt** is defined as the energy expended per second by an unvarying electric current of one International Ampere under an electric pressure of one International Volt.

DIMENSIONS OF UNITS

The dimensions in terms of length, mass, and time are denoted by the indices given under L, M, and T. Thus the dimensions of power are L^2MT^{-3}.

MECHANICAL AND HEAT UNITS

Quantity.	L.	M.	T.	Quantity.	L.	M.	T.	Quantity.	L.	M.	T.
Length . .	1	0	0	Momentum .	1	1	−1	Strain . . .	0	0	0
Mass . .	0	1	0	Moment of momentum . .	2	1	−1	Elasticity . .	−1	1	−2
Time . .	0	0	1	Moment of inertia . .	2	1	0	Compressibility	1	−1	2
Angle . .	0	0	0	Angular momentum . .	2	1	−1	Viscosity . .	−1	1	−1
Surface . .	2	0	0	Force . .	1	1	−2	Diffusion . .	2	0	−1
Volume . .	3	0	0	Couple, Torque	2	1	−2	Capillarity . .	0	1	−2
Density . .	−3	1	0	Work, Energy	2	1	−2	Temperature .	0	0	0
Velocity . .	1	0	−1	Power . . .	2	1	−3	Heat * . . .	2	1	−2
Angular vel. .	0	0	−1	Pressure, Stress	−1	1	−2	Thermal Conductivity * .	1	1	−3
Acceleration .	1	0	−2								
Angular acceleration . .	0	0	−2					Entropy * . .	2	1	−2

ELECTRICAL AND MAGNETIC UNITS

v, the ratio of the electromagnetic to the electrostatic unit of quantity, is usually taken as 3×10^{10}, and is a pure number. (p. 69). (See Rücker, *Phil. Mag.*, 22, 1889.)

Unit.	Symbol.	E.S. Unit. L	M	T	k	E.M. Unit. L	M	T	μ	E.S.U. / E.M.U.	Practical Unit.	EMU	ESU
Electrical													
Charge or quantity . . .	e	$\frac{3}{2}$	$\frac{1}{2}$	−1	$\frac{1}{2}$	$\frac{1}{2}$	$\frac{1}{2}$	0	$-\frac{1}{2}$	$1/v$	coulomb	$= 10^{-1}$	$= 3 \times 10^9$
Resistance . .	R	−1	0	1	−1	1	0	−1	1	v^2	ohm	$= 10^9$	$= \frac{1}{9} \times 10^{-11}$
Current . .	i	$\frac{3}{2}$	$\frac{1}{2}$	−2	$\frac{1}{2}$	$\frac{1}{2}$	$\frac{1}{2}$	−1	$-\frac{1}{2}$	$1/v$	ampere	$= 10^{-1}$	$= 3 \times 10^9$
Potential or E.M.F. . . .	E	$\frac{1}{2}$	$\frac{1}{2}$	−1	$-\frac{1}{2}$	$\frac{1}{2}$	$\frac{1}{2}$	−2	$\frac{1}{2}$	v	volt	$= 10^8$	$= 1/300$
Electric field	F	$-\frac{1}{2}$	$\frac{1}{2}$	−1	$-\frac{1}{2}$	$\frac{1}{2}$	$\frac{1}{2}$	−2	$\frac{1}{2}$	v	(volt/cm.)		
Conductivity . .	K	0	0	−1	1	−2	0	1	−1	$1/v^2$	"reciprocal ohm"	$= 10^{-9}$	$= 9 \times 10^{11}$
Capacity . . .	C	1	0	0	1	−1	0	2	−1	$1/v^2$	microfarad ‡	$= 10^{-15}$	$= 9 \times 10^5$
Self and mutual induction .	$L; M$	−1	0	2	−1	1	0	0	1	v^2	{ henry { cm.	$= 10^9$ $= 1$	$= \frac{1}{9} \times 10^{-11}$ $= \frac{1}{9} \times 10^{-20}$
Dielectric constant † . . .	k	0	0	0	1	−2	0	2	−1	$1/v^2$	—	—	—
Magnetic													
Pole strength .	m	$\frac{3}{2}$	$\frac{1}{2}$	0	$-\frac{1}{2}$	$\frac{1}{2}$	$\frac{1}{2}$	−1	$\frac{1}{2}$	v			
Flux (total lines)	N	$\frac{1}{2}$	$\frac{1}{2}$	0	$-\frac{1}{2}$	$\frac{1}{2}$	$\frac{1}{2}$	−1	$\frac{1}{2}$	$1/v$	maxwell	$= 1$	$= 3 \times 10^{10}$
Force; field strength .	H	$-\frac{1}{2}$	$\frac{1}{2}$	−2	$\frac{1}{2}$	$-\frac{1}{2}$	$\frac{1}{2}$	−1	$-\frac{1}{2}$	$1/v$	gauss	$= 1$	$= 3 \times 10^{10}$
Induction . . .	B	$\frac{1}{2}$	$\frac{1}{2}$	0	$-\frac{1}{2}$	$-\frac{1}{2}$	$\frac{1}{2}$	−1	$\frac{1}{2}$	v	gauss	$= 1$	$= \frac{1}{3} \times 10^{-10}$
Intensity of magnetization . .	I	$-\frac{3}{2}$	$\frac{1}{2}$	0	$-\frac{1}{2}$	$-\frac{1}{2}$	$\frac{1}{2}$	−1	$\frac{1}{2}$	v	—	—	—
Permeability . .	μ	−2	0	2	−1	0	0	0	1	v^2	—	—	—

* In dynamical units.   † Specific inductive capacity.   ‡ 10^{-6} farad.

Example :—To find the number (n) of ergs per sec. in a horse-power (33,000 ft.-lbs. per min.).

Dimensions of power $= L^2MT^{-3} = LT^{-1}$ [Force].

$$n = 33,000 \frac{\text{ft.}}{\text{cm.}} \left(\frac{\text{min.}}{\text{sec.}}\right)^{-1} \frac{\text{lb. weight}}{\text{dyne}} = \frac{33,000 \times 30 \cdot 48}{60} \times 453 \cdot 6 \times 981$$

$$= 7 \cdot 46 \times 10^9 \text{ ergs per sec.} = 746 \text{ watts.}$$

ABSOLUTE DETERMINATIONS OF ELECTRICAL UNITS

See Baillehache, "Unités Électriques," Paris, 1909, and the "Report of the London Conference" (p. 6). The appendix to this report (issued separately, 9d.) gives full particulars as to the realization of the ampere and ohm, together with the specification of the Weston normal (cadmium) cell.

THE OHM

The **mean value 106·25** cms. of Hg of 1 sq. mm. cross-section at 0° C. may be taken as a measure of the present experimental value of the true ohm, which is equal to 10^9 E.M. (c.g.s.) units. Compare the international ohm (p. 6).

cm./0°.	Method.	Observer.	cm./0°.	Method.	Observer.
106·28	Spinning disc	Rayleigh, 1882	106·29	Induced discharge	Glazebrook, '88
106·22	„ „	Rayleigh and Mrs. Sedgwick, 1883	106·32	Spinning disc	V. Jones, 1894
106·32	Mean result	Rowland, 1887	106·27	„ „	Ayrton and V. Jones, 1897
			106·24$_5$	„ „	Smith, N.P.L., '14

The 1884 "**legal**" ohm = ·9972 intl. ohm; the **B.A. ohm** = ·9866 intl. ohm.

THE AMPERE

The electrochemical equivalent of silver is given in milligrams per coulomb (1 ampere for 1 sec.) = 10^{-1} E.M. unit of quantity. **Mean = ·0011182**7 **gm./coulomb.** Compare the international ampere (p. 6).

mg. Ag.	Method.	Observer.	mg. Ag.	Method.	Observer.
1·11828	Dynamometer	Kohlrausch, '84 Corrected 1908	1·11821	Dynamometer	Janet, Laporte, de la Gorce, 1909
1·11827	Current weigher	Smith, Mather, and Lowry, 1907	1·11829	„	Do., 1910

E.M.F. OF WESTON CELL

The electromotive force (E) of the Weston (cadmium) cell in volts (10^8. E.M. units) as realized from one of the accepted specifications. The present accepted international value of E is **1·0183 international volts** (see p. 6) at 20° C.

Temperature coefficient.—Over the range 0° to 40°, Wolff (1908) obtained for the E.M.F. at t°—

$$E_t = E_{20} - \cdot0000406(t - 20) - 9 \cdot 5 \times 10^{-7}(t - 20)^2.$$

E at 20°.	Method.	Observer.	E at 20°.	Method.	Observer.
1·0185	Intl. ohm and dynamometer	Guthe, 1906	1·01820	Intl. ohm and current weigher	Ayrton, Mather, and Smith, 1908
1·01822		Guillet, 1908			
1·01841		Pellat, 1908	1·01822	„ „	Dorsey, 1911
1·01869	Intl. ohm and current weigher	Janet, Laporte, Jouaust, 1908	1·01834	Intl. ohm and intl. ampere	Jaeger and v. Steinwehr, 1909

The E.M.F. of the **Clark cell** = 1·433 volts at 15° C. It diminishes by about 1·2 parts in 1000 for 1° C. rise of temp.

BRITISH INTO METRIC CONVERSION FACTORS

Conversion factors based on the relations given on p. 4. g is taken as 981 cm.-sec.$^{-2}$. Reciprocals are given for converting metric into British measure.

British.	Metric.	(Reciprocal.)	British.	Metric.	(Reciprocal.)
Length—			**Force—**		
1 inch =	2·5400 cm.*	·3937 †	1 poundal =	13,825 dynes	7·233 × 10^{-5}
1 yard =	·9144 metre*	1·0936	1 pound wgt. =	4·45 × 10^5 dynes	2·247 × 10^{-6}
1 mile =	1·6093 km.	·6214	**Pressure—**		
Area—			1 lb./sq. inch =	68,971 dynes/cm.^{2}	1·45 × 10^{-5}
1 sq. inch =	6·4516 sq. cm	·1550 †	„ „ =	70·31 gm./cm.^{2}	·01422
Volume—			1 ton/sq. inch =	1·545 × 10^8 dynes/cm.^{2}	6·47 × 10^{-9}
1 cubic inch =	16·387 c.c.	·0610			
1 cubic foot =	28·317 litre	·03531	„ „ =	1·575 k. gm./mm.^{2}	·6349
1 pint =	·5682 litre	1·7598	**Work—**		
1 gallon =	4·5460 litre ‡	·2200 ‡	1 ft.-pound =	1·356 joules§	·7373
Mass—			**Power—**		
1 grain =	·0648 gram	15·432	1 horse-power =	·746 k.watt.	1·34
1 oz. (avoir.) =	28·350 grams	·03527	**Heat—**		
1 lb. „ =	·4536 k. gm.	2·2046	1 B. Th. unit (1 lb., 1° F.) } =	252·00 calories	·00397
1 ton „ =	1016 k. gm.‖	·0₉9842			
Density—					
1 lb./cub. ft. =	·01602 gm./cm.^{3}	62·43			
Velocity—					
1 mile/hour =	44·70 cm./sec.	·02237			

MISCELLANEOUS DATA

CONVENIENT APPROXIMATE RELATIONS

1 yard = 1 metre, less 10%

2 lbs. = 1 k. gram, „

2 galls. = 10 litres, „

1 ton = { 1 tonne / (1000 k. gm.) } less 2%

	British.	U.States.	
	Stnd.) yd. at 62°F.) =	(Stand. yd. at 59°·6F.	1 mm. = 10^{-3} metre
			1 micron, μ = 10^{-6} „
			$\mu\mu$ = 10^{-9} „
	1 lb. =	1 lb.	1 Å.Ū. = 10^{-10} „
	1 gal. =	1·20 gal.	1 mil = 10^{-3} inch

SOME BRITISH WEIGHTS AND MEASURES

Useful in photography, etc.

The avoirdupois, troy, and apothecaries **grain** are the same in weight.

1 lb. (avoir.) = 7000 grains = 454 grams

1 oz. „ = 437½ „ = 28·3 „

1 oz. (troy) = }

1 oz. (apothe- } = 480 „ = 31·1 „

caries) }

1 fl. drachm = 60 minims = 3·55 c.cs.

1 fl. oz. ℥ = 8 fl. drachms = 28·41 „

1 pint = 20 fl. ozs. = 568 „

A 10% solution is

1 grain in 10 minims of solution

1 oz. (avoir.) „ 10 fl. ozs. „

2 oz. „ „ 1 pint „

MATHEMATICAL

	Number.	Log. of Number.
π	3·141592654	·49715
π^2	9·869604401	·99430
$1/\pi$	·318309886	1̄·50285
$\sqrt{\pi}$	1·772453851	·24857
1 radian	57°·29578	1·75812
1°	·017453 radian	2̄·24188
e	2·718281828	·43429
log, 10	2·302585	·36222

To convert	*Multiply by*
Common into hyperbolic logs,	2·3026
Hyperbolic „ common „	·4343

---

* Correct to 1 part in a million.   † Correct to 3 parts in a million.

‡ Owing to the definition of the gallon (see p. 4), this number is dependent on assumed buoyancy and temperature corrections.

§ 1 joule = 10^7 ergs.   ‖ 1 tonne = 1000 k. gm.

MISCELLANEOUS DATA—*continued*.

BRITISH COINAGE			NAUTICAL
Coin.	Weight.	Diameter.	
sovereign	8 grams less ·15%	2·18 cm.	1 nautical mile $= 6082.66$ feet
penny	$\frac{1}{3}$ oz. (avoir.)	1·2 inch	1 admiralty mile $= 6080$ feet
halfpenny	$\frac{1}{5}$,, ,,	1·0 ,,	1 knot $= 1$ nautical mile/hour
farthing	$\frac{1}{10}$,, ,,	·8 ,,	1 fathom $= 6$ feet
			1 point $= 11\frac{1}{4}°$

10° **Centigrade** = 50° **Fahrenheit**,		British.	Continental.
whence the following is convenient for transforming room temperatures :— $5(t° F. - 50) = 9(t° C. - 10)$	Million . . .	10^6	10^6
	Billion . . .	10^{12}	10^9
	Trillion . .	10^{18}	10^{12}

VOLUME OF A KILOGRAMME OF PURE WATER

At 4° C. and 760 mm.   Values recalculated by Benoît.   (*Trav. et Mém. Bur. Intl.*, 14, 1910.) (See p. 4.)

Observer.	c.cs.	Observer.	c.cs.
Lefévre-Geneau and Fabbroni, 1799 .	1000·030	Chaney, 1893	1000·150
Schuckburgh and Kater, 1798 and 1821	999·525	Guillaume, 1904	1000·029
Svanberg and Berzélius, 1825 . . .	999·710	Chappuis, 1907	1000·027
Stampfer, 1831	1000·250	de Lépinay, Benoît, and Buisson,	
Kupffer, 1842	1000·069	1907	1000·028

DENSITIES OF GASES

Supplementary to p. 26.   Densities in grams per litre at 0° C., 760 mm., sea-level, and lat. 45°.

Gas.	gms./litre.	Observer.	Gas.	gms./litre.	Observer.
He .	·1782	Watson, *J.C.S.*, 1910	Ra, Em.	9·727	Gray & Ramsay, *P.R.S.* 1910
Ne .	·9002	,, ,, ,,	CH₄	·7168	Baume & Perrot, *C.R.*,
Kr .	3·708	Moore ,, 1908			1909
Xe .	5·851	,, ,, ,,			

C.R., *Compt. Rend.*; *J.C.S.*, *Journ. Chem. Soc.*; *P.R.S.*, *Proc. Roy. Soc.*

PRESSURE COEFFICIENTS OF PV

Pressure coefficient, m, of pv for gases at 1 atmosphere and constant temperature ; p is the pressure in atmospheres, and v is the volume. $m = \frac{\delta(pv)}{pv} \cdot \frac{1}{\delta p}$; m is a measure of the deviation of the gas from Boyle's law.

Air, $m = -·00191$, Regnault.

N, $m = -·000559$ ⎫
H, $m = +·000772$ ⎬ Chappuis, Rayleigh, Leduc, and Sacerdote.

VALUES OF GRAVITY ("g") LONGITUDE AND LATITUDE

Helmert's formula connecting "gravity" with latitude and height is $g = 980·617 - 2·593 \cos 2\lambda - ·0003086H$, where λ is the latitude, H is the height in metres above sea-level, and 980·617 cms./sec.² is the value of g attributed to lat. 45° and sea-level. The values of g calculated by this formula are for most places in fair agreement with the observed values. Some discrepancy is found in the vicinity of large mountain ranges, such as the Himalayas.

No absolute standard determination of g has been made in England for many years, but comparisons have been made with Potsdam and Sèvres. For relative measurements, the relation $dg = ·0226 \, dN$ is useful, where N is the number of vibrations which a pendulum makes in a mean solar day of 86,400 mean time seconds. The length (l) of the "seconds" pendulum (i.e. 2 secs. period) $= g/\pi^2 = ·101321g$. l varies from 99·094 cms. at the equator to 99·620 cms. at the pole.

See Helmert's "Höhere Geodäsie," "Die Grösse der Erde," 1906, and "Die Schwerkraft im Hochgebirge," Clarke's "Geodesy," 1880, Sir Geo. Darwin's "Tides and Kindred Phenomena," Fisher's "Physics of the Earth's Crust," and for recent aspects of the subject, the reports to the triennial International Geodetic Conferences (...1906, 1909...), and the reports of the U.S. Geodetic Survey. (See also p. 13.)

Place.	Longitude E. or W. of Greenwich.			Latitude (λ).			Height (H) above Sea-level.	"g" (calculated).
	°	′	″	°	′	″	metres.	cms./sec.²
Pole	—			90	0	0	——	983·210 *
Equator	—			0	0	0	——	978·024 *
British Isles—								
Aberdeen (Univ.)‡	2	6	38 W	57	8	58 N	21	981·68
Aberystwith	4	4	W	52	25	N	——	981·28 *
Bangor	4	8	W	53	13	N	——	981·35 *
Belfast	5	56	W	54	37	N	——	981·47 *
Birmingham	1	54	W	52	28	N	——	981·28 *
Bristol	2	35	W	51	28	N	——	981·20 *
Cambridge (Univ. Obs.) . .	0	5	41 E	52	12	52 N	28	981·254
Cardiff	3	10	W	51	28	N	——	981·20 *
Dublin (Trin. Coll.) . . .	6	15	W	53	20	35 N	7 †	981·36
„ (R.C.S)	6	40	32 W	53	23	13 N	15	981·36
Dundee (Univ. Col.) ‡ . .	2	58	45 W	56	27	26 N	27 †	981·62
Durham	1	34	56 W	54	46	6 N	——	981·48 *
Edinburgh	3	11	3 W	55	55	28 N	134	981·54
Eskdalemuir (Obs.) . . .	3	12	18 W	55	18	48 N	244	981·45
Glasgow (Univ.) ‡	4	17	12 W	55	52	31 N	46	981·56
Greenwich (Obs.)	0	0	0	51	28	38 N	47	981·184
Kew (Obs.)	0	18	46 W	51	28	6 N	5	981·200
Leeds (Univ.) ‡	1	33	15 W	53	48	30 N	81	981·38
Liverpool (Univ.) ‡ . . .	2	57	37 W	53	24	19 N	51	981·35
London (Natl. Phys. Lab.) §	0	20	11 W	51	25	20 N	10	981·19₅
„ (Univ., S. Kens.) .	0	10	23 W	51	29	54 N	14	981·19
„ (Univ. Coll.)‡ . .	0	7	57 W	51	31	27 N	28	981·19
Manchester (Univ.)‡ . . .	2	14	2 W	53	27	53 N	39	981·37
Newcastle (Armstrong Coll.)	1	36	53 W	54	58	50 N	55	981·48
Nottingham (Univ. Coll.) ‡ .	1	8	45 W	52	57	10 N	58 ‖	981·31
Oxford (Radcliffe Obs.) . .	1	15	39 W	51	45	34 N	65	981·20
Plymouth	4	9	W	50	22	N	——	981·10 *
Portsmouth	1	6	12 W	50	48	3 N	5	981·14
St. Andrews (Univ.) . . .	2	48	W	56	20	N	——	981·62 *
Sheffield (Univ. Obs.) . . .	0	5	50 E	53	23	2 N	——	981·36 *
Stonyhurst (Obs.)	2	28	10 W	53	50	40 N	114	981·37
Africa—								
Bloemfontein	26	40	E	29	0	S	——	979·24 *

* No correction has been applied for height above sea-level.   † Ground floor.
‡ Physics laboratory.   § Teddington.   ‖ Second floor.

GRAVITY

Place.	Longitude E. or W. of Greenwich.	Latitude (λ).	Height (H) above Sea-level.	"g" (calculated).
	o ′ ″	o ′ ″	metres.	cms./sec.²
Africa (*contd.*)—				
Cairo (Observatory)	31 17 14 E	30 4 38 N	33	979·32
Cape Town	18 29 E	33 56 S	12	979·64
Durban	30 40 E	29 40 S	—	979·29 *
Johannesburg (Univ. Coll.)	28 7 E	26 11 S	1753	978·49
Mauritius (Roy. Alf. Obs.)	57 33 9 E	20 5 39 S	55	978·63
America—				
Baltimore (Meteorol. Stn.)	76 37 W	39 18 N	23	980·10
Boston (Meteorol. Stn.)	71 4 W	42 21 N	38	980·37
Chicago (Meteorol. Stn.)	87 38 W	41 52 N	251	980·26
Harvard, Camb. (Obs.)	71 7 46 W	42 22 48 N	24	980·37
Jamaica (Montego Bay Obs.)	77 52 22 W	18 24.51 N	69	978·52
Montreal (McGill Obs.)	73 34 39 W	45 30 17 N	57	980·64
New York (Ruthfd. Obs.)	73 59 9 W	40 43 49 N	96	980·20
Philadelphia (Obs.)	75 9 37 W	39 57 8 N	36	980·15
Princeton (N.J.)	74 39 22 W	40 20 58 N	65	980·20
Quebec (Obs.)	71 13 8 W	46 48 21 N	70	980·76
St. Louis (Obs.)	90 12 17 W	38 38 4 N	171	979·99
Toronto (Obs.)	79 23 40 W	43 39 36 N	107	980·46
Washington (Bur. of Stands.)	77 3 59 W	38 56 32 N	102	980·097
Yale, New Haven (Obs.)	72 55 8 W	41 19 22 N	32	980·28
Asia—				
Bombay (Obs.)	72 48 56 E	18 53 45 N	10	978·57
Calcutta (Surv. Office)	88 21 30 E	22 32 54 N	6	978·76
Hong Kong (Obs.)	114 10 28 E	22 18 13 N	33	978·76
Madras (Obs.)	80 14 54 E	13 4 8 N	7	978·29
Australasia—				
Adelaide (Obs.)	138 35 8 E	34 55 39 S	430	979·68
Brisbane (Obs.)	153 1 36 E	27 28 S	42	979·12
Melbourne (Obs.)	144 58 32 E	37 49 53 S	28	979·97
Perth	115 52 E	31 57 S	14	979·47
Sydney (Obs.)	151 12 23 E	33 51 41 S	44	979·63
Wellington (Obs.), N.Z.	174 46 37 E	41 18 1 S	43	980·27
Europe—				
Berlin (Reichsanstalt) †	13 19 E	52 31 N	30	981·287
Christiania (Obs.)	10 43 23 E	59 54 44 N	25	981·90
Copenhagen (Obs.)	12 34 40 E	55 41 13 N	14	981·56
Geneva (Obs.)	6 9 11 E	46 11 59 N	374	980·61
Leyden (Obs.)	4 29 3 E	52 9 20 N	6	981·26
Paris (Obs.)	2 20 14 E	48 50 11 N	59	980·95
„ (Bureau Intl.) ‡	2 13 10 E	48 49 53 N	70	980·951
Potsdam (Astron. Inst.)	13 3 59 E	52 22 56 N	94	981·249
Rome (Coll. Obs.)	12 28 53 E	41 53 54 N	59	980·32
St. Petersburg (Acad. Obs.)	30 18 22 E	59 56 30 N	3	981·91
Vienna (Impl. Obs.)	16 20 21 E	48 12 47 N	—	980·91 *
Zurich (Poly. Obs.)	8 33 4 E	47 22 40 N	468	980·69

* No correction applied for height above sea-level.   † Charlottenburg.   ‡ Sèvres.

DISTANCES ON THE EARTH'S SURFACE
(See Ball's "Spherical Astronomy," 1909.)

At Lat.	Miles per degree of Longitude.	Latitude.	At Lat.	Miles per degree of Longitude.	Latitude.	At Lat.	Miles per degree of Longitude.	Latitude.
0	69·15	68·69	40	53·05	69·00	60	34·66	69·21
10	68·11	68·70	45	48·99	69·05	70	23·73	69·32
20	65·01	68·77	50	44·54	69·10	80	12·05	69·38
30	59·94	68·88	55	39·75	69·16	90	0	69·39

SIZE AND SHAPE OF THE EARTH

The spheroid of revolution which most nearly approximates to the earth, has the following dimensions :— [1 kilom. = ·6214 mile.]

Observer.	Equatorial radius, a.	Polar radius, b.	Ellipticity, $(a-b)/a$.
Bessel, 1841 . .	6,377,397 metres	6,356,079 metres	1/299·2
Clarke, 1866 . . .	8,206 ,,	584 ,,	1/295·0
,, 1880 . .	8,249 ,,	515 ,,	1/293·5
Helmert, 1906 * .	8,200 ,,	818 ,,	1/298·3
U.S. Survey, 1906 †	8,388 ,, ‡	909 ,, ‖	1/297·0

* " Die Grosse der Erde."
† " The Figure of the Earth," 1909, and Supplement, 1910 ; U.S. Coast and Geodetic Survey.
‡ 3963·339 miles.   ‖ 3949·992 miles.

MEAN DENSITY OF THE EARTH

(See Poynting's " Mean Density of the Earth," 1893.)

Observer.	Density.
Common Balance Method.	
Poynting, 1878	5·493
Richarz and Krigar-Menzel, 1898	5·505
Torsion Balance Method.	
Cavendish, 1798	5·45
Boys, *Phil. Trans.*, 1895 . .	5·527
Braun, 1896	5·527
Eötvos, 1896	5·534
Mean density of surface . . .	2·65

Mean polar quadrant $\Big\} = 10,002,100$ metres *
Volume of earth $= 1·083 \times 10^{21}$ metres³ *
Mass of earth $= 5·98 \times 10^{27}$ grams †
$= 5·87 \times 10^{21}$ tons
Area of land $= 1·45 \times 10^{18}$ cm.²
Area of ocean $= 3·67 \times 10^{18}$ cm.²
Mean depth of ocean (Murray) $\Big\} = 3·85 \times 10^{5}$ cm.
Volume of ocean $= 1·41 \times 10^{24}$ cm.²
Mass of ocean $= 1·45 \times 10^{24}$ grms.

* Mean of Helmert and U.S. Survey.
† Using Boys' and Braun's result for density.

SUN

The mean equatorial solar parallax (Hinks, 1909) $\Big\} = 8''·807$

Whence mean distance from earth to sun $\Big\} = \begin{cases} 1·494 \times 10^{11} \text{ metres} \\ 9·282 \times 10^{7} \text{ miles} \end{cases}$

Mean time taken by light to travel from sun to earth $\Big\} = 498·2$ secs.

MOON

Mean distance from earth to moon $\Big\} = \begin{cases} 60·27 \times \text{earth's radius} \end{cases}$
Mass of the moon (Hinks, 1909) $\Big\} = \begin{cases} (1/81·53) \times \text{earth's mass} \end{cases}$
Inclination of moon's orbit to ecliptic $\Big\} = 5° 8' 43''$

Constant of Gravitation (G in law of attraction) $= 6·658 \times 10^{-8}$ c.g.s.

Obliquity of the Ecliptic to the equator $= 23° 27' 4''·04$ in 1909, subject to a small fluctuation by nutation, and a slow continuous decline of $46''·84$ per century.

Constant of aberration of a star is theoretically equal to (Earth's orbital velocity)/(velocity of light) $= 20''·43 \pm ''·03$ (Renan and Ebert, 1905).

Constant of precession, *i.e.* annual precessional increase of the longitude of a star $= 50''·2564 + ''·00022225t$, where t is the interval in years from 1900 (**Newcomb**).

ELEMENTS OF THE SOLAR SYSTEM

8″·806 is taken as the equatorial horizontal solar parallax from the observations of the as eroid Eros in 1900–1 ; 5·527 is adopted as the Earth's mean density (Boys, 1895 ; Braun, 1896). The constants for Mercury are those adopted by Stroobant and Backland (1909). The value of the mass of Jupiter is that obtained by Cookson (1908). The time of rotation of Venus is that suggested by Hansky and Stefánik (1907). (See Newcomb's "Spherical Astronomy" and Ball's "Spherical Astronomy.")

Name.	Equatorial Semi-diameter. Angular.*	Equatorial Semi-diameter. Miles.	Equatorial Semi-diameter. Earth = 1	Mass Earth = 1	Mean Density. Earth = 1	Mean Density. Water = 1	Gravity at Surf. Earth = 1	No. of Satellites.‡
Sun . .	16 1·18	432,890	109·2	329,390	·25	1·39	27·61	——
Mercury .	3·08	1387	·350	·34	·88	4·86	·28	o
Venus .	8·40	3783	·955	>·818	>·94	>5·20	>·91	o
Earth .	8·80	3963·3	1·000	1·000	1·00 ,	5·527	1·00	1 (D)
Mars . .	4·68	2108	·532	·106	0·71	3·90	·38	2 (D)
Jupiter .	1 37·36	43850	11·06	314·50	·25	1·36	2·57	8(7 D; 1 R)
Saturn .	1 24·75	38170	9·63	94·07	·12	·63	1·01	10(9 D; 1 R)
Uranus .	34·28	15440	3·90	14·40	·24	1·34	·95	4 (R)
Neptune .	36·56	16470	4·15	16·72	·23	1·28	·97	1 (R)

Name.	Inclination of Equator to Orbit.	Time of Axial Rotation.	Semi-major Axis of Orbit. Earth = 1.	Semi-major Axis of Orbit.	Semi-major Axis of Orbit. Millions of Miles.	Sidereal Period. Mean Solar Days.	Sidereal Period. Julian Years.
Sun . .	7 15†	d h m 25 9 7		Bode's Law			
Mercury.	?	h m s ?	·3870986	4 = (0 + 4)	36·0	87·9693	·24
Venus .	?	23 40 (?)	·7233315	7 = (3 + 4)	67·2	224·7008	·62
Earth .	23 27 8	23 56 4·09	1·00000000	10 = (6 + 4)	92·9	365·2564	1·00
Mars . .	24 52	24 37 22·74	1·523688	16 = (12 + 4)	141·6	686·9797	1·88
Asteroids			2·55 to 2·85	28 = (24 + 4)	237 to 265		
Jupiter .	3 5	9 56 ±	5·202803	52 = (48 + 4)	483·3	4332·588	11·86
Saturn .	26 49	10 15 ±	9·538844	100 = (96 + 4)	886·2	10759·20	29·46
Uranus .	?	13 ?	19·19098	196 = (192 + 4)	1782·8	30586·29	83·74
Neptune .	27 ?	?	30·07067		2793·5	60187·65	164·78

Name.	Ellipticity of Planet.§	Mean Daily Motion in Orbit.	Longitude of Perihelion. ‖	Longitude of Ascending Node. ¶	Inclination of Orbit to Ecliptic.	Eccentricity of Orbit.**
Mercury.	?	4 5 32·4	75 53 59	47 8 45	7 0 16	·205614
Venus .	?	1 36 7·7	130 9 50	75 46 47	3 23 37	·006821
Earth .	1/298·3	59 8·2	101 13 15	0 0 0	0 0 0	·016751
Mars . .	1/270 ?	31 26·5	334 13 7	48 47 9	1 51 1	·093309
Jupiter .	1/17	4 59·1	12 36 20	99 26 42	1 18 42	·048254
Saturn .	1/9	2 0·5	90 48 32	112 47 12	2 29 39	·056061
Uranus .	1/95 ?	42·2	169 2 56	73 29 25	0 46 22	·047044
Neptune .	?	21·5	43 45 20	130 40 44	1 46 45	·008533

* This is the angle subtended by the semi-diameter at a distance equal to the Earth's mean distance from the Sun.

† The inclination of the plane of the Sun's equator to the plane of the ecliptic.

‡ D means direct ; R, retrograde.

§ The ellipticity = $(a−b)/a$, where a is the major axis and b the minor axis of the spheroid of revolution. The value given for the Earth is Helmert's (p. 13).

‖ Perihelion is the point in the orbit nearest the Sun. Longitude is the angular distance from the first point of Aries (see p. 3), measured along the ecliptic.

¶ A node is one of the two points at which a planet's orbit intersects the plane of the ecliptic. At the ascending node the planet passes from south to north of the ecliptic.

** The eccentricity = $\sqrt{(a^2−b^2)}/a$, where a and b are the major and minor axes of the orbit.

EQUATION OF TIME

(+) means that the equation of time has to be added to the apparent solar time (*i.e.* sundial time) to give the mean solar or clock time (see p. 3). (M) = maximum or minimum. The values below vary by a few seconds from year to year. $C = D + E$, where C = clock time, D = dial time, and E = equation of time.

Date.	Equation of time.	Date.	Equation of time.	Date.	Equation of time.	Date.	Equation of time.
	m. s.		m. s.		m. s.		m. s.
Jan. 1	+ 3 11	April 1	+ 4 1	July 1	+ 3 32	Oct. 16	− 14 20
„ 16	+ 9 33	„ 16	0 0	„ 26	+ 6 18 (M)	Nov. 3	− 16 21 (M)
Feb. 1	+ 13 37	May 1	− 2 57	Aug. 16	+ 4 11	„ 16	− 15 10
„ 12	+ 14 25 (M)	„ 14	− 3 49 (M)	Sept. 1	0 0	Dec. 1	− 10 56
Mar. 1	+ 12 34	June 1	− 2 27	„ 16	− 5 6	„ 12	− 6 15
„ 16	+ 8 51	„ 15	0 0	Oct. 1	− 10 16	„ 25	0 0

PARALLAXES OF STARS

The **proper motion** of a star is its real change of place arising from the actual motion of the star itself.

The **annual parallax** is the angle between the direction in which a star appears as seen from the earth and the direction in which it would appear if it could be observed from the centre of the sun.

A **light-year** is the distance that light travels in one year (see p. 69).

Star and Magnitude.	Proper motion per year.	Annual parallax.	Distance.	
			Sun's dist. = 1	Light-years.
	"	" "		
α Centauri (·2)	3·7	·75 ± ·01	·28 × 10⁶	4·4
21185 Lalande (7·5)	7·3	·48 ± ·02	·43 „	6·8
61 Cygni (4·8)	5·2	·37 ± ·02	·56 „	8·8
Sirius (− 1·4)	1·3	·37 ± ·01	·56 „	8·8
Procyon (·5)	1·3	·31	·69 „	11
Altair (·9)	·7	·28 ± ·02	·74 „	12
Aldebaran (1·1)	·2	·17 ± ·02	1·4 „	22
Capella (·2)	·4	·12 ± ·02	1·7 „	27
Vega (·1)	·4	·12 ± ·02	1·7 „	27
1830 Groombridge (6·4)	7·0	·10 ± ·02	2·0 „	33
Polaris (2·1)	0·0	·07 ± ·02	3·0 „	47
Arcturus (·2)	2·3	·024	8·7 „	149

SYSTEMATIC MOTIONS OF THE STARS

The apparent proper motions of the stars show drifts in two directions. The assigned positions of the apices of these directions are:—

Computer.	Stream I.		Stream II.	
	R.A.	Dec.	R.A.	Dec.
Kapteyn, 1904	85°	− 11°	260°	− 48°
Eddington	90°	− 19°	292°	− 58°
Dyson	94°	− 7°	240°	− 74°

STANDARD TIMES

Referred to Greenwich time.

Gt. Britain, France, Portugal, Belgium, Spain	Greenwich time
Ireland	„
Austria, Denmark, Germany, Italy, Norway, Switzerland	1 hour fast
British South Africa, Egypt, Turkey	1½ or 2 hours fast
Japan	9 hours fast
Australia	8, 9, or 10 hours fast
New Zealand	11½ „
Canada and United States	4, 5, 6, 7, or 8 hours slow

SCREWS

It is customary for British metal screws, cf $\frac{1}{4}$-inch diameter and above, to have a Whitworth thread, for smaller sizes a British Association thread. In the Whitworth thread the angle between the slopes is 55°, in the B.A. thread 47·5°.

The **pitch** is the distance between adjoining crests (say) of the same thread measured parallel to the axis of the screw. It is the reciprocal of the number of turns per inch or mm. as the case may be. The **full diameter** is the maximum over-all diameter.

Micrometer screws are made with some multiple or sub-multiple of 100 threads to the inch or mm.

"**Woodscrews**" of iron or brass are numbered as follows: No. 0 has a diameter of ·05 inch, each succeeding number adding ·014 inch to the diameter of the screw: this applies to all lengths. The length of countersunk screws is measured over all; that of round-headed screws, from under the head.   [1 inch = 25·4 mm.]

STANDARD WHITWORTH.				BRITISH ASSOCIATION.								
Full diameter.	Threads to inch.	Full diameter.	Threads to inch.	No.	Full diameter.	Pitch.	No.	Full diameter.	Pitch.	No.	Full diameter.	Pitch.
inch.		inch.			mm.	mm.		mm.	mm.	—	mm.	mm.
$1\frac{3}{4}$	5	$\frac{3}{4}$	10	0	6·0	1·0	9	1·9	·39	18	·62	·15
$1\frac{5}{8}$	5	$1\frac{1}{8}$	11	1	5·3	·9	10	1·7	·35	19	·54	·14
$1\frac{1}{2}$	6	$\frac{5}{8}$	11	2	4·7	·81	11	1·5	·31	20	·48	·12
$1\frac{3}{8}$	6	$\frac{9}{16}$	12	3	4·1	·73	12	1·3	·28	21	·42	·11
$1\frac{1}{4}$	7	$\frac{1}{2}$	12	4	3·6	·66	13	1·2	·25	22	·37	·10
$1\frac{1}{8}$	7	$\frac{7}{16}$	14	5	3·2	·59	14	1·0	·23	23	·33	·09
1	8	$\frac{3}{8}$	16	6	2·8	·53	15	·9	·21	24	·29	·08
$\frac{7}{8}$	9	$\frac{5}{16}$	18	7	2·5	·48	16	·79	·19	25	·25	·07
$1\frac{3}{16}$	10	$\frac{1}{4}$	20	8	2·2	·43	17	·70	·17			

MOMENTS OF INERTIA

M = mass of body.   (See A. M. Worthington, "Dynamics of Rotation." London.)

Body.	Axis of rotation.	Moment of inertia.
Uniform thin rod (length l)	(1) Through centre, perpendicular to length	$M\dfrac{l^2}{12}$
	(2) Through end, perpendicular to length	$M\dfrac{l^2}{3}$
Rectangular lamina (sides a and b)	(1) Through centre of gravity, perpendicular to plane	$M\dfrac{a^2 + b^2}{12}$
	(2) Through centre of gravity, parallel to side b	$M\dfrac{a^2}{12}$
Circular lamina (radius r)	(1) Through centre, perpendicular to plane	$M\dfrac{r^2}{2}$
	(2) Any diameter	$M\dfrac{r^2}{4}$
Solid cylinder (radius r; length l)	(1) Axis of cylinder	$M\dfrac{r^2}{2}$
	(2) Through centre of gravity, perpendicular to axis of cylinder	$M\left(\dfrac{l^2}{12} + \dfrac{r^2}{4}\right)$
Hollow cylinder (external and internal radii R and r; length l)	(1) Axis of cylinder	$M \cdot \dfrac{R^2 + r^2}{2}$
	(2) Through centre of gravity, perpendicular to axis	$M\left(\dfrac{l^2}{12} + \dfrac{R^2 + r^2}{4}\right)$
Solid sphere (radius r)	Through centre	$M \cdot \dfrac{2r^2}{5}$
Hollow sphere (external and internal radii R and r)	Through centre	$M\left(\dfrac{2}{5} \cdot \dfrac{R^5 - r^5}{R^3 - r^3}\right)$
Anchor ring (mean radius of ring R; radius of cross-section r)	(1) Through centre, perpendicular to plane of ring	$M\left(R^2 + \dfrac{3r^2}{4}\right)$
	(2) Any diameter	$M\left(\dfrac{R^2}{2} + \dfrac{5r^2}{8}\right)$

VOLUME CALIBRATION OF VESSELS BY WATER OR MERCURY

Volume content of vessel at $t°$ C. $=$ V$_t$ $=$ W$_t v_t \equiv w_t(f)$, where—

w_t = observed weight in grams (against brass weights in air) of contained water (or mercury) at $t°$ C.

W$_t$ = weight of such liquid *in vacuo* (*i.e.* corrected for buoyancy in air).

v_t = volume of 1 gram of liquid at $t°$ C.

(f) is a factor which introduces the buoyancy and specific volume corrections.

The following table of values of the factor (f) is based on tables on pp. 19 and 22.

Temp. (t) of weighing	10° C.	11°	12°	13°	14°	15°	16°	17°
Value of {H$_2$O . factor (f)|Hg .	1·00133 ·073683	1·00143 ·073697	1·00154 ·073710	1·00166 ·073724	1·00179 ·073737	1·00193 ·073750	1·00209 ·073764	1·00226 ·073777

Temp. (t) of weighing	18°	19°	20°	21°	22°	23°	24°	25°
Value of {H$_2$O . factor (f)|Hg .	1·00244 ·073790	1·00263 ·073804	1·00283 ·073817	1·00305 ·073831	1·00327 ·073844	1·00350 ·073857	1·00375 ·073871	1·00400 ·073884

The above gives the volume content V$_t$ of the vessel at the temperature of weighing, $t°$ C. At any other temperature, t', the volume V$_{t'}$ $=$ V$_t$ $\{1 + \gamma(t' - t)\} \equiv$ V$_t$(F), where γ is the coefficient of cubical expansion of the material of the vessel. Values of the factor (F) for **glass vessels** ($\gamma = ·000025$) are tabulated below.

$(t' - t)$	2° C.	4°	6°	8°	$-2°$ C.	$-4°$	$-6°$	$-8°$
Value of factor (F)	1·00005	1·00010	1·00015	1·00020	·99995	·99990	·99985	·99980

Example.—Weight of water contained in a vessel at 10° C. $=$ 10 grams : thence volume of vessel at 10° C. $=$ 10 × 1·00133 c.cs. The same vessel, if of glass, would contain at 16° C., 10 × 1·00133 × 1·00015 = 10·0148 c.cs.

CAPILLARITY CORRECTIONS OF MERCURY COLUMNS

The height of the meniscus and the value of the capillary depression depend on the bore of the tubing, on the cleanliness of the mercury, and on the state of the walls of the tube. The correction is negligible for tubes with diameters greater than about 25 mms. The table below gives the amount of the correction (which has to be added to the height) for various diameters of glass tubing and meniscus heights. (Mendeléeff and Gutkowsky, 1877. See also Scheel and Heuse, *Ann. d. Phys.*, 33, 1910.)

Bore of tube.	Height of meniscus in mms.								Bore of tube.	Height of meniscus in mms.					
	·4	·6	·8	1·0	1·2	1·4	1·6	1·8		·8	1·0	1·2	1·4	1·6	1·8
mm.	mm.	mm.	mm.	mm.	mm.	mm.	mm.	mm.	mm.	mm.	mm.	mm.	mm.	mm.	mm.
4	·83	1·22	1·54	1·98	2·37	—	—	—	9	·21	·28	·33	·40	·46	·52
5	·47	·65	·86	1·19	1·45	1·80	—	—	10	·15	·20	·25	·29	·33	·37
6	·27	·41	·56	·78	·98	1·21	1·43	—	11	·10	·14	·18	·21	·24	·27
7	·18	·28	·40	·53	·67	·82	·97	1·13	12	·07	·10	·13	·15	·18	·19
8	—	·20	·29	·38	·46	·56	·65	·77	13	·04	·07	·10	·12	·13	·14

REDUCTION OF BAROMETER READINGS TO 0° C.

Corrected height $H_0 = H\left\{1 - \dfrac{(\beta - a)t}{(1 + \beta t)}\right\}$, where H and t are the observed height and temperature of the barometer, $\beta = \cdot0001818$ (Regnault), the coefficient of cubical expansion of mercury; $a = \cdot0000085$, the coefficient of linear expansion of glass, or $\cdot0000184$ for brass. Hydrogen temperature scale. (After Broch, Inter. Bur. Weights and Measures.)

(In standard English barometry the mercury is reduced to 32° F., and the scale to 62° F. In the table below, both are reduced to the ice point.)

Correction in mms. to be subtracted.

Temp. (t).	GLASS SCALE. Uncorrected height in mms.					BRASS SCALE. Uncorrected height in mms.				
	700	720	740	760	780	700	720	740	760	780
	mm.					mm.				
2° C.	·24	·25	·26	·26	·27	·23	·24	·24	·25	·25
4	·48	·49	·51	·53	·54	·46	·47	·48	·50	·51
6	·73	·75	·77	·79	·81	·69	·71	·72	·74	·76
8	·97	·99	1·02	1·05	1·08	·91	·94	·97	·99	1·02
10	1·21	1·25	1·28	1·31	1·35	1·14	1·17	1·21	1·24	1·27
12	1·45	1·49	1·53	1·58	1·62	1·37	1·41	1·45	1·49	1·53
14	1·69	1·74	1·79	1·84	1·89	1·60	1·64	1·69	1·73	1·78
16	1·94	1·99	2·05	2·10	2·16	1·82	1·88	1·93	1·98	2·03
18	2·18	2·24	2·30	2·36	2·43	2·05	2·11	2·17	2·23	2·29
20	2·42	2·49	2·56	2·62	2·69	2·28	2·34	2·41	2·47	2·54
22	2·66	2·73	2·81	2·89	2·96	2·51	2·58	2·65	2·72	2·79
24	2·90	2·98	3·06	3·15	3·23	2·73	2·81	2·89	2·97	3·05
26	3·14	3·23	3·32	3·41	3·50	2·96	3·04	3·13	3·21	3·30
28	3·38	3·47	3·57	3·67	3·77	3·19	3·28	3·37	3·46	3·55
30	3·62	3·72	3·83	3·93	4·03	3·41	3·51	3·61	3·71	3·80
32	3·86	3·97	4·08	4·19	4·30	3·64	3·74	3·85	3·95	4·05
34	4·10	4·21	4·33	4·45	4·57	3·87	3·98	4·09	4·20	4·31

REDUCTION OF BAROMETER READINGS TO LAT. 45° AND SEA-LEVEL

It is a convention to take "g" at lat. 45° and sea-level as the standard value for "gravity." The corrections below result from the variation of "g" with latitude and height above sea-level (see p. 11). The barometer correction for **latitude** $= \dfrac{H_0}{760}(\text{c})$, has to be subtracted from the temperature—corrected barometer reading H_0 for latitudes between 0° and 45°; and added for latitudes from 45° to 90°.

Latitude	0° 90°	5° 85°	10° 80°	15° 75°	20° 70°	25° 65°	30° 60°	35° 55°	40° 50°	45° 45°
	mm.									
c	1·97	1·94	1·85	1·70	1·51	1·27	·98	·67	·34*	·00

The correction of the barometer due to diminution of gravity with increasing **height** above sea-level amounts to about ·24 mm. of mercury per 1000 metres above sea-level. The correction has to be subtracted from the observed reading.

* London, ·45.

REDUCTION OF WEIGHINGS TO VACUO

The buoyancy correction $= M\sigma(1/\Delta - 1/\rho) = Mk$, where M is the apparent mass in grams of the body in air, σ is the density of air ($= \cdot 0012$) in grams per c.c., Δ is the density of the body, ρ is the density of the weights. The correction is true to 4% for the following limits: 740 mm. press., 1° to 22°; 760 mm., 8° to 29°; 780 mm., 15° to 35°. If the correction is required more accurately, multiply the value of k given below by $\sigma'/\cdot 0012$, where σ' is the true density of the air for the temp. and press. at the time of the weighing (for σ', see p. 25). The corrections for quartz weights are the same as for Al. $+$ means cor^n. to be added to observed weight.

Density of Body weighed Δ.	Correction Factor (k) in Milligms.			Density of Body weighed Δ.	Correction Factor (k) in Milligms.		
	Brass wgts. $\rho = 8\cdot4$.	Pt wgts. $\rho = 21\cdot5$.	Al wgts. $\rho = 2\cdot65$.		Brass wgts. $\rho = 8\cdot4$.	Pt wgts. $\rho = 21\cdot5$.	Al wgts. $\rho = 2\cdot65$.
$\cdot5$	$+ 2\cdot26$	$+ 2\cdot34$	$+ 1\cdot95$	$1\cdot6$	$+ \cdot61$	$+ \cdot69$	$+ \cdot30$
$\cdot55$	$+ 2\cdot04$	$+ 2\cdot13$	$+ 1\cdot73$	$1\cdot7$	$+ \cdot56$	$+ \cdot65$	$+ \cdot25$
$\cdot6$	$+ 1\cdot86$	$+ 1\cdot94$	$+ 1\cdot55$	$1\cdot8$	$+ \cdot52$	$+ \cdot62$	$+ \cdot21$
$\cdot65$	$+ 1\cdot70$	$+ 1\cdot79$	$+ 1\cdot39$	$1\cdot9$	$+ \cdot49$	$+ \cdot58$	$+ \cdot18$
$\cdot7$	$+ 1\cdot57$	$+ 1\cdot66$	$+ 1\cdot26$	2	$+ \cdot46$	$+ \cdot54$	$+ \cdot15$
$\cdot75$	$+ 1\cdot46$	$+ 1\cdot55$	$+ 1\cdot15$	$2\cdot5$	$+ \cdot34$	$+ \cdot43$	$+ \cdot03$
$\cdot8$	$+ 1\cdot36$	$+ 1\cdot44$	$+ 1\cdot05$	3	$+ \cdot26$	$+ \cdot34$	$- \cdot05$
$\cdot85$	$+ 1\cdot27$	$+ 1\cdot36$	$+ \cdot96$	$3\cdot5$	$+ \cdot20$	$+ \cdot29$	$- \cdot11$
$\cdot9$	$+ 1\cdot19$	$+ 1\cdot28$	$+ \cdot88$	4	$+ \cdot16$	$+ \cdot24$	$- \cdot15$
$\cdot95$	$+ 1\cdot12$	$+ 1\cdot21$	$+ \cdot81$	5	$+ \cdot10$	$+ \cdot19$	$- \cdot21$
1	$+ 1\cdot06$	$+ 1\cdot14$	$+ \cdot75$	6	$+ \cdot06$	$+ \cdot14$	$- \cdot25$
$1\cdot1$	$+ \cdot95$	$+ 1\cdot04$	$+ \cdot64$	8	$+ \cdot01$	$+ \cdot09$	$- \cdot30$
$1\cdot2$	$+ \cdot86$	$+ \cdot94$	$+ \cdot55$	10	$- \cdot02$	$+ \cdot06$	$- \cdot33$
$1\cdot3$	$+ \cdot78$	$+ \cdot87$	$+ \cdot47$	15	$- \cdot06$	$+ \cdot03$	$- \cdot37$
$1\cdot4$	$+ \cdot71$	$+ \cdot80$	$+ \cdot40$	20	$- \cdot08$	$+ \cdot004$	$- \cdot39$
$1\cdot5$	$+ \cdot66$	$+ \cdot75$	$+ \cdot35$	22	$- \cdot09$	$- \cdot001$	$- \cdot40$

REDUCTION OF GASEOUS VOLUMES TO 0° AND 760 MMS. PRESSURE

Corrected volume $v_0 = \{v/(1 + \cdot00367t)\} \cdot p/760$, where v, t, and p are the observed volume, temp., and pressure (in mms. of mercury) of the gas respectively. $g = 980\cdot62$ cms. per sec^2. The coefficient $\cdot00367$ observed by Regnault.

Values of $(1 + \cdot00367t)$.

Temp. (t).	0	1	2	3	4	5	6	7	8	9
0° C.	1·0000	1·0037	1·0073	1·0110	1·0147	1·0183	1·0220	1·0257	1·0294	1·0330
10	0367	0404	0440	0477	0514	0550	0587	0624	0661	0697
20	0734	0771	0807	0844	0881	0917	0954	0991	1028	1064
30	1101	1138	1174	1211	1248	1284	1321	1358	1395	1431
40	1468	1505	1541	1578	1615	1651	1688	1725	1762	1798
50	1835	1872	1908	1945	1982	2018	2055	2092	2129	2165
60	2202	2239	2275	2312	2349	2385	2422	2459	2496	2532
70	2569	2606	2642	2679	2716	2752	2789	2826	2863	2899
80	2936	2973	3009	3046	3083	3119	3156	3193	3230	3266
90	3303	3340	3376	3413	3450	3486	3523	3560	3597	3633
100	3670	3707	3743	3780	3817	3853	3890	3927	3964	4000
110	4037	4074	4110	4147	4184	4220	4257	4294	4331	4367

Values of $p/760$

Press. (p).	0	1	2	3	4	5	6	7	8	9
700 mm.	·9211	·9224	·9227	·9250	·9263	·9276	·9289	·9303	·9316	·9329
710	·9342	·9355	·9368	·9382	·9395	·9408	·9421	·9434	·9447	·9461
720	·9474	·9487	·9500	·9513	·9526	·9539	·9553	·9566	·9579	·9592
730	·9605	·9618	·9632	·9645	·9658	·9671	·9684	·9697	·9711	·9724
740	·9737	·9750	·9763	·9776	·9789	·9803	·9816	·9829	·9842	·9855
750	·9868	·9882	·9895	·9908	·9921	·9934	·9947	·9961	·9974	·9987
760	1·0000	1·0013	1·0026	1·0039	1·0053	1·0066	1·0079	1·0092	1·0105	1·0118
770	1·0132	1·0145	1·0158	1·0171	1·0184	1·0197	1·0211	1·0224	1·0237	1·0250

DENSITIES OF THE ELEMENTS

Average densities of liquid and solid elements in grams per c.c. at ordinary temperature unless otherwise stated. For gaseous densities see p. 26. The density of a specimen ma depend considerably on its state and previous treatment, *e.g.* the density of a cast metal i increased by drawing, rolling, or hammering.

Element.	Density.	Element.	Density.	Element.	Density.
Aluminium . . .	2·70	Indium . . .	7·12	Samarium . .	7·8
Antimony . . .	6·62	Iodine . . .	4·95	Scandium . . .	(?)
Argon (liq). . .	1·4/−185°	Iridium . . .	22·41	Selenium, amorph.	4·8
Arsenic	5·73	Iron (pure) . .	7·86	„ cryst.	4·5
Barium	3·75	Krypton (liq.) . .	2·16	„ liq. . .	4·27
Beryllium . . .	1·93	Lanthanum . .	6·12	Silicon	c. 2·3
Bismuth	9·80	Lead	11·37	Silver	10·5
Boron	2·5 (?)	Lithium . . .	·534	Sodium . . .	·971
Bromine . . .	3·102/25°	Magnesium . .	1·74	Strontium . .	2·54
Cadmium . . .	8·64	Manganese . .	7·39	Sulphur, rhombic	2·07
Cæsium . . .	1·87	Mercury (see p. 22)	13·56/15°	„ monoclinic	1·96
Calcium . . .	1·55/29°	Molybdenum . .	10·0	„ amorphous	1·92
Carbon—		Neodymium . .	6·96	„ liquid 113°	1·81
Diamond . . .	3·52	Neon (liq.) . . .	(?)	Tantalum . . .	16·6
Graphite . . .	2·3	Nickel	8·9	Tellurium . . .	6·25
Cerium . . .	6·92	Niobium . . .	12·75	Terbium	(?)
Chlorine (liq.) . .	2·49/0°	Nitrogen (liq.) . .	·79/−196°	Thallium . . .	11·9
Chromium . . .	6·50	Osmium	22·5	Thorium . . .	11·3
Cobalt	8·6	Oxygen (liq.) . .	1·27/−235°	Tin	7·29
Copper	8·93	Palladium . . .	11·4	Titanium . . .	3·54
Erbium	4·77 (?)	Phosphorus, red .	2·20	Tungsten . . .	18·8
Fluorine (liq.) . .	1·11/−187°	„ yellow	1·83	Uranium . . .	18·7
Gadolinium . . .	(?)	Platinum . . .	21·50	Vanadium . . .	5·5
Gallium . . .	5·95	Potassium . . .	·862	Xenon (liq) . .	3·5
Germanium . . .	5·47	Praseodymium .	6·48	Ytterbium . . .	(?)
Gold	19·32	Radium . . .	(?)	Yttrium	3·8 (?)
Helium (liq.) . .	·12/B.P.	Rhodium . . .	12·44	Zinc	7·1
Hydrogen (liq.) . .	·07/B.P.	Rubidium . . .	1·532	Zirconium . . .	4·15
„ „	·086/M.P.	Ruthenium . . .	12·3		

The densities of the alkali metals Li, Na, K, Rb, Cs are due to Richards and Brink, 1907; of He a −268°·6, Onnes, 1908; of W, Gin, 1908; of Ta, Nb, and Th, von Bolton, 1905, 1907, 1908; of Ca, Goodwin, 1904; of Rh and Ir, Holborn, Henning, and Austin, 1904; of Br, Andrews and Carlton, 1907.

DENSITIES OF COMMON SUBSTANCES

Average densities in grams per c.c. at ordinary temperatures. For densities of acids, alkalies, and other solutions, see pp. 23 *et seq*.; of "chemical compounds," p. 109; of gases, p. 26; of other minerals, p. 126.

Substance.	Density.	Substance.	Density.	Substance.	Density.
Metals & Alloys.		Coins (English)		**Woods** (seasoned).	
Iron, cast . .	7·1–7·7	„ silver §	10·31	Ash ; mahogany .	·6–·8
„ wrought . .	7·8–7·9	Constantan (Eureka)‖ . . . }	8·88	Bamboo	c. ·4
„ wire . .	7·7	German silver ¶	8·5–8·9	Beach ; oak ; teak	·7–·9
Steel	7·7–7·9	Gunmetal . .	8·0–8·4	Box	·9–1·1
Brass (ordy.) * .	8·4–8·7	Magnalium ** . .	c. 2	Cedar	·5–·6
Brass weights . .	c. 8·4	Manganin †† . .	8·5	Ebony	1·1–1·3
Bronze (Cu, Sn) .	8·7–8·9	Phosphor bronze ‡‡	8·7–8·9	Lignum vitæ . .	1·2–1·3
Coins (English)		Platinoid §§ . .	c. 9	Pitchpine ; walnut	·6–·7
„ bronze † . .	8·96	Pt (90), Ir (10). .	21·62	Red pine (deal) .	·5–·7
„ gold ‡ . .	17·72			White pine . .	·4–·5

* c. 66 Cu, 34 Zn. † 95 Cu, 4 Sn, 1 Zn. ‡ 91⅔ Au, 8⅓ Cu. § 92½ Ag, 7½ Cu. ‖ 60 Cu, 40 Ni.
¶ 60 Cu, 15 Ni, 25 Zn. ** c. 70 Al, 30 Mg. †† 84 Cu, 12 Mn, 4 Ni. ‡‡ 92½ Cu, 7 Sn, ½ P.
§§ Described as German silver with a little tungsten.

DENSITIES

DENSITIES OF COMMON SUBSTANCES (contd.)

Substance.	Density.	Substance.	Density.	Substance.	Density.
Minerals, etc.		**Liquids.**		Gelatine	1·27
				Glass, flint . . .	2·9–4·5
Agate ; slate . .	2·5–2·7	Glycerine . . .	1·26	„ crown ; }	2·4–2·6
Asbestos	3·0	Methylated spirit .	·83	window }	
„ board .	1·2	Milk	c. 1·03	„ Jena . . .	(see p. 74.)
Carbon (see above)		Naphtha . . .	·85	Ice (Roth, 1908), 0°	·9168
Charcoal	·3–·6	Oil, castor . . .	·97	„ (Vincent,'02),0°	·9160
Coal	1·2–1·5	„ linseed . .	·91–·93	Indiarubber (pure)	·91–·93
„ anthracite	1·4–1·8	„ lubricating .	·90–·92	Ivory	1·8–1·9
Coke	1·0–1·7	„ olive ; palm .	·91–·93	Leather	·85–1
Gas carbon . . .	1·9	„ paraffin . .	c. ·8	Paper	·7–1·1
Emery	4·0	Petrol	·63–·72	Pitch	c. 1·1
Granite	2·5–3	Sea-water . . .	1·01–1·05	Porcelain . . .	2·2–2·4
Marble	2·5–2·8	Turpentine . . .	·87	Resin	c. 1·1
Masonry	c. 2	Vinegar	1·02	Red fibre	1·45
Pumice (natural) .	·4–·9	**Miscellaneous.**		Snow (loose) . .	c. ·12
Quartz	2·66	Amber	1·1	Tar	1·02
Silica, fused		Bone	1·8–2·0	Wax, soft paraffin .	·87–·88
„ transparent	2·21	Butter, lard . . .	·92–·94	„ hard „	·88–·93
„ translucent .	2·07	Celluloid	1·4	„ white ; bees-	·95–·96
Sand (silver) . .	2·63	Cork	·22–·26	„ sealing . .	c. 1·8
Sandstone ; kaolin	2·2–2·3	Ebonite	1·8	„ soft red . .	c. 1·0

DENSITY DETERMINATION CORRECTIONS

In the determination of the density of a body by weighing in water, the true density (corrected for air buoyancy and water density) is given by $\Delta(D - \sigma) + \sigma$, where Δ is the uncorrected density of the body, D is the density of the water, and σ is the density of the air. The table below gives the correction to be applied to Δ. D is taken as ·9992 (correct to 1 part in 2000 between 10° and 18° C., see p. 22) and σ as ·0012 (see p. 25). — means that the correction has to be subtracted from Δ. (See Stewart and Gee, "Practical Physics," vol. i.)

Δ	Corr.	Δ	Corr.	Δ	Corr.	Δ	Corr.	Δ	Corr.	Δ	Corr.
0·5	+·0002	4·0	—·0068	7·5	—·0138	8·4	—·0156	9·5	—·0178	16·0	—·0308
1·0	—·0008	4·5	—·0078	7·8	—·0144	8·5	—·0158	10·0	—·0188	17·0	—·0328
1·5	—·0018	5·0	—·0088	7·9	—·0146	8·6	—·0160	11·0	—·0208	18·0	—·0348
2·0	—·0028	5·5	—·0098	8·0	—·0148	8·7	—·0162	12·0	—·0228	19·0	—·0368
2·5	—·0038	6·0	—·0108	8·1	—·0150	8·8	—·0164	13·0	—·0248	20·0	—·0388
3·0	—·0048	6·5	—·0118	8·2	—·0152	8·9	—·0166	14·0	—·0268	21·0	—·0408
3·5	—·0058	7·0	—·0128	8·3	—·0154	9·0	—·0168	15·0	—·0288	22·0	—·0428

DENSITY OF DAMP AIR

The density of damp air may be derived from the expression $\sigma = \sigma_d(H - 0·378p)/H$, where σ_d is the density of dry air at a pressure H mms. (see p. 25), H is the barometric height, and p is the pressure of water-vapour in the air.

HYDROMETERS

Common : Density = degrees/1000.

Baumé : Density at 15° = 144·3/(144·3 — Baumé degrees).

Twaddell : Density = 1 + (Twaddell degrees/200).

Sikes : One degree = a density interval of ·002 on the average.

DENSITIES

DENSITY OF WATER

In grams per millilitre.* Pure air-free water under 1 atmos. Temps. on const.-vol. H.scale. Water has a **maximum density** at 3°·98 (Chappuis, 1897 ; Thiesen, Scheel and Diesselhorst ; De Coppet, 1903). The temp. (t_m) of maximum density at different pressures (p), measured in atmos., is given by $t_m = 3·98 - ·0225(p - 1)$.

The **specific volume** is the reciprocal of the density. For reciprocals, see p.136. (See Chappuis, *Trav. et Mém. Bur. Intl*, 13, 1907.)

For density of ice see p. 21 ; of steam, p. 26.　　[* 1 litre = 1000·027 c.cs.]

Density of water at −10° = ·99815 ; at −5° = ·99930.

Temp.	0	2	4	6	8	10	12	14	16	18
0°C.	·99987	·99997	1·00000	·99997	·99988	·99973	·99953	·99927	·99897	·99862
20	·99823	·99780	·99732	·99681	·99626	·99567	·99505	·99440	·99371	·9930
40	·9922	·9915	·9907	·9898	·9890	·9881	·9872	·9862	·9853	·9843
60	·9832	·9822	·9811	·9801	·9789	·9778	·9767	·9755	·9743	·9731
80	·9718	·9706	·9693	·9680	·9667	·9653	·9640	·9626	·9612	·9598
100	·9584	—	—	—	—	·951	—	—	—	—

Density at 150° = ·917 ; at 200° = ·863 ; at 250° = ·79 ; at 300° = ·70.

DENSITY OF MERCURY

In grams per c.c. Hydrogen scale of temp. For reciprocals, see p. 136. (See Chappuis, *Trav. et Mém. Bur. Intl.*, 13, 1907.)

Temp.	0	2	4	6	8	10	12	14	16	18
−20°C.	13·6450	13·6400	13·6351	13·6301	13·6251	13·6202	13·6152	13·6103	13·6053	13·6004
0	·5955	·5905	·5856	·5806	·5757	·5708	·5659	·5609	·5560	·5511
20	·5462	·5413	·5364	·5315	·5266	·5217	·5168	·5119	·5070	·5022
40	·4973	·4924	·4875	·4826	·4778	·4729	·4680	·4632	·4583	·4534
60	·4486	·4437	·4389	·4340	·4292	·4243	·4195	·4146	·4098	·4050
80	·4001	·3953	·3904	·3856	·3808	·3759	·3711	·3663	·3615	·3566

	0	20	40	60	80	100	120	140	160	180
100	13·3518	13·304	13·2·7	13·209	13·152	13·115	13·068	13·021	12·974	12·927
300	12·881	12·834	12·787	12·740	—	—	—	—	—	—

DENSITY OF ETHYL ALCOHOL, C₂H₅OH . Aq

In grams per c.c. % indicates grams of C_2H_6OH in 100 grams of aqueous solution. Hydrogen scale of temp. (Calculated by E. W. Morley from Mendeléeff's Observations, *Jour. Am. Chem. Soc.*, Oct. 1904.)

At 17° C.

%	0	1	2	3	4	5	6	7	8	9
0	·9988	·9969	·9951	·9933	·9916	·9899	·9884	·9869	·9854	·9840
10	·9826	·9813	·9800	·9787	·9775	·9762	·9750	·9737	·9725	·9713
20	·9700	·9687	·9674	·9661	·9647	·9633	·9619	·9604	·9589	·9573
30	·9557	·9540	·9524	·9506	·9489	·9470	·9452	·9433	·9414	·9394
40	·9375	·9354	·9334	·9313	·9292	·9271	·9250	·9228	·9207	·9185
50	·9163	·9140	·9118	·9096	·9073	·9051	·9028	·9005	·8982	·8959
60	·8936	·8913	·8890	·8867	·8843	·8820	·8797	·8773	·8749	·8726
70	·8702	·8678	·8655	·8631	·8607	·8582	·8558	·8534	·8510	·8485
80	·8461	·8436	·8411	·8386	·8361	·8336	·8310	·8285	·8259	·8232
90	·8206	·8179	·8152	·8124	·8096	·8068	·8039	·8010	·7980	·7950
100	·7919	—	—	—	—	—	—	—	—	—

For other temperatures, interpolate from the above and the following :—

At 22° C.

0%, ·9978 ; 10%, ·9813 ; 20%, ·9678 ; 30%, ·9526 ; 40%, ·9338 ; 50%, ·9122 ; 60%, ·8895 ; 70%, ·8660 ; 80%, ·8417 ; 90%, ·8162 ; 100%, ·7876.

DENSITY OF HYDROCHLORIC ACID, HCl. Aq

Grams per c.c. at 15° C.   (Lunge and Marchlewski, 1891.)

Dens.	Grams HCl in 100 gm.	Grams HCl in 1 litre	Dens. Change for ± 1°	Dens.	Grams HCl in 100 gm.	Grams HCl in 1 litre	Dens. Change for ± 1°	Dens.	Grams HCl in 100 gm.	Grams HCl in 1 litre	Dens. Change for ± 1°
	of Solution.				of Solution.				of Solution.		
1·01	2·14	22	·00016	1·08	16·15	174	·00035	1·15	29·6	340	·00052
1·02	4·13	42	·00019	1·09	18·1	197	·00038	1·16	31·5	366	·00054
1·03	6·15	64	·00021	1·10	20·0	220	·00040	1·17	33·5	392	·00056
1·04	8·16	85	·00024	1·11	21·9	243	·00043	1·18	35·4	418	·00058
1·05	10·17	107	·00027	1·12	23·8	267	·00045	1·19	37·2	443	·00059
1·06	12·19	129	·00030	1·13	25·7	291	·00048	1·20	39·1	469	·00060
1·07	14·17	152	·00032	1·14	27·7	315	·00050				

DENSITY OF NITRIC ACID, HNO₃. Aq

Grams per c.c. at 15° C.   % N$_2$O$_5$ = ·857 × % HNO$_3$—by weight.   (Lunge and Rey, 1891.)

Dens.	Grams HNO₃ in 100 gm.	Grams HNO₃ in 1 litre	Dens. Change for ± 1°	Dens.	Grams HNO₃ in 100 gm.	Grams HNO₃ in 1 litre	Dens. Change for ± 1°	Dens.	Grams HNO₃ in 100 gm.	Grams HNO₃ in 1 litre	Dens. Change for ± 1°
	of Solution.				of Solution.				of Solution.		
1·02	3·70	38	·00022	1·22	35·3	430	·00080	1·42	69·8	991	·00137
1·04	7·26	75	·00028	1·24	38·3	475	·00086	1·44	74·7	1075	·00143
1·06	10·7	113	·00034	1·26	41·3	521	·00091	1·46	80·0	1168	·00149
1·08	13·9	151	·00040	1·28	44·4	568	·00097	1·48	86·0	1274	·00154
1·10	17·1	188	·00045	1·30	47·5	617	·00103	1·50	94·1	1411	·00160
1·12	20·2	227	·00051	1·32	50·7	669	·00109	1·504	96·0	1444	·00161
1·14	23·3	266	·00057	1·34	54·1	725	·00114	1·508	97·5	1470	·00162
1·16	26·4	306	·00062	1·36	57·6	783	·00120	1·512	98·5	1490	·00163
1·18	29·4	347	·00068	1·38	61·3	846	·00126	1·516	99·2	1504	·00164
1·20	32·4	388	·00074	1·40	65·3	914	·00132	1·520	99·7	1515	·00166

DENSITY OF SULPHURIC ACID, H₂SO₄. Aq

Grams per c.c. at 15° C.   % SO$_3$ = ·816 × % H$_2$SO$_4$ —by weight.   (Lunge and Isler, 1895.)

Density.	Grams H₂SO₄ in 100 gm.	Grams H₂SO₄ in 1 litre	Density.	Grams H₂SO₄ in 100 gm.	Grams H₂SO₄ in 1 litre	Density.	Grams H₂SO₄ in 100 gm.	Grams H₂SO₄ in 1 litre
	of Solution.			of Solution.			of Solution.	
1·02	3·03	31	1·44	54·1	779	1·822	90·4	1647
1·04	5·96	62	1·46	56·0	817	1·824	90·8	1656
1·06	8·77	93	1·48	57·8	856	1·826	91·2	1666
1·08	11·60	125	1·50	59·7	896	1·828	91·7	1676
1·10	14·35	158	1·52	61·6	936	1·830	92·1	1685
1·12	17·01	191	1·54	63·4	977	1·832	92·5	1695
1·14	19·61	223	1·56	65·1	1015	1·834	93·0	1706
1·16	22·19	257	1·58	66·7	1054	1·836	93·8	1722
1·18	24·76	292	1·60	68·5	1096	1·838	94·6	1739
1·20	27·3	328	1·62	70·3	1139	1·840	95·6	1759
1·22	29·8	364	1·64	72·0	1181			
1·24	32·3	400	1·66	73·6	1222	1·8405	95·9	1765
1·26	34·6	435	1·68	75·4	1267	1·8410	97·0	1786
1·28	36·9	472	1·70	77·2	1312	1·8415	97·7	1799
1·30	39·2	510	1·72	78·9	1357	1·8410	98·2	1808
1·32	41·5	548	1·74	80·7	1404	1·8405	98·7	1816
1·34	43·7	586	1·76	82·4	1451	1·8400	·99·2	1825
1·36	45·9	624	1·78	84·5	1504	1·8395	99·4	1830
1·38	48·0	662	1·80	86·9	1564	1·8390	99·7	1834
1·40	50·1	702	1·81	88·3	1598	1·8385	99·9	1838
1·42	52·1	740	1·82	90·0	1639			

DENSITIES : ALKALIES

DENSITY OF AMMONIA, NH₄HO . Aq
Grams per c.c. at 15° C.

Dens.	Grams NH₃ in 100 gm.	1 litre	Dens. Change for ± 1°.	Dens.	Grams NH₃ in 100 gm.	1 litre	Dens. Change for ± 1°.	Dens.	Grams NH₃ in 100 gm.	1 litre	Dens. Change for ± 1°.
	of Solution.				of Solution.				of Solution.		
·996	·91	9·1	·00019	·956	11·03	105·4	·00031	·916	23·03	210·9	·00049
·992	1·84	18·2	·00020	·952	12·17	115·9	·00033	·912	24·33	221·9	·00051
·988	2·80	27·7	·00021	·948	13·31	126·2	·00035	·908	25·65	232·9	·00053
·934	3·80	37·4	·00022	·944	14·46	136·5	·00037	·904	26·98	243·9	·00055
·980	4·80	47·0	·00023	·940	15·63	146·9	·00039	·900	28·33	255·0	·00057
·976	5·80	56·6	·00024	·936	16·82	157·9	·00041	·896	29·69	266·0	·00059
·972	6·80	66 1	·00025	·932	18·03	168·1	·00042	·892	31·05	277·0	·00060
·968	7·82	75·7	·00026	·928	19·25	178·6	·00043	·888	32·50	288·6	·00062
·964	8 84	85·2	·00027	·924	20·49	189·3	·00045	·884	34·10	301·4	·00064
·960	9·91	95·1	·00029	·920	21·75	200·1	·00047	·880	35·70	314·2	·00066

DENSITY OF SODIUM HYDROXIDE, NaHO . Aq
Grams per c.c. at 18° C. The percentages indicate grams of NaOH in 100 grams of solution. (Bousfield and Lowry, 1905.)

%	Density.	%	Density.	%	Density.	%	Density.	%	Density.
0	·9986	10	1·1098	20	1·2202	30	1·3290	40	1·4314
1	1·0100	11	1·1208	21	1·2312	31	1·3396	41	1·4411
2	1·0213	12	1·1319	22	1·2422	32	1·3502	42	1·4508
3	1·0324	13	1·1429	23	1·2532	33	1·3605	43	1·4604
4	1·0435	14	1·1540	24	1·2641	34	1·3708	44	1·4699
5	1·0545	15	1·1650	25	1·2751	35	1·3811	45	1·4794
6	1·0656	16	1·1761	26	1·2860	36	1·3913	46	1·4890
7	1·0766	17	1·1871	27	1·2968	37	1·4014	47	1·4985
8	1·0877	18	1·1982	28	1·3076	38	1·4115	48	1·5080
9	1·0987	19	1·2092	29	1·3184	39	1·4215	49	1·5174

DENSITY OF SODIUM CARBONATE, Na₂CO₃ . Aq
Grams per c.c. at 15° C. (Lunge.)

Density.	Grams Na₂CO₃ in 100 gm.	1 litre	Density.	Grams Na₂CO₃ in 100 gm.	1 litre	Density.	Grams Na₂CO₃ in 100 gm.	1 litre
	of Solution.			of Solution.			of Solution.	
1·007	·67	6·8	1·060	5·71	60·5	1·116	10·95	122·2
1·014	1·33	13·5	1·067	6·37	68·0	1·125	11·81	132·9
1·022	2·09	21·4	1·075	7·12	76·5	1·134	12·61	143·0
1·029	2·76	28·4	1·083	7·88	85·3	1·142	13·16	150·3
1·036	3·43	35·5	1·091	8·62	94·0	1·152	14·24	164·1
1·045	4·29	44·8	1·100	9·43	103·7			
1·052	4·94	52·0	1·108	10·19	112·9			

Change of density per 1° C. (0° to 30°), 6 to 7 % = ·0002 ; 11 to 20 % = ·0004.

DENSITY OF CALCIUM CHLORIDE, CaCl₂ . Aq
Grams per c.c. at 17·9° C. The percentages indicate grams of anhydrous CaCl₂ in 100 grams of solution. (Pickering, 1894.)

%	Density.	%	Density.	%	Density.	%	Density.	%	Density.
1	1·007	11	1·094	21	1·189	31	1·294	41	1·406
3	1·024	13	1·112	23	1·209	33	1·316	43	1·429
5	1·041	15	1·131	25	1·229	35	1·338		
7	1·058	17	1·150	27	1·250	37	1·361		
9	1·076	19	1·169	29	1·272	39	1·384		

DENSITIES OF SOME AQUEOUS SOLUTIONS

Grams per c.c. at 18° C.   The indicated % is the number of grams of anhydrous substance in 100 grams of solution.   (Kohlrausch, " Prakt. Phys.")

Substance.	5%	10%	15%	20%	25%	Substance.	5%	10%	15%	20%
NaCl .	1·034	1·071	1·109	1·148	1·190	MgSO₄.	1·050	1·104	1·160	1·220
NaNO₃	1·033	1·068	1·105	1·144	1·185	BaCl₂ .	1·044	1·093	1·147	1·204
NaA .	1·025	1·051	1·078	1·105	1·132	NH₄Cl.	1·014	1·029	1·043	1·057
H₂PO₄ .	1·027	1·054	1·083	1·114	1·145	CuSO₄ .	1·051	1·107	1·167	1·230
ZnSO₄ .	1·051	1·107	1·167	1·232	1·305	KCl . .	1·031	1·064	1·098	1·133
FeCl₃ .	1·130	1·175	1·226	1·278	1·331	KNO₃ .	1·030	1·063	1·097	1·133
SrCl₂ .	1·044	1·093	1·146	1·202	1·256	K₂SO₄ .	1·039	1·081	—	—
MgCl₂ .	1·042	1·086	1·130	1·176	1·225	K₂Cr₂O₇	1·035	1·072	1·109	—

Substance.	5 %	10 %	15 %	20 %	25 %	30 %	35 %	40 %	45 %	50 %
KBr . .	1·035	1·073	1·114	1·157	1·204	1·254	1·307	1·365	1·429	—
KI . .	1·036	1·076	1·120	1·168	1·218	1·273	1·332	1·397.	1·468	1·545
K₂CO₃ .	1·044	1·091	1·140	1·191	1·244	1·299	1·356	1·415	1·477	1·541
LiCl. .	1·027	1·056	1·085	1·115	1·147	1·181	1·217	1·255	—	—
CdSO₄ .	1·049	1·103	1·161	1·224	1·295	1·372	1·457	—	—	—
AgNO₃.	1·042	1·089	1·140	1·196	1·255	1·321	1·394	1·477	1·570	1·674
PbA₂ .	1·036	1·075	1·118	1·163	1·212	1·265·	1·322	1·386	—	—
Sugar *.	1·018	1·039	1·060	1·081	1·104	1·128	1·152	1·177	1·203	1·230

* **60**%, 1·287 ; [**75**%, 1·380 (supersaturated)].

DENSITY OF DRY AIR AT DIFFERENT TEMPERATURES AND PRESSURES

Grams per c.c. ; pressures in mm. of mercury at 0° C. lat. 45° ; $g = 980\cdot62$ cms. per sec.², These densities are calculated by the expression $\dfrac{\cdot001293}{(1 + \cdot00367t)} \cdot \dfrac{H}{760}$, where ·001293 is due to Leduc, 1898, and Rayleigh, 1893 (p. 26) ; and ·00367 to Regnault.   For density of damp air, see p. 21.

Temp. (t).	Pressure in Millimetres (H).							
	710	720	730	740	750	760	770	780
0° C.	·001208	·001225	·001242	·001259	·001276	·001293	·001310	·001327
2	·001199	·001216	·001233	·001250	·001267	·001284	·001300	·001317
4	·001190	·001207	·001224	·001241	·001258	·001274	·001291	·001308
6	·001182	·001199	·001215	·001232	·001248	·001265	·001282	·001298
8	·001173	·001190	·001207	·001223	·001240	·001256	·001273	·001289
10	·001165	·001182	·001198	·001214	·001231	·001247	·001264	·001280
12	·001157	·001173	·001190	·001206	·001222	·001238	·001255	·001271
14	·001149	·001165	·001181	·001197	·001214	·001230	·001246	·001262
16	·001141	·001157	·001173	·001189	·001205	·001221	·001237	·001253
18	·001133	·001149	·001165	·001181	·001197	·001213	·001229	·001245
20	·001125	·001141	·001157	·001173	·001189	·001205	·001220	·001236
22	·001118	·001133	·001149	·001165	·001181	·001196	·001212	·001228
24	·001110	·001126	·001141	·001157	·001173	·001188	·001204	·001220
26	·001103	·001118	·001134	·001149	·001165	·001180	·001196	·001211
28	·001095	·001111	·001126	·001142	·001157	·001173	·001188	·001203
30	·001088	·001103	·001119	·001134	·001149	·001165	·001180	·001195

DENSITIES OF GASES

Only those gases for which accurate density determinations have been made are included in this table (see also p. 10). Other gases will be found in the table below. For density of air under different temperatures and pressures, see p. 25.

Densities are in grams per litre (1000·027 c.cs. ; see p. 10) at 0° C. under 760 mm. of mercury at 0° C. and lat. 45° ($g = 980·62$), *i.e.* under a pressure of 1·01323 × 10⁶ dynes per sq. cm. (After P. A. Guye, *Chem. News*, 1908.)

Gas.	Density and Observer.	Accepted density.	Density rel. to O
		Grams/litre.	
Air	1·2927 L. ; 1·2928 R.	1·2928	0·90469
Oxygen, O_2	1·4288 L. ; 1·42905 R. ; 1·42900 M. ; 1·42896 Gr. ; 1·4292 J.P.	1·42900	**1·00000**
Hydrogen, H_2 . . .	0·08982 L. ; 0·08998 R. ; 0·089873 M.	0·08987	0·06289
Nitrogen, N_2	1·2503 L. ; 1·2507 R. ; 1·2507 Gr.	1·2507	0·87523
Argon, A	1·7809 R. ; 1·7808 Ra.	1·7809	1·2463
Nitrous oxide, N_2O .	1·9780 L. ; 1·9777 R. ; 1·9774 G.P.	1·9777	1·3840
Nitric oxide, NO .	1·3429 L. ; 1·3402 Gr. ; 1·3402 G.D.	1·3402	0·93786
Ammonia, NH_3 . . .	0·7719 L.; 0·77085 P.D.; 0·7708 G.P.	0·7708	0·5394
Carbon monoxide, CO .	1·2501 L. ; 1·2504 R.	1·2504	0·87502
Carbon dioxide, CO_2 . .	1·9763 L. ; 1·9769 R. ; 1·9768 G.P.	1·9768	1·3833
Hydrochloric acid, HCl	1·6407 L. ; 1·6397 Gr. ; 1·6398 G.G.	1·6398	1·1475
Sulphur dioxide, SO_2 .	2·9266 L.; 2·9266 J.P.; 2·9266 B.	2·9266	2·0480

B., Berthelot ; G.D., Guye & Davila ; G.G., Guye & Gazarian ; G.P., Guye & Pintza ; Gr., Gray ; J.P., Jacquerod & Pintza ; L., Leduc ; M., Morley ; P.D., Perman & Davies ; R., Rayleigh ; Ra., Ramsay.

The densities below are all **experimental values,** and are relative to that of oxygen ($O_2 = 16$) at 0° and 760 mms. at lat. 45° (see above).

Gas.	Rel. dens.	Gas.	Rel. dens.	Gas.	Rel. dens.
Acetylene, C_2H_2 . .	13·32	Helium, He . . .	1·98	Nitrogen oxychloride,	
Arsine, AsH_3 . .	39·02	Hydrobromic acid,		NOCl	33·45
Boron fluoride, BF_3 .	33·48	HBr	39·24	Nitrogen peroxide—	
Bromine, Br_2 c.**228**°C.	79·99	Hydrofluoric acid, HF	10·32	(N_2O_4) **26°·7** C.	38·37
Butane, C_4H_{10} . . .	29·10	Hydriodic acid, HI	63·36	,, ,, **39°·8**	35·62
Carbon oxychloride,		Hydrogen selenide,		,, ,, **60°·2**	30·12
$COCl_2$	50·75	H_2Se	40·47	,, ,, **80°·6**	26·06
,, oxysulphide,COS	30·47	,, sulphide, H_2S	17·22	,, ,, **100°·1**	24·33
Chlorine, Cl_2 . .	36·07	,, telluride, H_2Te	65·00	,, ,, **121°·5**	23·46
,, monoxide, Cl_2O .	43·54	Krypton, Kr . . .	41·5	,, (NO_2)**154°·0**	22·88
,, dioxide, ClO_2 .	33·74	Methane, CH_4 (1909)	8·03	,, ,, **183°·2**	22·73
Cyanogen, C_2N_2 . .	26·16	Methylamine,		Phosphine, PH_3 . .	17·58
Ethane, C_2H_6 . . .	15·57	CH_3NH_2 . .	15·64	Phosphorus chloro-	
Ethylamine,		Methyl chloride,		fluoride, PCl_2F_3	78·19
$C_2H_5NH_2$ . . .	22·77	CH_3Cl	25·06	,, oxyfluoride, POF_3	53·29
Ethyl chloride,		Methyl ether, C_2H_6O	23·41	,, pentafluoride, PF_5	65·01
C_2H_5Cl	32·13	,, fluoride, CH_3F	17·67	,, trifluoride, PF_3	43·76
Ethyl fluoride,C_2H_5F	24·62	Methylene fluoride,		Propylene, C_3H_6 . .	21·69
Ethylene, C_2H_4 . .	14·27	CH_2F_2	26·21	Silicon fluoride, SiF_4	52·13
Fluorine, F_2 . . .	18·97	Neon, Ne (1910) . .	10·82	Xenon, Xe	65·35

DENSITY OF SATURATED WATER VAPOUR

Densities in grams per litre under different pressures. (Zeuner, 1890.)

Atmos.	0	0·5	1	1·5	2	2·5	3	3·5	4	4·5
0	—	0·315	0·606	0·887	1·16	1·43	1·70	1·97	2·23	2·49
5	2·75	3·01	3·26	3·52	3·77	4·02	4·27	4·52	4·77	5·02
10	5·27	5·52	5·76	6·01	6·25	6·50	6·74	6·99	7·23	—

ELASTICITIES

Young's Modulus, or Longitudinal Elasticity, E in dynes per sq. cm.
Rigidity, Torsion Modulus, or Shear Modulus, n in dynes per sq. cm.
Volume Elasticity, Cubic Elasticity, or Bulk Modulus, k in dynes per sq. cm.
Compressibility (cubic), $C = 1/k$.
Poisson's Ratio, $\sigma =$ lateral contraction per unit breadth/longitudinal extension per unit length. For a homogeneous isotropic substance—

$$n = \frac{E}{2(1 + \sigma)} \ldots (a); \quad \sigma = \frac{E}{2n} - 1 \ldots (b); \quad k = \frac{E}{3(1 - 2\sigma)} \ldots (c)$$

For an isotropic solid Poisson's Ratio must lie between $+\frac{1}{2}$ and -1, but for some materials it may, when deduced from E and n, exceed $+1$. (See Searle's "Elasticity.")

1 megabar = 10^6 dynes per sq. cm. = ·987 atmos. = $1/1·013$ atmos. = the pressure measured by 750·15 mms. of mercury at 0° C. sea-level, and latitude 45°=749·66 mms. at 0° in London.

The elasticities of a substance depend considerably upon its history. The extent of the agreement between the calculated and observed values of n and of σ below gives an indication of the degree of isotropy of the metals used. (Grüneisen, Reichsanstalt, *Ann. d. Phy.*, 1908.)

ELASTICITIES OF METALS

Metal at 18° C. (see also below and pp. 28, 29).	Young's Modulus, E. By static method or longl. vibns.	Rigidity, n.		Poisson's Ratio, σ.		Vol. Elast. k.	Compress'y C. per megabar (calculated).
		By oscilln. method.	Calcd. by formula (a).	Ob-served.	Calcd. by formula (b).	Calcd. by formula (c).	
Aluminium (W) * .	$7·05 \times 10^{11}$	$2·67 \times 10^{11}$	$2·63 \times 10^{11}$	·339	·310	$7·46 \times 10^{11}$	$1·33 \times 10^{-6}$
Bismuth (C), pure .	3·19	—	1·20	·33	—	3·14	3·2
Cadmium (C), pure	4·99	—	1·92	·30	—	4·12	2·4
Copper (W), pure .	12·3	4·55	4·55	·337	·356	13·1	·74
Gold (W), pure .	8·0	2·77	2·80	·422	·495	16·6	·60
Iron (W), ·1%C.	21·3	—	8·31	·280	—	16·1	·63
Steel (W), 1%C.	20·9	8·12	8·12	·287	·287	16·4	·62
Lead (C), pure .	1·62	—	·562	·446	—	5·00	2·0
Nickel (W) † . .	20·2	—	7·70	·309	—	17·6	·57
Palladium (C), pure	11·3	5·11	4·04	·393	·101	17·6	·57
Platinum (C), pure	16·8	6·10	6·04	·387	·368	24·7	·41
Silver (W), pure .	7·90	2·87	2·86	·379	·369	10·9	·92
Tin (C), pure .	5·43	—	2·04	·33	—	5·29	1·9
Bronze (C)‡ . . .	8·08	3·43	2·97	·358	·177	9·52	1·05
Constantan (W) § .	16·3	6·11	6·11	·325	·329	15·5	·65
Manganin (W)‖ .	12·4	4·65	4·65	·329	·329	12·1	·83

(C) means cast ; (W) worked. * ·5% Fe, ·4% Cu. † 97% Ni, 1·4% Co, 1% Mn.
‡ 85·7% Cu, 7·2% Zn, 6·4% Sn. § 60% Cu, 40% Ni. ‖ 84% Cu, 12% Mn, 4% Ni.

The (experimental) results below are mostly for **ordinary laboratory materials**, chiefly wires.

Substance.	Young's Modulus, E.	Rigidity, n.	Volume Elast. k.	Poisson's Ratio, σ.
Copper	$12·4–12·9 \times 10^{11}$ S.	$3·9–4 \times 10^{11}$ S.	$14·3 \times 10^{11}$ M.	·26 S.
Iron (wrought) . .	19–20	7·7–8·3	14·6	c. ·27
„ (cast) . . .	10–13 G.	3·5–5·3	9·6	·23–·31
Steel	19·5–20·6	7·9–8·9	18·1 M.	·25–·33
Zinc (1 % Pb) . . .	8·7 § G.	3·8		·21
Brass (c. 66 Cu, 34 Zn) .	9·7–10·2	c. 3·5	10·65 M.	·34–·40
German silver * . . .	11·6 S.	4·3–4·7	—	·37
Platinoid †	13·6 S.	3·60 S.	—	·37
Phosphor bronze ‡ . .	12·0 S.	4·36 S.	—	·38 S.
Quartz fibre	5·18	3·0 H.	1·4	
Indiarubber	·048–·052	·00016		·46–·49 Sc.
Jena Glasses, Crowns .	6·5–7·8	2·6–3·2	4·0–5·9	·20–·27
„ „ Flints .	5·0–6·0	2·0–2·5	3·6–3·8	·22–·26

(G.) Grüneisen, 1907. (H.) Horton, 1905. (M.) Mallock, 1905. (S.) Searle, 1900.
(Sc.) Schiller, 1906. * 60 Cu, 15 Ni, 25 Zn. † German silver with a little tungsten.
‡ 92·5 Cu, 7 Sn, ·5 P. § Pure Zn, $12·5 \times 10^{11}$ dynes/cm².

TENSILE STRENGTHS

ELASTICITIES (contd.)

Substance.	Young's Modulus, E. dynes/cm.²	Temperature coefficient α in Elast$_t$ = Elast$_{15}${1 − α (t − 15)} :			Compressibility C. per megabar (i.e. 10⁶ dynes/cm.²) (Buchanan, *Proc. R. Soc.*, 1904).	
		At 15° C.	α for E.*	α for n†	7–11° C.; 200–300 megabars (see also pp. 27, 29).	
Iridium ‖	5·2 × 10¹¹ (G.)	Aluminium	21·3 × 10⁻⁴	13·5 × 10⁻⁴	Aluminium	1·7 × 10⁻⁶
Rhodium ‖	3·2 (G.)	Copper . .	3·64	4·0	Copper . .	·88
Tantalum	18·6 (Bo.)	Gold . .	4·8	3·3	Gold . .	·80
Invar .	14·1	Iron . .	2·3	7·3	Lead . .	2·8 (A.)
90 Pt, 10 Ir	21·0	Steel . .	2·4	2·6	Magnesium	3·2
Silk fibre	·65 ‡	Platinum .	·98	1·0	Platinum .	·56
Spider thread .	·3 (B.)§	Silver . .	7·5	4·5	Flint glass	3·0
Catgut .	·32	Tin . . .	—	5·9	Germ. glass tubing	2·57
Ice (− 2°)	·28	Brass . .	3·7	4·6	Steel . .	·51 (Br.)
Quartz (crystal)	6·8	German silver . .		6·5		
Marble .	2·6	Phosphor-bronze . .		c. 3		
Oak . .	1·3	Quartz fibre	—	−1·2		
Deal . .	·9					
Mahogany	·88					
Teak . .	1·66					

(A.) Amagat. (B.) Benton, 1907 and 1908. (Bo.) v. Bolton, 1905. (Br.) Bridgman, 1909. (G.) Grüneisen, 1907. * Wassmuth, 1906, and Schaefer, 1902. † Horton, 1904 and 1905. ‡ Diminishes rapidly with increasing load. § Shows marked elastic fatigue. ‖ Pure.

TENSILE STRENGTHS OF MATERIALS

Tenacities or breaking stresses in dynes per sq. cm. The elastic limit is always exceeded before the breaking stress is reached. The process of drawing into wire seems to strengthen the material, and the finer the wire the greater is the breaking stress. (See Poynting and Thomson's "Properties of Matter.")

For crushing and shearing strengths, see Ewing's "Strength of Materials" or one of the Engineering "Pocket-books." For bursting strengths of tubing, see p. 39; for tensile strengths of liquids, see p. 39.

To reduce to kilogrammes per sq. mm., it is sufficient to divide by 10⁸; to lbs. per sq. inch, divide by 7 × 10⁴; to tons per sq. inch, divide by 1·5 × 10⁸. * Along the grain.

Substance.	Tenacity.	Substance.	Tenacity.
	dynes/cm.²		dynes/cm.²
Aluminium, cast	·6–·9 × 10⁹	White or yellow pine * . . .	·2–·5 × 10⁹
" rolled	·9–1·5	Leather belt	c. ·3
Copper, cast	1·2–1·9	Hemp rope	·6–1·0
" rolled	2·0–2·5	Catgut	4·2
Iron, (a) cast	·8–2·3	Spider thread	1·8
(b) wrought	2·9–4·5	Silk fibre	2·6
(c) steel castings . .	2·3–7·0	Quartz fibre	c. 10
Mild steel (·2 % C) .	4·3–4·9	WIRES.	
High carbon annld. .	7·0–7·7	Aluminium	1·7–2·0
(for springs) {temprd.	9·3–10·8	Copper, hard drawn . .	4·0–4·6
Tungsten or chrome	11–12	" annealed . .	2·8–3·1
Ni steel, 5 %; 12 %	6·2 ; 14	Gold	2·6
Lead	c. ·16	Iron (charcoal), hard drawn	5·4–6·2
Tin	·16–·38	" " annealed	c. 4·6
Zinc, rolled	1·1–1·5	Steel; (1) ordinary; (2) tempd.	c. 11; 15·5
Brass (ordinary), {66 Cu} cast	1·5–1·9	" pianoforte . . .	18·6–23·3
" " {34 Zn} rolled	2·3–3·7	Nickel	5·3
Phosphor-bronze . . .	2·5–2·8	Platinum	3·3
Gun-metal (90 Cu, 10 Sn) . .	1·9–2·6	Silver	2·9
Soft solder	c. ·5	Tantalum	4·2
Glass	·3–·9	Brass	3·1–3·9
Ash, beech, oak, teak, mahogany*	·6–1·1	Phosphor-bronze, hard drawn	6·9–10·8
Fir, pitch-pine *	·4–·8	German silver	4·6
Red or white deal *	·3–·7		

COMPRESSIBILITIES OF ELEMENTS

Coefficient of compressibility $C = \frac{1}{V} \cdot \frac{\delta V}{\delta p}$, where δV is the change in volume of a volume V under a change of pressure δp (temp. constant).

The values of C below are per megabar (*i.e.* 10^6 dynes per sq. cm.). To express as compressibility per atmosphere, increase C by $\frac{1}{80}$ of its value. Room temp. Pressure range, 100–500 megabars. Based on compressibility of mercury = ·0,371 per megabar. The results show a periodic relation with atomic weight. See also pp. 27, 28. (Richards, *Zeit. Phys. Chem.*, 61, 1907, and *Journ. Chem. Soc.*, 1911.)

Element.	C	Element.	C	Element.	C	Element.	C
Al.	$1·3 \times 10^{-6}$	Cl (liq.).	95×10^{-6}	Hg	$3·71 \times 10^{-6}$	Si .	$·16 \times 10^{-6}$
Sb.	2·2 ,,	Cr.	·7 ,,	Mo	·26 ,,	Ag	·84 ,,
As.	4·3 ,,	Cu	·54 ,,	Ni	·27 ,,	Na	15·4 ,,
Bi.	2·8 ,,	Au	·47 ,,	Pd	·38 ,,	S	12·5 ,,
Br.	51·8 ,,	I	13 ,,	P, red	9·0 ,,	Tl	2·6 ,,
Cd	1·9 ,,	Fe	·40 ,,	white	20·3 ,,	Sn	1·7 ,,
Cs	61 ,,	Pb	2·2 ,,	Pt	·21 ,,	Zn	1·5 ,,
Ca.	5·5 ,,	Li	8·8 ,,	K	31·5 ,,		
C,diamond	·5 ,,	Mg	2·7 ,,	Rb	40 ,,		
graphite	3 ,,	Mn	·67 ,,	Se	11·8 ,,		

COMPRESSIBILITIES OF LIQUIDS

C = compressibility per megabar (*i.e.* 10^6 dynes per cm.^{2}). To express as compressibility per atmosphere, increase C by $\frac{1}{80}$ of its value.

As the pressure increases C becomes less. In general a rise in temperature increases the compressibility of a liquid; but water, however, shows a minimum value of C at about 50° C. (Amagat). The compressibility of a solution diminishes as the concentration increases (see Poynting and Thomson's "Properties of Matter.").

Where the limits of pressure are not given, they are—for Amagat, 8–37 atmos.; for Röntgen, 8 atmos.; for Richards, 100–200 atmos.

Liquid.	Temp.	Comp. C per megabar.	Liquid.	Temp.	Comp. C per megabar.
Water, 1–25 atmos. (A.)	15° C.	$48·9 \times 10^{-6}$	Carbon tetrachloride		
900–1000 ,, (A.)	15	36·3 ,,	(Ri.)	20° C.	$89·6 \times 10^{-6}$
900–1000 ,, (A.)	198	55·4 ,,	Carbon bisulphide (A.)	15·6	85·9 ,,
2500–3000 ,, (A.)	14·2	25·8 ,,	Ether, 1–50 atmos. (A.)	0	145·2 ,,
Sea-water (Grassi, 1851)	—	43·1 ,,	900–1000 ,, (A.)	0	64·2 ,,
Mercury . . . (A.)	20	3·82 ,,	,, ,, (A.)	198	142·2 ,,
,, . . . (Ri.)	15	3·71 ,,	Methyl acetate (A.)	14·3	95·8 ,,
Methyl alcohol,CH_3OH			Ethyl acetate . . (A.)	13·3	102·7 ,,
(A.)	14·7	102·7 ,,	,, bromide . (A.)	99·3	291·3 ,,
Ethyl alcohol—			,, chloride . (A.)	15·2	151·1 ,,
1–500 atm. (A.)	0	76 ,,	Acetic acid, 1–16 atm.		
150–200 atm. (Ba.)	310	4147 ,,	(C. & S.)	0	40·2 ,,
Propyl alcohol,			Glycerine, $C_3H_5(OH)_3$		
C_3H_7OH . . (R.)	17·7	95·8 ,,	(Q.)	20·5	24·8 ,,
Propyl alcohol iso- (R.)	17·8	101·7 ,,	Olive oil . . . (Q.)	20·5	62·5 ,,
Butyl alcohol,C_4H_9OH			Paraffin oil (de Metz,		
(R.)	17·4	83·9 ,,	1890)	14·8	61·9 ,,
Butyl alcohol iso- (R.)	17·9	96·8 ,,	Petroleum (Martini) .	16·5	68·7 ,,
Amyl alcohol,			Pentane, C_5H_{12} . (A.)	20	314 ,,
$C_5H_{11}OH$ . . (R.)	17·7	89·4 ,,	Benzene, C_6H_6 . (R.)	17·9	90·8 ,,
Chloroform . . (Ri.)	20	9·4 ,,	Turpentine,$C_{10}H_{16}$ (Q.)	19·7	78·14 ,,

(A.) Amagat, *Comptes Rendus*, 1884–93; (B.) Bartoli, 1896; (Ba.) Barus, 1891; (C. & S.), Colladon and Sturm, 1827; (G.) Grimaldi, 1886; (Q.) Quincke, *Wied. Ann.*, 19, 1883; (R.) Röntgen, *Wied. Ann.*, 44, 1891; (Ri.) Richards, 1907.

VISCOSITIES OF LIQUIDS

If two parallel planes are at unit distance apart in a fluid, and one of them is moving in its own plane with unit velocity relatively to the other plane, then the tangential force exerted per unit area on each of the planes is equal to the viscosity. The dimensions of a viscosity are $ML^{-1}T^{-1}$.

For the capillary-tube method of determining viscosities, Poiseuille's formula is, Viscosity $\eta = \dfrac{\pi p r^4 t}{8 l V}$, where p is the pressure difference between the two ends of the tube, r the radius of the tube, l its length, V the volume of liquid delivered in a time t.

VISCOSITY OF WATER

Determined by an efflux method and corrected for kinetic energy of outflow. (Hosking, *Phil. Mag.*, 1909, 1, 502 ; 2, 260.)

Temp.	Viscosity.	Temp.	Viscosity.	Temp.	Viscosity.	Temp.	Viscosity.
	c.g.s.						
0° C.	·01793	20° C.	·01006	50° C.	·00550	90° C.	·00316
5	·01522	25	·00893	60	·00469	100	·00284
10	·01311	30	·00800	70	·00406	124 *	·00223
15	·01142	40	·00657	80	·00356	153 *	·00181

* de Haas, 1894.

VISCOSITY OF MERCURY   (Koch, 1881.)

Temp.	−20° C.	0°	20°	50°	100°	200°	300°
Viscosity (c.g.s.)	·0186	·0169	·0156	·0141	·0122	·0101	·0093

VISCOSITIES OF VARIOUS LIQUIDS

Substance.	0° C.	10°	20°	30°	40°	50°	60°	70°
	c.g.s.							
Methyl alcohol, CH_4O	·00813	·00686	·00591	·00515	·00450	·00396	·00349	——
Ethyl „ C_2H_6O	·0177	·0145	·0119	·00989	·00827	·00697	·00591	·00504
Propyl „ C_3H_8O	·0388	·0292	·0225	·0178	·0140	·0113	·00919	·00757
Isopropyl „	·0456	·0324	·0237	·0175	·0133	·0103	·00804	·00642
Ether $(C_2H_5)_2O$	·00286	·00258	·00234	·00212	——	——	——	——
Chloroform, $CHCl_3$	·00700	·00626	·00564	·00511	·00465	·00426	·00390	——
Carbon tetrachloride	·0135	·0113	·00969	·00841	·00738	·00653	·00583	·00524
„ bisulphide	·00429	·00396	·00367	·00342	·00319	——	——	——
„ dioxide (liq.)	——	·00085	·00071	·00053	——	——	——	——
Benzene, C_6H_6	·00902	·00759	·00649	·00562	·00492	·00437	·00390	·00351
Aniline, $C_6H_5NH_2$	——	·0655	·0440	·0319	·0241	·0189	·0156	——
Glycerine, $C_3H_5(OH)_3$	46·0	21·0	8·5	3·5	——	——	——	——
Bromine	·0126	·0111	·00993	·00898	·00817	·00746	——	——
Turpentine, dens. = ·87	·0225	·0178	·0149	·0127	·0107	·00926	·00821	·00728
Pentane (n), C_5H_{12}	·00283	·00255	·00232	·00212	——	——	——	——
Hexane (n), C_6H_{14}	·00396	·00355	·00320	·00290	·00264	·00241	·00221	——
Formic acid, HCO_2H	——	·0224	·0178	·0146	·0122	·0103	·0089	·0077
Acetic acid, CH_3CO_2H	——	——	·0122	·0104	·0090	·0079	·0070	·0062
Propionic acid, $C_3H_6O_2$	·0152	·0129	·0110	·0096	·0084	·0075	·0067	·0060
Butyric „ $C_4H_8O_2$	·0228	·0185	·0154	·0130	·0112	·0097	·0085	·0076
Isobutyric „ „	·0188	·0157	·0131	·0113	·0098	·0086	·0076	·0068
Methyl formate	·00429	·00384	·00347	·00317	——	——	——	——
Ethyl „	·00505	·00448	·00402	·00362	·00328	·00299	——	——
Methyl acetate	·00478	·00425	·00381	·00344	·00312	·00284	——	——

Machine oil, c. 1/19° ; olive oil, ·99/15° ; paraffin oil, c. ·02/19° ; rape oil, 1·6/20°.

RELATIVE VISCOSITIES OF SOME AQUEOUS SOLUTIONS

Strength of solutions 1 normal. Viscosities relative to that of water at same temp. For a complete list, see Stöckl in L.B.M., and Moore, *Phys. Rev.*, 1896.

Substance.	Temp.	Relative Viscosity.	Substance.	Temp.	Relative Viscosity.
Ammonia	25° C.	1·02	Potassium chloride .	17°·6 C.	·98
Ammonium chloride	17·6	·98	Potassium iodide . .	17·6	·91
Calcium chloride .	20	1·31	Sodium hydrate . .	25	1·24
Hydrochloric acid .	25	1·07.	Sulphuric acid . . .	25	1·09

VISCOSITIES OF SOLIDS

Venice turpentine * at 17°·3, 1300, c.g.s.   **Shoemaker's wax** † at 8°, 4·7 × 10⁶. c.g.s.
Pitch † at 0°, 51 × 10¹⁰; at 15°, 1·3 × 10¹⁰.   **Soda glass** † at 575°, 11 × 10¹³.
Glacier ice, ‡ 12 × 10¹³.   **Golden Syrup** (Lyle), 1400/12°.

* R. Ladenburg, 1906.   † Trouton and Andrews, 1904.   ‡ Deeley, 1908.

VISCOSITIES OF GASES AND VAPOURS

Clerk Maxwell showed in 1860 that, on the basis of the kinetic theory, the coefficient of viscosity of a gas would be independent of the pressure, and would vary as the square root of the absolute temperature. The first relation is true except at very low pressures; the second deduction is not supported by experiment.

Of the formulæ connecting gaseous viscosity (η) and temperature (t), there are the convenient but only approximate relation of O. E. Meyer, $\eta_t = \eta_0 (1 + at)$, where a is a const.; and the less manageable but accurate formula of Sutherland (*Phil. Mag.*, 31, 1893), who, by taking account of the effects of molecular forces in bringing about collisions which otherwise would have been avoided, derived the expression $\eta_t = \eta_0 \dfrac{273 + C}{\theta + C} \cdot \left(\dfrac{\theta}{273}\right)^{\frac{3}{2}}$, where θ is the absolute temperature, and C is Sutherland's constant. The formula only holds for temps. above the critical, and for pressures such that Boyle's law is approximately obeyed. Sutherland's relation is thus of the form (which lends itself to graphical treatment), $\theta = \dfrac{K\theta^{3/2}}{\eta} - C$, where K is a constant. (See Fisher, *Phys. Rev.*, 1907, 1909 *et seq.*; O. E. Meyer's "Kinetic Theory of Gases." For a bibliography of gaseous viscosity, see Pedersen, *Phys. Rev.*, 25, 1907.) The values below are for dry gases.

Gas or Vapour.	Temp.	η. ×10⁻⁶	Observer.	Gas or Vapour.	Temp.	η. ×10⁻⁶	Observer.
Air . . .	−21° C.	164	Breitenbach	Nitrogen	0° C.	166	v.Obermayer
	0	173	,, (1901)	(*contd.*)	11	171	,, (1876)
	0	171	Hogg, 1905		54	190	,, ,,
	0	170	G. & G.*1908	Helium .	0	189	Schultze, '01
	0	171	Fisher, 1909		15	197	,,
	15	181	Markowski		185	270	,,
	99·6	221	,, (1904)	Neon . .	15	312	Rankine, '10
	302	299	Breitenbach	Argon . .	0	210	Schultze, '01
Hydrogen	−21	82	,, (1901)		15	221	,,
	0	86	,, ,,		184	322	,,
	15	89	,, ,,	Krypton .	15	246	Rankine, '10
	99	106	,, ,,	Xenon . .	15	222	,,
	302	139	,, ,,	Chlorine .	0	129	Graham, '46
Oxygen .	0	187	v.Obermayer		20	147	,,
	15	195	,, (1876)	Water(vap.)	0	90	Puluj, 1878
	54	216	,, ,,		15	97	K.&W.‡1875
Nitrogen .	−21	157	,, ,,		100	132	M.&S.§1881

* Grindley and Gibson.   ‡ Kundt and Warburg.   § Meyer and Schumann.

VISCOSITIES

VISCOSITIES OF GASES AND VAPOURS (contd.)

Gas or Vapour.	Temp.	η.	Observer.	Gas or Vapour.	Temp.	η.	Observer.
		$\times 10^{-6}$				$\times 10^{-6}$	
Mercury	0° C.	162 *	S. Koch, '83	Carbon	99° C.	186	Breitenbach
(vap.)	300	532	,,	dioxide	302	268	,, (1901)
	380	656	,,	Methane,	0	104	Graham, '46
Nitrous	−21	125	v.Obermayer	CH₄	20	120	,,
oxide	0	135	,, (1876)	Ethylene,	−21	89	Breitenbach
	100	183	,, ,,	C₂H₄	0	97	,, (1901)
Nitric	0	165	Graham, '46		15	102	,, ,,
oxide	20	186	,,		99·3	128	,, ,,
Sulphur	0	123	,,	Alcohol	0	83	Puluj, 1878
dioxide	20	138	,,	(vap.)	17	89	,, ,,
Sulphuretd	0	115	,,		78	142	,, ,,
hydrogen	20	130	,,	Ether (vap.)	0	69	,, ,,
Cyanogen .	0	95	,,		16	73	,, ,,
	20	107	,,		36	79	,, ,,
Carbon	0	163	v.Obermayer	Chloroform	0	99	Breitenbach
monoxide	20	184	,, (1876)	(vap.)	17·4	103	,, (1901)
Carbon	−21	129	Breitenbach		61	189	,, ,,
dioxide	0	139	,, (1901)	Benzene	0	69	Schumann
	15	146	,, ,,	(vap)	19	79	,, (1884)
					100	118	,, ,,

* Extrapolated.

TEMPERATURE COEFFICIENTS OF VISCOSITY

Based largely on W. J. Fisher's computations (ref. above).

Gas or Vapour.	Sutherland's Consts.		Meyer's Const. a	Gas or Vapour.	Sutherland's Consts.		Meyer's Const. a
	C	K			C	K	
Air	124	150 × 10⁻⁷	·00273	Xenon	252	246 × 10⁻⁷	—
Hydrogen . . .	72	66 ,,	—	Water (vap.) . .	72	—	—
Oxygen	127	175 ,,	·00283	Carbon monoxide	102	135 ,,	·00269
Nitrogen . . .	110	143 ,,	·00269	,, dioxide .	240	158 ,,	·00350
Helium	80	148 ,,	—	Nitrous oxide . .	313	172 ,,	·00345
Neon	56	220 ,,	—	Ethylene . . .	226	106 ,,	·00350
Argon	170	207 ,,	—	Chloroform (vap.)	454	159 ,,	—
Krypton	188	240 ,,					

SIZE, VELOCITY, AND FREE PATH OF MOLECULES

ρ = density of gas in gms./c.c. at 0° C. and 76 cms.

p = 1 atmos. = $1·0132 \times 10^6$ dynes/cm.²

θ = absolute temperature.

R = gas constant.

b = b of Van der Waals' equation (p. 34).

k = thermal conductivity of gas (p. 52).

c_v = specific heat at const. volume (p. 58).

η = viscosity of gas (p. 31).

N = number of molecules of gas per c.c. at 0° C. and 76 cms.

σ = molecular diameter in cms.

m = mass of a single molecule (in grams).

G = square root of mean square molecular vel. (cm./sec. at 0° C.).

Ω = mean molecular velocity (cm./sec.).

L = length of mean free path in cms.

Assuming a Maxwell-Boltzmann distribution of velocities—

$$G = \sqrt{3p/(Nm)} = \sqrt{3p/\rho} = \sqrt{3R\theta}$$

$$\Omega = 4G/\sqrt{6\pi} = ·921G$$

$$L = \eta/(·31\rho\Omega) = 2·02\eta/\sqrt{p\rho}$$

Collision frequency = $\Omega/L = 5 \times 10^9$ per sec. for O_2

SIZE, VELOCITY, AND FREE PATH OF MOLECULES (contd.)

MOLECULAR SIZE

The molecular diameter σ has been calculated by the following formulæ :—

1. The **viscosity** η of a gas is a function of the size of its molecules.

$$\eta = \cdot 44\rho\Omega/(\sqrt{2}N\pi\sigma^2) \quad . \quad . \quad \text{Jeans} \quad \therefore \sigma = \{\cdot 0912\rho G/(N\eta)\}^{\frac{1}{2}}$$

2. The **thermal conductivity**, $k = 1\cdot 6\eta c_v = \cdot 158\rho\Omega c_v/N\sigma^2$

$$\therefore \sigma = \{\cdot 146\rho G c_v/(N k)\}^{\frac{1}{2}}$$

3. **Van der Waals'**, $b = 2\pi N\sigma^3/3 \qquad \therefore \sigma = \{3b/(2\pi N)\}^{\frac{1}{3}}$

4. **Limiting density**, i.e. density D of densest known form. $\sigma = \{6\rho/(\pi DN)\}^{\frac{1}{3}}$

The values of ρ and η used in calculating G and L below are given on pp. 26, 31. The values of σ tabulated are mostly taken from Jeans' "Dynamical Theory of Gases," or Rudorf (*Phil. Mag.*, 1909, p. 795). Jeans takes $N = 4 \times 10^{19}$, while in the table following, the more recent value $2\cdot75 \times 10^{19}$ has been used.

Gas.	G at 0° C.	Mean free path, L.	Molecular diameter σ deduced from			
			η	k	b	Lt. ρ [= D]
	cm./sec.	cm.	cm.	cm.	cm.	cm.
Hydrogen, H_2 .	$18\cdot39\times10^4$	$18\cdot3\times10^{-6}$	$2\cdot47\times10^{-8}$	$2\cdot40\times10^{-8}$	$2\cdot32\times10^{-8}$	$2\cdot92\times10^{-8}$
Helium, He .	$13\cdot11$,,	$28\cdot5$,,	$2\cdot18$,,	——	$2\cdot30$,,	$4\cdot31$,,
Nitrogen, N_2 .	$4\cdot93$,,	$9\cdot44$,,	$3\cdot50$,,	$3\cdot31$,,	$3\cdot53$,,	$2\cdot97$,,
Oxygen, O_2 .	$4\cdot61$,,	$9\cdot95$,,	$3\cdot39$,,	$3\cdot11$,,	——	$2\cdot79$,,
Neon, Ne . .	$5\cdot61$,,	$19\cdot3$,,	——	——	——	——
Argon, A . .	$4\cdot13$,,	$10\cdot0$,,	$3\cdot36$,,	——	$2\cdot86$,,	$4\cdot43$,,
Krypton, Kr .	$2\cdot86$,,	$9\cdot49$,,	——	——	$3\cdot14$,,	$4\cdot93$,,
Xenon, Xe . .	$2\cdot28$,,	$5\cdot61$,,	——	——	$3\cdot42$,,	$4\cdot88$,,
Chlorine, Cl .	$3\cdot07$,,	$4\cdot57$,,	$4\cdot96$,,	——	——	——
Methane, CH_4	$6\cdot48$,,	$7\cdot79$,,	——	——	——	——
Ethylene, C_2H_4	$4\cdot88$,,	$5\cdot47$,,	$4\cdot55$,,	$4\cdot68$,,	——	$5\cdot26$,,
Carbon mon-oxide, CO	$4\cdot93$,,	$9\cdot27$,,	$3\cdot50$,,	$3\cdot31$,,	——	——
Carbon di-oxide, CO_2 .	$3\cdot92$,,	$6\cdot29$,,	$4\cdot18$,,	$4\cdot32$,,	$3\cdot40$,,	$4\cdot42$,,
Ammonia, NH_3	$6\cdot28$,,	$6\cdot95$,,	——	——	——	——
Nitrous oxide, N_2O . .	$3\cdot92$,,	$6\cdot10$,,	$4\cdot27$,,	$4\cdot20$,,	——	$4\cdot58$,,
Nitric oxide, NO . . .	$4\cdot76$,,	$9\cdot06$,,	$3\cdot40$,,	$3\cdot40$,,	——	——
Sulph. hydro-gen, H_2S .	$4\cdot44$,,	$5\cdot90$,,	——	——	——	——
Sulph. dioxide, SO_2 . .	$3\cdot22$,,	$4\cdot57$,,	——	——	——	——
Hydrochloric acid, HCl .	$4\cdot30$,,	$6\cdot86$,,	——	——	——	——
Water, H_2O .	$7\cdot08$,,	$7\cdot22$,,	$4\cdot09$,,	——	——	$3\cdot45$,,

The formulæ above assume the molecules to be spherical. Sutherland (*Phil. Mag.*, 1910), adopting his formula (see p. 31) for the variation of η with temp., obtains the following values of σ. Unit, 10^{-8} cm.

H	He	A	O_2	N_2	N_2O	NO	CO	CO_2	C_2H_4	Cl_2
$2\cdot17$	$1\cdot92$	$2\cdot66$	$2\cdot71$	$2\cdot95$	$3\cdot33$	$2\cdot59$	$2\cdot74$	$2\cdot90$	$3\cdot31$	$3\cdot76$

CRITICAL DATA AND VAN DER WAALS' CONSTANTS

Critical temperature, θ_c, is the highest temperature at which a gas can be liquefied by subjecting it to pressure.

Critical pressure, p_c, is the pressure (of gas and liquid) at the critical temperature.

Critical volume, v_c, is here defined as the ratio of the volume that a gas has at the critical temp. and press. to that which it would have at o° C. and 760 mms., *i.e.* it is the volume of gas at θ_c and p_c which at N.T.P. would have unit volume. Some writers take the critical volume to be the specific volume (c.cs. per gram) at θ_c and p_c.

Most of the characteristic equations of state which have been proposed for gases take the form $(p + a/v^2)(v - b) = R\theta$, where p is the pressure, v the volume, θ the absolute temperature of the gas, and R is the "gas constant." a expresses the mutual attraction of the molecules. The "covolume" b is proportional to the space occupied by the molecules : O. E. Meyer takes $b = 4\sqrt{2}$ (volume of molecule). Van der Waals assumes a is constant : if this were true the constant volume and thermodynamic scales of temperatures would agree—they do not, however (see p. 44). Joule and Thomson, Clausius, Amagat, and Berthelot, among others, regard a as a function of θ (*e.g.* $a \propto 1/\theta$), and b as constant.

Assuming with Van der Waals that a and b are constants, the equation can be regarded as a cubic in v, which has its three roots equal at the critical point, whence $a = 27R^2\theta_c^2/(64p_c)$, and $b = R\theta_c/(8p_c)$.

Taking pressures in atmos., and the volume of the gas at o° C. and 1 atmos. as 1, $R = pv/\theta = 1/273$. In these units, b is in terms of the volume of the gas at o° C. and 1 atmos.

Example.—For CO_2 $p_c = 73$ atmos. and $\theta_c = 273 + 31\cdot1 = 304\cdot1$, whence $b = 304\cdot1/(8 \times 273 \times 73) = \cdot00191$ of the volume of the gas at o° C. and 1 atmos.

See Preston's "Heat," Nernst's "Theoretical Chemistry," Young's "Stoichiometry," Berthelot (*Trav. et Mém. Bur. Intl.*, 1907). * Indicates calculated values.

Substance.	Critical			Van der Waals'		Observer.
	Temp. θ_c	Press. p	Vol. v_c	a.	b.	
		atmos.				
Hydrogen	−234°·5 C.	20	·00264*	·00042	·00088	Olszewski, '95
Oxygen.	−118	50	·00426*	·00273	·00142	v. Wroblewski, '85
Nitrogen	−146	33	·00517*	·00250	·00165	,, ,,
Air	−140	39	·00468*	·00257	·00156	Olszewski, '84
Helium.	−268	2·3	·00299*	·0000615	·000995	Onnes, 1908
Neon	< −210					
Argon	−117·4	52·9	·00404*	·00259	·00135	Ramsay and
Krypton	−62·5	54·3	·00532*	·00462	·00178	Travers, 1900
Xenon	14·7	57·2	·0069*	·00818	·00230	
Chlorine	146	93·5	·00615*	·01063	·00205	Knietch, '90
Bromine	302	131*	·00605	·01434	·00202	Nadejdine, '85
Water	365	194·6	·00386	·0118	·00150	Battelli, '90
Hydrochloric acid . .	52·3	86	·0052*	·00697	·00173	Dewar, 1884
Carbon monoxide . .	−141·1	35·9	·00505*	·00275	·00168	v. Wroblewski, '83
Carbon dioxide . .	31·1	73	·0066	·00717	·00191	Andrews, 1869
Carbon bisulphide . .	273	72·9	·0090	·02316	·00343	Batelli, 1890
Ammonia, NH_3 . .	130	115·0	·00481*	·00798	·00161	Dewar, 1884
Nitrous oxide, N_2O .	38·8	77·5	·00436	·00710	·00184	Villard, 1894
Nitric oxide, NO . .	−93·5	71·2	·00347*	·00257	·00116	Olszewski, '85
Nitrogen tetroxide, NO_2	171·2	144*	·00413	·00756	·00138	Nadejdine, '85
Sulphuretted hydrogen	100	88·7	·00578*	·00888	·00193	Olszewski, '90
Sulphur dioxide. . .	155·4	78·9	·00745*	·01316	·00249	Sajotschewsky,'78
Methane, CH_4 . . .	−95·5	50	·00488*	·00357	·00162	Dewar, 1884
Acetylene, C_2H_2 . .	36·5	61·6	·0069*	·00880	·00230	Mackintosh, '07
Ethylene, C_2H_4 . .	10	51·7	·00752*	·00877	·00251	Olszewski, '95
Ethane, C_2H_6. . . .	34	50·2	·00839*	·01060	·0028	,, ,, ['86
Ethyl alcohol, C_2H_5OH	243	62·7	·0071	·02467	·00377	Ramsay & Young,
Ether $(C_2H_5)_2O$ . .	197	35·8	·0158	·03496	·00602	Battelli, '92
Chloroform, $CHCl_3$.	260	54·9	·0133	·0293	·00445	Sajotschewsky,'78
Aniline, $C_6H_5NH_2$. .	425·6	52·3	·0183*	·05282	·00611	Guye & Mallet, '02
Benzene, C_6H_6 . . .	288·5	47·9	·0161*	·03726	·00537	Young, 1900

DIFFUSION OF GASES

The Coefficient of diffusion, D, is the mass of the "diffusing" gas which crosses unit area in unit time under unit concentration gradient : the dimensions of the coefficient are $cm.^2$ $sec.^{-1}$. D is inversely proportional to the total pressure of the two gases, and roughly proportional to the square of their absolute temperature. Total pressure 1 atmosphere. H_2—O_2 implies that H_2 is diffusing into O_2.
(See Meyer's "Kinetic Theory of Gases.")

Gases.	t° C.	D	Gases.	t° C.	D	Gas (Winkelmann).	t° C.	D into		
								Air.	CO_2	H_2
H_2—O_2 .	0°	·677, O.	CO—H_2 .	0°	·642, L.	Formic acid .	0°	·131	·088	·513
H_2—O_2 .	0	·681, O.	CO—C_2H_4	0	·101, O.	Acetic . . .	0	·106	·071	·404
H_2—CH_4 .	0	·625, O.				Propionic acid	0	·082	·058	·326
H_2—CO .	0	·649, O.	CO_2—CO	0	·131, O.	Butyric acid .	0	·053	·037	·201
H_2—CO_2.	0	·538, O.	CO_2—CO	0	·141, L.	Isobutyric acid	0	·07	·047	·271
H_2—C_2H_4	0	·483, O.	CO_2—Air	0	·142, L.	Me. alcohol .	0	·132	·088	·500
H_2—N_2O	0	·535, O.	CO_2—CH_4	·0	·146, O. ; ·16, L.	Et. „	0	·102	·068	·378
			CO_2—O_2.	0	·18, L.	Propyl alcohol	0	·080	·058	·315
O_2—N_2.	0	·171, O.	CO_2—N_2O	0	·1, L. ; ·15, O.	Butyl „ .	0	·068	·048	·272
O_2—H_2.	0	·722, L.	CO_2—H_2	0	·55, L.	„ „ .	99	·126	·088	·504
H_2O—CO_2	18	·155, G.	Air—O_2 .	0	·178, O.	Benzene . .	0	·075	·053	·294
H_2O—Air	8	·239, G.	Air—H_2 .	17	·66, Sc.	Me. acetate .	0	·084	·056	·328
H_2O—Air	15	·246, G.				Et. formate .	0	·085	·057	·336
H_2O—Air	18	·248, G.	CS_2—Air	0	·1, S.	Et. acetate .	0	·071	·049	·273
H_2O—Air	0	·203, H.				Et. butyrate .	0	·057	·041	·224
						Et.iso-butyrate	0	·055	·040	·224

G., Guglielmo, 1884 ; H., Houdaille, 1896 ; L., Loschmidt, 1870 ; O., v. Obermayer, 1887 ; S., Stefan, 1879 ; Sc., Schulze, 1897.

DETERMINATION OF ALTITUDES BY THE BAROMETER

Babinet's formula (*Compt. Rend.*, 1850) is, Altitude $= \dfrac{C(H_1 - H_2)}{H_1 + H_2}$, where $H_1 =$ barometer reading at lower station, H_2 at upper station. If altitudes are in metres, and barometric heights in mms.,

$$C = 32(500 + t_1 + t_2)$$

where t_1 and t_2 are the corresponding station temperatures (° C.).
In the table below the mean temperature, $(t_1 + t_2)/2$, is taken as 10° C., and the barometric height at sea-level as 760 mm., so that altitudes are in metres above sea-level. The values are of course only approximate. Babinet's formula is not applicable to very great altitudes.

Altitude	0	100	200	300	400	500	600	700	800	900
metres.	mm.	mm.	mm.	mm.	mm.	mm.	mm.	mm.	mm.	mm.
0	760	751	742	733	724	716	707	699	690	682
1000	674	666	658	650	642	635	627	620	612	605

THICKNESS OF THIN METAL FOIL

Approximate thickness of the thinnest beaten metal leaf at present commercially obtainable. Unit 10^{-6} cm.

Metal.	Al	Cu	Au	Pt	Ag	Dutch metal.	(Cigarette paper.)
Thickness	20	34	8	25	21	70	2500

SURFACE TENSIONS

In dynes per cm. (A) indicates liquid in contact with air, (V) indicates liquid in contact with its vapour. The surface tension of a liquid varies somewhat with the age (and contamination) of the surface.

Temperature variation. It follows from Eötvos' rule, that the surface tension T at temp. t is approximately proportional to $(t_c - t)$, where t_c is the critical temp., the constant of proportionality being much the same for chemically similar substances. The surface tension at t_c is zero. (For critical temps. see p. 34.)

See Poynting and Thomson's "Properties of Matter."

WATER ($t_c = 365°$ C.)

Surf. Tens. T at 15° C.	Method.	Observer.	Temp. (t).	T_t/T_{15}	Temp. (t).	T_t/T_{15}
dynes per cm.						
72·8 (A)	Vibrating jet	Bohr., *Phil. Trans.*, '09	0° C.	1·030	60° C	·901
74·3 (A)	Vibrating jet	Pedersen, *P. Trans.*, '07	10	1·010	70	·876
74·2 (A)	Capillary waves	Kalähne, *Ann. d. Phy.*,	15	1·000	80	·851
73·8 (A)	Hanging drop	Sentis, 1897	['02			
			20	·990	90	·827
73·3 (A)	Tension of film	Hall, 1893	['93			
			30	·970	100	·80
74·3 (A)	Capillary waves	Rayleigh, *Phil. Mag.*,	40	·947	120	·75
73·3 (A)	Capillary tube	Volkmann, 1895	50	·925	140	·70
71·4 (V)	Capillary tube	Ramsay & Shields, '93				
77·6 (A)	Pull on ring	Weinberg, 1892	Ramsay & Shields, '93; Volkmann & Brunner			

Substance.		Temp. (t).	Surf. Tens.	Method.	Observer.
INORGANIC.			dynes cm.		
Cadmium	CO_2	Molten	693	Weight of drop	Quincke
Gold	A	1070°C.	612	Curvature of drop	Heydweiller, '98
Lead	CO_2	335	473	Capillary waves	Grunmach
Mercury ($T_t = T_0 - ·379t$)	A	17·5	547	Capillary tube	Quincke
Potassium	CO_2	58	364	Weight of drop	,,
Sodium	CO_2	90	520	,, ,,	,,
Sulphur (M.P. 115°) . .	A	160	59	Press. reqd. to bubble air from cap. tube thro' liquid	Zickendraht, '06; and Quincke, '08
,, . . .	A	250	118		
,, . . . (B.P.)	A	445	44		
Liquid oxygen	A	−183	13·1	Capillary waves	Grunmach, 1906
,, nitrogen . .	A	−196	8·5	,, ,,	,, 1906
,, nitrous oxide . .	A	−89·4	26·3	,, ,,	,, 1904
Nickel carbonyl, Ni(CO)₄	V	19·8	14·2	Capillary tube	Ramsay and Shields, 1893
Ammonia soln. ($d = ·96$)	A	15	64·7	Vibrating jet	Pedersen, 1907
Sulph. acid sol. ($d = 1·14$)	A	15	74·4	,, ,,	,, 1907
Other solns. (see below)					
CARBON COMPOUNDS.					
Acetone, $(CH_3)_2CO$ . .	V	16·8	23·3	Capillary tube	Ramsay and Shields, 1893
	V	78·3	15·9	,, ,,	
Acetic acid, CH_3CO_2H .	V	20	23·5	,, ,,	,, ,,
	V	300	1·16	,, ,,	,, ,,
Alcohol—methyl, CH_4O	V	20	23	,, ,,	,, ,,
	V	200	5·2	,, ,,	,, ,,
—ethyl, C_2H_5OH	V	20	22·0	,, ,,	,, ,,
($T_t = T_0 - ·092t$) . .	V	150	9·5	,, ,,	,, ,,
—propyl (n),	V	16·4	23·8	,, ,,	,, ,,
C_3H_7OH	V	78·3	18·7	,, ,,	,, ,,
Aniline, $C_6H_5.NH_2$ . .	A	15	43·0	Vibrating jet	Pedersen, 1907
Benzene, C_6H_6.	A	17·5	29·2	Capillary tube	Volkmann
($T_t = T_0 - ·146t$)					

Substance.		Temp. (t).	Surf. Tens.	Method.	Observer.
CARBON COMPOUNDS.— *(contd.)*			dynes cm.		
Butyric acid, $C_3H_7CO_2H$	V	**15°.** C.	26·7	Capillary tube	{ Ramsay and
	V	**132**	16·4	,, ,,	Shields, 1893
Carbon bisulphide . .	V	**19·4**	33·6	,, ,,	,, ,,
	V	**46·1**	29·4	,, ,,	,, ,,
Carbon tetrachloride. .	V	**20**	25·7	,, ,,	,, ,,
	V	**250**	1·93	,, ,,	,, ,,
Chloroform, $CHCl_3$ . .	A	**15**	27·2	,, ,,	Kaye, 1905
Ether (ethyl), $(C_2H_5)_2O$.	V	**20**	16·5	,, ,,	Jaeger, 1892
($T_t = T_0 - ·115t$) .	V	**150**	2·9	,, ,,	,,
Ethyl acetate,	V	**20**	23·6	,, ,,	,,
$CH_3CO_2C_2H_5$	V	**100**	14	,, ,,	,,
Formic acid, $HCOOH$.	V	**17**	37·5	,, ,,	{ Ramsay and
	V	**80**	30·8	,, ,,	Shields, 1893
Olive oil ($d/20° = ·91$) .	A	**20**	32	Curvature of drop	Magie, 1888
Paraffin oil ($d = ·847$) .	A	**25**	26·4	Capillary tube	Frankenheim, '47
Propionic acid, $C_3H_6O_2$	V	**16·6**	26·6	,, . ,,	{ Ramsay and
	V	**132**	15·5	,, ,,	Shields, 1893
Pyridine, C_5H_5N . .	V	**17·5**	36·7	,, ,,	{ Dutoit and Fri-
	V	**91**	26·5	,, ,,	derich, 1900
Toluene, $C_6H_5 . CH_3$.	A	**15**	28·8	Vibrating jet	Pedersen, 1907
Turpentine, $C_{10}H_{16}$ . .	A	**15**	27·3	Capillary tube	Kaye, 1905

SURF. TENSIONS OF SOLUTIONS

The surface tension of aqueous salt solutions is generally greater than that of pure water. Dorsey (*Phil. Mag.*, 1897) has shown

$$T_n = T + A . n$$

T_n is the surf. tens. of a sol. of n gram – equivalents per litre, T that of water at same temp.

Salt.	A.
NaCl	1·53
KCl 4	1·71
$\frac{1}{2}(Na_2CO_3)$	2·00
$\frac{1}{2}(K_2CO_3)$	1·77
$\frac{1}{2}(ZnSO_4)$	1·86

SURFACE TENSIONS AT INTER-LIQUID BOUNDARIES

Liquids at 20° C.	Surface Tension T.	Observer.
	dynes/cm.	
Water-benzene . . .	33·6	Pockels, 1899
,, chloroform † .	29·5	Quincke
,, ether . .	12·2	,,
,, olive oil ‡ . .	20·6	,,
,, paraffin oil . .	48·3	Pockels, 1899
Mercury-water . . .	427 *	Gouy, 1908
,, alcohol § . .	399	Quincke
,, chloroform † .	399	,,

* Diminishes with time. † Density = 1·49.
‡ Density = ·91. § Density = ·79.

ANGLES OF CONTACT BETWEEN GLASS AND LIQUIDS

Angles of contact vary largely with the freshness of the surfaces in contact.

Liquid.	Angle.	Observer.	Liquid.	Angle.	Observer.
Mercury . . .	52° 40′ *	Quincke	Acetic acid . .	20°	Magie, '88
Water	8°–9°	,,	Benzene . . .	0°	,,
Water	~ 0° †	Wilberforce	Paraffin oil . .	26°	,,
Methyl alcohol .	0°	Magie, '88	Turpentine . .	17°	,,
Ethyl alcohol . .	0°	,,			
Ether . . .	16°	,,	* For freshly formed drop, 41° 5′.		
Chloroform . .	0°	,,	† Glass quite clean.		

The angle of contact of water against different **metals** varies between 3° and 11°.

SIZE OF DROPS AND THICKNESS OF LIQUID FILMS

Reference may be made to the writings of J. J. Thomson ("Conduction of Electricity through Gases"), C. T. R. Wilson, Laby (*Phil. Trans. A*, 1908), Reinold & Rücker (*Phil. Trans.*, 1886), Lord Rayleigh, and Johonnot (*Phil. Mag.*, 1906).

HYGROMETRY

RELATIVE HUMIDITY AND DEW-POINT

Relative humidity $= \dfrac{[p]_t}{[p]'_t} \cdot 100$, where $[p]_t$ is the actual pressure of water-vapour

at temperature $t°$, and is equal to $[p]'_{t_{dp}}$ the saturated vapour-pressure at the dew-point (dp); $[p]'_t$ is the pressure of saturated vapour at $t°$. For a table of saturated water-vapour pressures, see p. 40. (See "Smithsonian Meteorological Tables.")

Percentage relative humidities for different dew-points and dew-point depressions are tabulated below.

Dew-point (dp).	Depression of dew-point $= t° - (dp)°$.															
	0°C.	1°	2°	3°	4°	5°	6°	7°	8°	9°	10°	12°	14°	16°	18°
− 15° C.	100	92	85	79	73	67	62	58	53	49	46	39	34	29	26
0	100	93	87	81	75	70	65	61	57	53	50	44	38	34	30
+ 10	100	94	88	82	77	72	68	64	60	56	53	47	41	37	33
20	100	94	89	83	78	74	70	66	62	58	55	49	44	39	35
30	100	94	89	84	80	75	71	68	64	61	57	52	46	42	38

WET AND DRY BULB HYGROMETER

Apjohn (1835), August (1825), and others, by making various assumptions (some of doubtful legitimacy), have derived formulæ of the type—

$$[p]'_w - [p]_t = AH(t - t_w)\,[1 + B(t - t_w)]$$

where t is the temperature of the dry bulb, t_w that of the wet, $[p]_t$ is the actual pressure of water-vapour in the air (at temperature t), $[p]'_w$ is the saturated vapour pressure of water at the temperature (t_w) of the wet bulb, H is the barometric height, and A and B are constants. (See Love & Smeal, 1911.)

The indications of this hygrometer are so dependent on its environment that for most purposes B may be taken as zero, and H as constant, say 760 mms.

If H is measured in millimetres, and temperatures in Centigrade degrees, the following values of A are suitable for the conditions mentioned :—

A = ·00068 for moving air, as in a ventilated hygrometer.

A = ·00075 in a Stevenson screen as used by Meteorological Office.

A = ·0008 in open air with slight wind.

A = ·00084 in open air with no wind.

A = ·001 in a small closed room.

Rizzo (1897) takes A = ·00075 and B = − ·003, and the table below is derived by employing these values. $[p]'_w$ can be got from the table of saturated vapour pressures on p. 40, and thus the desired vapour pressure $[p]_t$ can be determined.

VALUES OF $[p]'_w - [p]_t$ (Rizzo)

Barom. Press. H.	Difference of temperature of dry and wet bulb thermometers $(t - t_w)$.									
	1° C.	2°	3°	4°	5°	6°	7°	8°	9°	10°
mm.	mm.	mm.	mm.	mm.	mm.	mm.	mm.	mm.	mm.	mm.
770	·57	1·13	1·69	2·23	2·78	3·30	3·81	4·32	4·87	5·31
760	·56	1·12	1·67	2·20	2·74	3·25	3·76	4·27	4·75	5·24
750	·55	1·11	1·65	2·17	2·71	3·21	3·71	4·21	4·69	5·17
730	·54	1·08	1·60	2·12	2·63	3·12	3·61	4·10	4·56	5·03
700	·52	1·03	1·54	2·03	2·52	3·00	3·46	3·93	4·37	4·82
670	·50	·99	1·47	1·94	2·42	2·87	3·32	3·76	4·19	4·62
	11° C.	12°	13°	14°	15°	16°	17°	18°	19°	20°
770	5·78	6·26	6·72	7·17	7·62	8·06	8·47	8·89	9·30	9·69
760	5·71	6·18	6·63	7·08	7·52	7·95	8·36	8·77	9·18	9·56
750	5·63	6·09	6·54	6·98	7·42	7·84	8·25	8·66	9·06	9·44
730	5·48	5·93	6·37	6·79	7·22	7·63	8·03	8·43	8·82	9·18
700	5·26	5·69	6·11	6·52	6·93	7·32	7·70	8·08	8·46	8·82
670	5·03	5·44	5·84	6·24	6·63	7·01	7·37	7·73	8·08	8·43

WET AND DRY·BULB HYGROMETER (contd.)
GLAISHER'S FACTORS

Mr. Glaisher, in 1841-5, took many thousands of observations with the wet and dry bulb hygrometer in Greenwich, India, and Toronto, and from simultaneous readings of a Daniell's hygrometer (now recognized as being an untrustworthy instrument) drew up a table of "factors."

The factor (f) at any dry-bulb reading is defined by

$$\text{depression of dew-point} = t - t_{dp} = f(t - t_w)$$

the notation being as above. Glaisher's factors are employed by the Meteorological Office and the Meteorological stations in this country. The hygrometer readings are taken in a Stevenson screen, which is essentially a box with double louvred sides.

The factors for a range of dry-bulb temperatures are tabulated below. The formula above yields the dew-point; and the saturated vapour pressure at the dew-point gives the actual vapour pressure at $t°$. For a table of saturated vapour pressures, see p. 40. (See "The Observers' Handbook," Meteorological Office.)

Dry Bulb Temp. (t).	0	1	2	3	4	5	6	7	8	9
− 10° C.	8·76	8·73	8·55	8·26	7·82	7·28	6·62	5·77	4·92	4·04
0	3·32	2·81	2·54	2·39	2·31	2·26	2·21	2·17	2·13	2·10
+ 10	2·06	2·02	1·99	1·95	1·92	1·89	1·87	1·85	1·83	1·81
20	1·79	1·77	1·75	1·74	1·72	1·70	1·69	1·68	1·67	1·66
30	1·65	1·64	1·63	1·62	1·61	1·60	1·59	1·58	1·57	1·56

CHEMICAL HYGROMETER

The values below are grams of water vapour contained in a cubic metre (10^6 c.cs.) of saturated air at 760 mms. total pressure. Calculated from Regnault's observations.

Temp.	0	1	2	3	4	5	6	7	8	9
0° C.	4·84	5·18	5·54	5·92	6·33	6·76	7·22	7·70	8·21	8·76
10	9·33	9·93	10·57	11·25	11·96	12·71	13·50	14·34	15·22	16·14
20	17·12	18·14	19·22	20·35	21·54	22·80	24·11	25·49	26·93	28·45
30	30·04	31·70	33·45	35·27	37·18	39·18	41·3	43·5	45·8	48·2

TENSILE STRENGTHS OF LIQUIDS

Liquids perfectly free from air can sustain considerable tension without rupture, e.g. water can withstand a tension of 5 atmospheres, alcohol 12, and strong sulphuric acid 12 atmospheres. Extensions of volume of 0·8 % for water, 1·1 % for alcohol, and 1·7 % for ether have been obtained. The volume elasticity (p. 29) of alcohol is the same for extension as for compression. (See Worthington, *Phil. Trans. A.*, 1892 ; Dixon, *Proc. Roy. Dub. Soc.*, 1909 ; Berthelot, *Ann. Chim. Phys.*, **30**, 1850 ; Poynting and Thomson's " Properties of Matter.")

BURSTING STRENGTHS OF GLASS TUBING

Bursting pressures in atmospheres for German soda glass tubing. Most glass-tubing is in a state of considerable strain, and a factor of safety of not less than two should usually be employed. (Roebuck, *Phys. Rev.*, 1909 ; and Onnes and Braak, *Kon. Ak. Wet.*, Amsterdam, 1908.) Ordinary boiler water-gauge glasses stand between 12 and 24 atmospheres.

Thickness of Wall.	Bore.						
	1 mm.	2	3	4	5	6	7
1 mm.	atmos. ——	310	280	230	220	150	190
2	570	——	340	——	330	240	220
3	560	420	460	400	——	——	230
4	——	450	——	400	310	320	280

VAPOUR PRESSURES

Inter- and Extrapolation of Vapour Pressures.—The Kirchhoff-Rankine-Dupré formula, $\log p = A + B/\theta + C \log \theta$, where p is the vapour pressure, θ the absolute temperature, and A, B, C are constants, is accurate and convenient (*e.g.* see p. 41). For values of A, B, C, see Juliusburger, *Ann. d. Phys.*, p. 618, 1900.

Ramsay and Young's Method.—If two liquids, one at absolute temperature θ and the other at θ', have the same vapour pressure, the ratio θ/θ', when plotted against θ, gives a straight line. This method may be used to find roughly the vap. press. of a substance at any temperature when only its boiling-point is known.

Interpolation by Logarithms.—The curve of vapour pressure (p) against temp. (t) is approximately hyperbolic, and thus $\log p$ plotted against t gives a graph of slight curvature, which over 10° intervals of t may, for approximate work, be regarded as a straight line : thus the following method of interpolation :—

Example.—Required vap. press. of water at 15°, given

t	p	$\log p$
10°	9·2	·964
20°	17·5	1·243

$$\frac{·964 + 1·243}{2} = 1·104 = \log 12·7 ; \text{ i.e. } p \text{ at } 15° = 12·7,$$

actually it is 12·8.

VAPOUR PRESSURE OF ICE

In mms. of mercury at 0° C. ; $g = 980·62$ cms. per sec.[2] ; hydrogen (const. vol.) scale of temps. (Scheel, and Heuse, Reichsanstalt *Ann. d. Phys.*, 1909.)

Temp. . .	−50° C.	−40°	−30°	−20°	−10°	−5°	−2°	0°
Vap. press.	·030 mm.	·096	·288	·784	1·963	3·022	3·885	4·579

(SATURATED) VAPOUR PRESSURE OF WATER

In mms. of mercury at 0° C. ; $g = 980·67$ cms. per sec.[2] Thermodynamic scale of temp. (see p. 44). From −20° to 0° the observations are due to Scheel and Heuse (*v.* ice); from 0° to 50°, to Thiesen and Scheel ; from 50° to 200°, to Holborn and Henning, Reichsanstalt (*Ann. d. Phys.*, 26, 833. 1908). For vapour pressures at temps. near 100° see also the table of boiling-points on next page.

Vap. press. at −20° C., ·960 mm.; −10°, 2·160 ; −5°, 3·171 ; −2°, 3·958 ; −1°, 4·258.

Temp.	0	1	2	3	4	5	6	7	8	9
0° C.	4·579	4·924	5·290	5·681	6·097	6·541	7·011	7·511	8·042	8·606
10	9·205	9·840	10·513	11·226	11·980	12·779	13·624	14·517	15·460	16·456
20	17·51	18·62	19·79	21·02	22·32	23·69	25·13	26·65	28·25	29·94
30	31·71	33·57	35·53	37·59	39·75	42·02	44·40	46·90	49·51	52·26

	0	2	4	6	8	10	12	14	16	18
40	55·13	61·30	68·05	75·43	83·50	92·30	101·9	112·3	123·6	135·9
60	149·2	163·6	179·1	195·9	214·0	233·5	254·5	277·1	301·3	327·2
80	355·1	384·9	416·7	450·8	487·1	525·8	567·1	611·0	657·7	707·3
100	760·0	815·9	875·1	937·9	1004	1074·5	1149	1227	1310	1397
120	1489	1586	1687	1795	1907	2026	2150	2280	2416	2560
140	2709	2866	3030	3202	3381	3569	3764	3968	4181	4402
160	4633	4874	5124	5384	5655	5937	6229	6533	6848	7175
180	7514	7866	8230	8608	8999	9404	9823	10256	10705	11168
200	11647	12142	12653	—	—	—	—	—	—	—

(Battelli, 1892.)

Temp. . .	220° C.	240°	260°	280°	300°	320°	340°	360°
Vap. Press.	17,380 mm.	25,170	35,760	50,600	67,620	88,340	113,830	141,870

Interpolate logs of vapour pressures as explained above.

BOILING-POINT OF WATER UNDER VARIOUS BAROMETRIC PRESSURES

Hydrogen scale of temps. Pressures in mms. of mercury at $0°$ C.; $g = 980·62$ cms. per sec.[2] (Regnault's measurements; reduced by Broch, 1881; recalculated by Wiebe, 1893.)

Barometric Height.	0	1	2	3	4	5	6	7	8	9
	°C.									
680 mm.	96·91	96·95	97·00	97·03	97·07	97·11	97·15	97·20	97·24	97·28
690	97·32	·36	·40	·44	·48	·52	·56	·59	·63	·67
700	97·71	·75	·79	·83	·87	·91	·95	·99	98·03	98·07
710	98·11	98·14	98·18	98·22	98·26	98·30	98·34	98·38	·42	·45
720	98·49	·53	·57	·61	·65	·69	·72	·76	·80	·84
730	98·88	·91	·95	·99	99·03	99·07	99·10	99·14	99·18	99·22
740	99·25	99·29	99·33	99·37	·41	·44	·48	·52	·56	·59
750	99·63	·67	·70	·74	·78	·81	·85	·89	·93	·96
760	100·00	100·03	100·07	100·11	100·15	100·18	100·22	100·26	100·29	100·33
770	100·37	·40	·44	·47	·51	·55	·58	·62	·66	·69
780	100·73	·76	·80	·84	·87	·91	·94	·98	101·01	101·05

VAPOUR PRESSURE OF MERCURY

In mms. of mercury at $0°$ C. Reduced from the observations of Hertz, Ramsay and Young, Callendar and Griffiths, Pfaundler, Morley, Gebhardt, Cailletet, Colardeau, Rivière. For interpolation from $15°$ to $270°$.

$$\log p = 15·24431 - 3623·932/\theta - 2·367233 \log \theta \quad . \quad . \quad . \quad . \quad . \quad (A)$$

From $270°$ to $450°$

$$\log p = 10·04087 - 3271·245/\theta - ·7020537 \log \theta$$

$\frac{\delta p}{\delta t}$ at the boiling-point $= 13·6$ mm. per degree (Laby, *Phil. Mag.*, Nov., 1908).

Temp.	Vap. Press.	Temp.	Vap. Press.	Temp.	Vap. Press.	Temp.	Vap. Press.	Temp.	Vap. Press.
	mm.		mm.		mm.		mm.		atmos.
0° C.	·00016*	25°	·00168	60°	·0246	250°	75·83	500°	8
5	·00026*	30	·00257	80	·0885	300	248·6	600	22·3
10	·00043*	35	·00387	100	·276	356·7	760	700	50
15	·00069	40	·00574	150	2·88	400	1566	800	102
20	·00109	50	·0122	200	17·81	450	3229	880	162

* Extrapolated by formula A.

VAPOUR PRESSURE OF ETHYL ALCOHOL

Vap. press. in mms. of mercury at $0°$ C. Calculated by Bunsen from Regnault's results (1862), which are in good agreement with the mean of those of Ramsay and Young (1886), and Schmidt (1891).

Regnault, Vapour press. at $-20°$, 3·34 mm.; at $-10°$, 6·47 mm.

Temp.	0	1	2	3	4	5	6	7	8	9
0° C.	12·73	13·65	14·6	15·59	16·62	17·7	18·84	20·04	21·31	22·66
10	24·08	25·59	27·19	28·9	30·7	32·6	34·6	36·8	39·0	41·4
20	44·0	46·7	49·5	52·5	55·7	59·0	62·5	66·2	70·1	74·1
30	78·4									

(Ramsay and Young, 1886.)

Temp.	30° C.	40°	50°	60°	70°	80°	100°	120°	140°	160°
Press.	78·1 mm.	133·4	219·8	350·2	541	812	1692	3220	5670	9370

Interpolate logs of vapour pressures as explained on p. 40.

VAPOUR PRESSURES

VAPOUR PRESSURES OF ELEMENTS

p = vapour pressure in mms. of mercury at $0°$ C. lat. $45°$ and sea-level ($g = 980.62$) (*i.e.* 1 mm. Hg = 1333·2 dynes per sq. cm). If followed by *at.*, p is in atmospheres; θ = absolute temp. (A.); t = temp. in $°$ C.; (s) solid; (l) liquid. The thermometry is in many cases somewhat dubious.

Interpolate logs of vapour pressures as explained on p. 40.

Argon (Olszewski, 1895)	t	-121° C.	-128·6	-129·6	-134·4	-135·1	-136·2	-138·3	-139·1	—
	p	50·6 at.	38·0	35·8	29·8	29·0	27·3	25·3	23·7	—
Argon (Ramsay & Travers)	θ	78°·9 A.	86·9	97·9	107·3	155·6	= crit. temp.		—	—
Krypton	θ	110°·5 A.	121·3	135·2	147·3	—	210·5	= crit. temp.	—	—
Xenon	θ	148°·9 A.	163·9	182·9	199·6	—	—	287·8	= crit. temp.	—
	p	300 mm.	760	2000	4000	40,200	41,240	43,500	—	—
Bromine (Ramsay & Young, 1886)	t	-16°·6 C.	-12·0	-5·0	8·2	16·9	23·4	40·5	51·9	58·7
	p	20 mm.	30	50	100	150	200	400	600	760
Chlorine (Knietsch, 1890)	t	-80° C.	-60°	-40	-33·6	-20	0	10	20	30
	p	62·5 mm.	210	560	760	1·84 at.	3·66	4·95	6·62	8·75
Iodine (Baxter, Hickey, & Holmes, 1907)	t	0° C.	15	30	55	85	117	137	160·9	185·3
	p	·03 mm.	·131	·469	3·08	20	100	200	400	760
Hydrogen (Travers & Jaquerod, 1902)	t	-258°·2C.	-256·7	-255·7	-255·0	-254·3	-253·7	-253·2	-252·9	H. Scale
	p	100 mm.	200	300	400	500	600	700·|	760	
Helium (Onnes, 1911)	θ	1°·2 A.	4·3	—	—	Neon (Travers & Jaquerod, '02)	θ	15°·65 A.(s)	20·4 (s)	He
	p	0·2 mm.	760	—	—		p	2·4 mm.	12·8	Scale
Mercury		See p. 41.				Ra. Emanation		See p. 103.		
Nitrogen (Baly, 1900; Fischer & Alt., 1902)	θ	62°·5 A.	67·8	72·4	77·3	80	83	86	89	91
	p	86 mm.	200	400	760	1013	1386	1880	2465	2916
Oxygen (Jaquerod, Travers, & Senter, 1902)	θ	79°·1 A.	82·1	84·4	86·3	87·9	89·3	90·1	90·6	H. Scale
	p	200 mm.	300	400	500	600	700	760	800	—
Phosphorus (Schrötter, 1848)	t	165° C.	170	180	200	209	219	226	230	287·3
	p	120 mm.	173	204	266	339	359	393	514	760
Sulphur (Ruff & Graff, '08; B., 1899; C., 1899)	t	50° C.	100	147	211	400	444·5	$\delta t/\delta p$ = 0°·09/mm. near		
	p	·0003 mm.	·0089	·192	3·14	c. 372	760	B.P. (see p. 50).		

VAPOUR PRESSURES OF COMPOUNDS

For a complete list, see Schenck in L.B.M.

Hydrochloric acid (F., 1845; Ansdell, 1880)	t	-73°·3 C.	-45·5	-23·3	-3·9	4·0	9·2	13·8	22·0	33·4
	p	1·8 at.	6·3	12·8	23·1	29·8	33·9	37·7	45·7	58·8
Sulphuretted hydrogen (R., 1862)	t	-25° C.	-15	-5	0	10	30	50	60	70
	p	4·93 at.	6·84	9·3	10·8	14·3	23·7	36·6	44·4	53·1
Sulphur dioxide (Regnault, 1862)	t	-30° C.	-20	-10	0	10	20	30	40	50
	p	·39 at.	·63	1·00	1·53	2·26	3·24	4·52	6·15	8·19
Ammonia, NH_3 (Brill, 1906)	t	-80° C.	-77·6	-70·4	-64·4	-60·8	-54·4	-46·2	-39·8	-33·0
	p	35·2 mm.	44·1	74·9	116·0	157·6	239·5	403·5	568·2	761
Nitrous oxide, N_2O (Cailletet, '78; R., '62)	t	-80° C.	-60	-40	-20	-10	0	10	20	40
	p	1·9 at.	5·05	11·0	23·1	28·9	36·1	44·8	55·3	83·4
Nitric oxide, NO (Olszewski, 1885)	t	-176·5°C.	-167	-138	-129	-119	-110	-105	-100·9	-97·5
	p	·024 at.	·182	5·4	10·6	20·0	31·6	41·0	49·9	57·8
Nickel carbonyl, $NiCO_4$ (D. & Jones, 1903)	t	-9° C.	-7	-2	0	10	16	20	30	—
	p	94·3 mm.	104·3	129·1	144·5	215·0	283·5	329·5	462	—

Interpolate logs of vapour pressures as explained on p. 40.

VAPOUR PRESSURES OF COMPOUNDS (contd.)

Interpolate logs of vapour pressures as explained on p. 40.

Carbon dioxide (Zeleny & Smith, 1906)	t	-130°C.(s)	-100(s)	-80(s)	-65(s)	-56.4‡	-65(l)	40(l)	-20(l)	-10(l)
	p	2.5 mm.	119	657	2100	3910	2508	7510	14,830	19,630
Carbon bisulphide (Regnault, 1862)	t	-20°C.	-10	0	10	20	40	60	80	100
	p	47.3 mm.	79.4	128	198	298	618	1164	2033	3325
Chloroform, CHCl₃ (Regnault, 1862)	t	20°C.	30	40	50	60	70	80	90	100
	p	160.5 mm.	248	369	535	755	1042	1408	1865	2429
Carbon tetrachloride, CCl₄ (R., 1862)	t	-20°C.	-10	0	10	20	40	60	80	100
	p	9.8 mm.	18.47	32.9	56	91	215	447	843	1467
Acetylene, C₂H₂ (Villard, 1895)	t	-90°C.(s)	-85(s)	-81	-70	-50	-23.8	0	20.2	36.5
	p	.69 at.	1.00	1.25	2.22	5.3	13.2	26.05	42.8	61.6 (M.)
Benzene, C₆H₆ (Young, 1889)	t	-10°C.	0	10	20	40	60	80	100	120
	p	14.8 mm.	26.5	45.4	74.6	181.1	389	754	1344	2238
Aniline, C₆H₅NH₂ (Kahlbaum, 1898)	t	101°.9 C.	119.4	138.7	151.5	161.1	168.7	175.0	180.8	183.9
	p	50 mm.	100	200	300	400	500	600	700	760
Bromnaphthalene C₁₀H₇Br (Ra. & Y., 1885)	t	215°C.	220	230	240	250	260	270	275	280.4
	p	158.9 mm.	181.8	236.0	303.4	386.4	487.4	608.8	677.9	760
Me. alcohol, CH₃OH (R.,'62; Ra. & Y.; Ri.,'86)	t	-10°C.	0	17	20	30	50	80	120	150
	p	14.8 mm.	28.5	78.3	88.7	150	381.7	1238	4312	9361
n. propyl alcohol, C₃H₇OH (Ra. & Y.; S.; Ri.,'86)	t	0°C.	10	17	30	40	60	80	100	120
	p	3.9 mm.	7.8	12.4	28.2	51.4	157	389	843	1668
Iso-butyl alcohol C₄H₉OH (Ri.,'86; S.,'91)	t	10°C.	17	20	40	60	80	100	108	120
	p	4.1 mm.	6.8	8.1	30.3	91.2	245	569	760	1195
Iso-amyl alcohol C₅H₁₁OH (Ri.,'86; S.,'91)	t	17°C.	30	40	50	60	80	100	120	130
	p	1.78 mm.	4.68	9.33	17.4	32.0	151	234	522	741
Formic acid, CH₂O₂ (S., 1891; K., 1898)	t	0°C.	10	17	20	30	40	70	80	101
	p	10.2 mm.	18.4	26.3	31.6	51.3	79.4	266	373	760
Acetic acid, C₂H₄O₂ (Ra. & Y.; Ri.,'86; S.,'91)	t	17°C.	30	50	70	90	110	130	150	200
	p	9.8 mm.	20.6	56.2	133	288	582	1058	1847	5905
Propionic acid, C₃H₆O₂ (Ri.,'86; S.,'91; K.,'98)	t	15°C.	17	20	30	40	60	70	80	140
	p	1.7 mm.	2.0	2.45	4.9	9.1	28.2	46.1	74.5	760
Butyric acid, C₄H₈O₂ (Ra.&Y.,'86; S.'91; K.'94)	t	17°C.	20	30	50	70	90	110	130	150
	p	.52 mm.*	.66*	1.4	5.2	16.2	44.9	111	245	497
Iso-butyric acid, C₄H₈O₂ (Ri.,'86; S.,'91; K.,'94)	t	17°C.	30	50	70	90	110	130	150	153.5
	p	.88 mm.*	1.9	8.2	25.1	67.6	162	347	684	760
Methyl formate CHO₂CH₃ (Y. & T.,'93)	t	-20°C.	-10	0	10	20	40	60	80	100
	p	67.7 mm.	117.6	195	309	476	1029	1990	3497	5782
Methyl butyrate C₄H₇O₂.CH₃ (Y. & T.,'93)	t	-10°C.	0	10	20	40	60	80	100	
	p	3.55 mm.	7.3	13.8	24.5	69.2	167.5	361	701	—
Methyl isobutyrate C₄H₇O₂.CH₃ (Y. & T.,'93)	t	-10°C.	0	10	20	40	60	80	100	120
	p	6.22 mm.	12.15	22.4	38.9	104.7	244	505	956	1660
Ethyl acetate C₂H₃O₂.C₂H₅ (Y. & T.,'93)	t	-20°C.	-10	0	10	20	40	60	80	100
	p	6.5 mm.	12.9	24.3	42.7	72.8	186	415	833	1515
Ethyl propionate C₂H₅O₂.C₂H₅ (Y. & T.,'93)	t	-10°C.	0	10	20	40	60	80	100	120
	p	4.05 mm.	8.3	15.5	27.7	77.9	188.0	403.6	785	1388
Propyl acetate C₂H₃O₂.C₃H₇ (Y. & T.,'93)	t	-10°C.	0	10	20	40	60	80	100	120
	p	3.6 mm.	7.4	15.9	25.1	70.8	172	373	724	1288
Ethyl ether, (C₂H₅)₂O (Young, 1910)	t	-10°C.	0	10	20	40	60	80	100	193.8‖
	p	112.3	184.9	290.8	439.8	921	1734	2974	4855	27,060

Interpolate logs of vapour pressure as explained on p. 40.

* Extrapolated.

† The vapour pressures here given have been graphically interpolated from the observers' values. B., Bodenstein ; C., Callendar ; D., Dewar ; F., Faraday ; K., Kahlbaum ; M., Mackintosh ; R., Regnault ; Ra. and Y., Ramsay and Young ; Ri., Richardson ; S., Schmidt ; Y. and T., Young and Thomas.

‡ Triple point.   ‖ Critical temp.

GAS THERMOMETRY

The standard thermometric scale of the International Committee of Weights and Measures (1887) is that of the constant-volume hydrogen thermometer, the hydrogen being taken at an initial pressure at 0° C. of 1000 mms. of mercury measured at 0° C. sea-level and lat. 45° (= 1·3158 standard atmosphere).

THERMODYNAMIC TEMPERATURE OF THE ICE-POINT

Method.	H_2	N_2	Air.	CO_2	Computer.
From Joule-Thomson effect	273·14°	273·09°	——	273·05°	Callendar, 1903
Extrapolation to zero pressure (see p. 54)	273·07	273·09	——	——	Berthelot and Chappuis, 1907
From Joule-Thomson effect	273·05	(273·17)	273·19°	273·10	Berthelot, 1907
,, ,, ,,	273·06	273·25	273·27	273·12	Buckingham,1908
,, ,, ,,	273·13	273·14	——	——	Rose-Innes, 1908

General mean = 273°·13.

THERMODYNAMIC CORRECTIONS TO GAS SCALES OF TEMPERATURE

The corrections to both the constant-pressure (C.P.) and the constant-volume (C.V.) scales are either (1) derived from characteristic equations of state (Callendar, 1903 ; Berthelot, 1907), or (2) in the case of the C.P. thermometer, computed from the Joule-Thomson effect ; whence from these C.P. corrections and a knowledge of the compressibility of the gas under different conditions the C.V. corrections can be calculated. Chappuis (1907)* has experimentally compared the C.P. and C.V. H. and N. thermometers each with mercury thermometers. The values below are based on computations by Callendar (*Phil. Mag.*, 1903), Berthelot * (from Chappuis' data 1907), Onnes and Braak (1907 and 1908), Rose-Innes (*Phil. Mag.*, 1908), and Buckingham (1908).† There is some divergence among the different computations for hydrogen ; the agreement is much better in the case of nitrogen. The thermodynamic correction to the C.V.H. thermometer is negligible, and with nitrogen also at extreme temps. the correction is less than the error of working in modern gas thermometry. The values for air are a little smaller than for nitrogen ; for helium they are slightly larger than for hydrogen except at the lowest temperatures, when the helium corrections are the smaller. New experiments on the Joule-Thomson effect are needed. ‡ (+) means that the correction has to be added to the gas scale temperature to give the thermodynamic temperature. The correction is proportional to the initial pressure of the gas in the thermometer.

* *Trav. et Mém. Bureau Intl.* 1907. † *Bull. Bureau of Standards*, 1908.
‡ See Dalton, *Proc. Konink. Akad. Weten. Amsterdam*, April, 1909.

$t°$ C.	Const. Pressure P = 1000 mm.		Const. Volume P at 0°=1000 mm.		$t°$ C.	Const. Pressure P = 1000 mm.		Const. Volume P at 0°=1000 mm.	
	H_2	N_2	H_2	N_2		H_2	N_2	H_2	N_2
−240°	+1°·2 (?)	—	+°·18	—	70°	−°·003	−°·019	−°·001	−°·004
−200	+ ·26	—	+ ·06	—	80	− ·002	− ·014	− ·000	− ·003
−150	+ ·10	+1°·3	+ ·033	+°·26 (?)	90	− ·001	− ·007	− ·000	− ·002
−100	+ ·04	+ ·40	+ ·010	+ ·10 (?)	100	0	0	0	0
− 50	+ ·02	+ ·12	+ ·005	+ ·03	200	+ ·014	+ ·12	+ ·004	+ ·04
0	0	0	0	0	300	+ ·034	+ ·28	+ ·011	+ ·10
10	− ·001	− ·009	− ·000	− ·002	400	+ ·07 (?)	+ ·46	+ ·018 (?)	+ ·17
20	− ·002	− ·017	− ·000	− ·004	450	+ ·09 (?)	+ ·56	+ ·02 (?)	+ ·19
30	− ·003	− ·021	− ·001	− ·005	600	—	+ ·87	—	+ ·3
40	− ·003	− ·023	− ·001	− ·006	800	—	+1·3	—	+ ·5
50	− ·003	− ·024	− ·001	− ·007	1000	—	+1·8	—	+ ·7
60	− ·003	− ·022	− ·001	− ·006	1200	—	+2·3	—	+1·0

MERCURY THERMOMETRY
CORRECTIONS TO REDUCE MERCURY-IN-GLASS SCALE TEMPS. TO GAS SCALE TEMPS.

The values for the English Kew glass (which is a lead potash silicate) are due to Harker (1906); the verre dur corrections are given by the International Bureau; those for the Jena glasses by Grützmacher. The method at Kew is to determine the ice-point correction before an observation is made. The other glasses have their ice-point or zero depressions determined immediately after each temperature reading. See Guillaume's "Thermométrie de Précision." Paris, 1889, and Chree's "Notes on Thermometry," *Phil. Mag.*, 1898. The French glass, verre dur, is used by Tonnelot of Paris. The normal glass, Jena 16''', may be known by the presence of a thin violet line near the surface. Jena 59''' is a borosilicate (p. 74).

Temp.	Kew Glass. $l_H - l_{K.G.}$	Verre Dur. $l_H - l_{V.D.}$	Jena 16''. $l_H - l_{16'''}$	Jena 59''. $l_H - l_{59'''}$	Temp.	Verre Dur. $l_N - l_{V.D.}$	Jena 16''. $l_N - l_{16''}$	Jena 59''. $l_N - l_{59'''}$
−20°	—	+°17	+°19	+°10	110°	+°04	+°03	— °00
0	0°	0	0	0	120	+ ·06	+ ·05	— ·02
10	— ·00	— ·05	— ·06	— ·02	130	+ ·07	+ ·07	— ·04
20	— ·00	— ·08	— ·09	— ·04	140	+ ·07	+ ·09	— ·08
30	+ ·005	— ·10	— ·11	— ·04	150	+ ·06	+ ·10	— ·13
40	+ ·01	— ·11	— ·12	— ·04	160	+ ·03	+ ·10	— ·19
50	+ ·01	— ·10	— ·11	— ·03	170	0	+ ·08	— ·28
60	+ ·01	— ·09	— ·10	— ·02	180	— ·04	+ ·06	— ·39
70	+ ·015	— ·07	— ·08	— ·01	190	— ·09	+ ·02	— ·52
80	+ ·02	— ·05	— ·06	— ·00	200	— ·13	— ·04	— ·67
90	+ ·025	— ·03	— ·03	— ·00	250	—	— ·63	— 1·7
100	0	0	0	0	300	—	— 1·91	— 4·1

DEPRESSION OF ZERO OF MERCURY THERMOMETERS

The values indicate the zero depressions after the thermometer has been heated to the temp. stated. They have been determined by Guillaume, Thiesen, Schloesser, and Böttcher because of the impossibility in practice of interrupting a series of temperature measurements to take a number of zero readings (see above).

Temp.	Verre Dur.	Jena 16'''.	Jena 59'''.	Temp.	Verre Dur.	Jena 16'''.	Jena 59'''.
10° C.	°008	°005	°005	60° C.	°060	°039	°024
20	·017	·011	·009	70	·071	·048	·027
30	·027	·017	·014	80	·084	·057	·030
40	·037	·024	·017	90	·097	·066	·033
50	·048	·031	·021	100	·111	·077	·035

STEM-EXPOSURE OR EMERGENT-COLUMN CORRECTION

The table below gives the (additive) "stem-exposure" correction for (1) the ordinary solid-stem thermometer, and (2) the German pattern sleeve-thermometer, which has a fine capillary in an outer glass tube. Both thermometers are of Jena 16''' glass, with degree intervals about 1 mm. long.

t is the indicated temperature, and t_{aux} the temperature of an auxiliary thermometer whose bulb is 10 cms. from and on a level with the mid-point of the exposed stem. The auxiliary thermometer must be shielded from the source of heat. (See Watson's "Practical Physics," and Rimbach, *Zeit. f. Inst.*, 10, 1890.)

No. of degree divs. of exposed thread	Solid Stem; Scale on Stem. $t - t_{aux}$						Sleeve Thermometer; Enclosed Scale.						No. of degree divs. of exposed thread
	70° C.	80°	100°	120°	140°	180°	70° C.	80°	100°	120°	140°	180°	
10	°02	°03	°07	·11	·17	·27	°01	°01	°04	°07	°10	°17	10
20	·13	·15	·22	·29	·38	·53	·08	·12	·19	·25	·28	·40	20
30	·24	·28	·39	·48	·59	·78	·25	·28	·36	·42	·48	·66	30
40	·35	·41	·56	·68	·82	1·04	·30	·35	·48	·60	·67	·92	40
60	·57	·66	·89	1·09	1·25	1·58	·52	·60	·79	·99	1·11	1·46	60
80	·80	·91	1·21	1·52	1·71	2·15	·75	·87	1·15	1·38	1·53	1·98	80
100	1·02	1·18	1·56	1·97	2·18	2·70	·98	1·12	1·47	1·82	2·03	2·55	100
120	—	—	1·98	2·43	2·69	3·26	—	—	1·88	2·28	2·49	3·13	120

ELECTRICAL THERMOMETRY

PLATINUM THERMOMETRY

TO REDUCE PT-SCALE TEMPS. (t_{pt}) TO CONST. VOL. N-SCALE TEMPS. (t)

Callendar's "difference formula" for the difference between the nitrogen-scale temp. (t) and the Pt-scale temp. (t_{pt}) is $t - t_{pt} = \delta \cdot t(t-100)10^{-4}$, where δ is close to $1\cdot5$. Pt-scale temps. result from assuming a linear relation $R_{pt} = R_0(1 + \alpha t_{pt})$ between temp. and the electrical resistance (R) of Pt ; α is the mean coefficient for the range $0°$ to $100°$. The "difference formula" gives the correction yielded by the truer parabolic relation $R_t = R_0(1 + \alpha t + \beta t^2)$. Pt thermometers should not be used above $1200°$ C. (See Callendar, *Phil. Mag.*, 1899. **1**, p. 191 ; **2**, p. 519. Camb. Sci. Inst. Co.'s list "Technical Thermometry;" and (for bibliography), Waidner and Burgess, *Bull. Bur. of Standards*, 1909.)

$$\delta = 1\cdot50.$$

(Harker, *Phil. Trans.*, 1904.)

Pt Temps. t_{pt}.	0	20	40	60	80	100	120	140	160	180
	t	t	t	t	t	t	t	t	t	t
$-200°$	—	$-172°\cdot9$	$-154°\cdot1$	$-135°\cdot2$	$-116°\cdot2$	$-97°\cdot13$	$-77°\cdot92$	$-58°\cdot61$	$-39°\cdot18$	$-19°\cdot65$
0	$0°$	$19\cdot76$	$39\cdot64$	$59\cdot64$	$79\cdot76$	100	$120\cdot4$	$140\cdot9$	$161\cdot5$	$182\cdot3$
$+200$	$203\cdot1$	$224\cdot2$	$245\cdot4$	$266\cdot7$	$288\cdot1$	$309\cdot8$	$331\cdot5$	$353\cdot4$	$375\cdot5$	$397\cdot8$
400	$420\cdot2$	$442\cdot8$	$465\cdot5$	$488\cdot5$	$511\cdot6$	$534\cdot9$	$558\cdot4$	$582\cdot1$	$606\cdot0$	$630\cdot1$
600	$654\cdot4$	$679\cdot0$	$703\cdot7$	$728\cdot7$	$754\cdot0$	$779\cdot4$	$805\cdot2$	$831\cdot2$	$857\cdot4$	$884\cdot0$
800	$910\cdot8$	$937\cdot9$	$965\cdot3$	$993\cdot0$	1021	1050	1078	1107	1137	1167
1000	1197	1228	1259	1290	1323	1355				

TO CALCULATE THE CHANGE Δt IN THE N-SCALE TEMP. (t) FOR A CHANGE OF $+\cdot01$ IN δ

t	Δt	t	Δt	t	Δt	t	Δt	t	Δt	t	Δt
$-200°$	$°\cdot060$	$-60°$	$°\cdot010$	$80°$	$-°\cdot002$	$250°$	$°\cdot038$	$600°$	$°\cdot30$	$950°$	$0\cdot8$
-180	$\cdot050$	-40	$\cdot006$	100	0	300	$\cdot060$	650	$\cdot36$	1000	$\cdot9$
-160	$\cdot042$	-20	$\cdot002$	120	$\cdot002$	350	$\cdot088$	700	$\cdot42$	1050	$1\cdot0$
-140	$\cdot034$	0	0	140	$\cdot006$	400	$\cdot120$	750	$\cdot49$	1100	$1\cdot1$
-120	$\cdot026$	20	$-\cdot002$	160	$\cdot010$	450	$\cdot158$	800	$\cdot56$	1150	$1\cdot2$
-100	$\cdot020$	40	$-\cdot002$	180	$\cdot014$	500	$\cdot20$	850	$\cdot64$	1200	$1\cdot3$
-80	$\cdot014$	60	$-\cdot002$	200	$\cdot020$	550	$\cdot25$	900	$\cdot72$	1250	$1\cdot4$

HIGH TEMPERATURES

(See Burgess and Le Chatelier's "High Temperature Measurements, 1912.")

For the measurement of high temperatures (say above $1200°$ C., which is about the present upper experimental limit of the gas scale) the instruments in general use are thermo-junctions and optical or radiation pyrometers. Both involve extrapolation. Thermo-couples have been used up to the temperature of the melting-point of platinum (*c.* $1750°$). At high temperatures thermo-junctions yield rather lower results than do optical pyrometers, *e.g.* see the M.P.'s of Pd and Pt on p. 49.

THERMO-ELECTRIC THERMOMETRY

Temperature readings with thermo-couples are reduced by one of the formulæ : (*a*) $E = a + bt + ct^2$, (*b*) $E = mt^n$, or $\log E = n \log t + m'$, E being the e.m.f. generated, and t the temperature of the hot junction, the cold junction being at $0°$. Up to about $1200°$ these formulæ with suitable constants agree to within $2°$ for the usual 10% (Pt, Pt – Rh) and (Pt, Pt – Ir) couples, but above $1200°$ formula (*b*) yields the higher results, *e.g.* see the melting-points of Pd and Pt on p. 49. The thermo-e.m.f.'s of these Pt couples gradually diminish with prolonged heating. The values of the constants below are only average values.

E IN MICRO VOLTS (10^{-6} VOLT)

	Couple.	a	b	c	n	m'
Cold junc- tion at $0°$ C.	Pt and (90 Pt, 10 Rh) .	-307*	$8\cdot1$*	$\cdot0017$*	$1\cdot19$	$\cdot52$
	Pt and (90 Pt, 10 Ir) .	-550*	$14\cdot8$*	$\cdot0016$*	$1\cdot10$	$\cdot89$
	Cu and Constantan † .	—	—	—	$1\cdot14$	$1\cdot34$
	Cu and Fe	0	$10\cdot34$	$-\cdot0183$		

* These constants are not suitable for temperatures below $300°$. † Eureka, 60 Cu, 40 Ni.

THERMO-ELECTRIC THERMOMETRY (contd.)

The following are the readings in 10^{-5} volt determined at the National Physical Laboratory for a Pt-Rh and a Pt-Ir couple, each having the cold junction at $0°$ C. The values only hold for the particular couples.

Couple.	Temp.	0	50	100	150	200	250	300	350	400	450
Pt and (90 Pt, 10 Rh)	0° C.	0	23	51	83	119	158	199	242	286	331
	500	377	423	470	518	567	617	668	720	773	826
	1000	880	935	991	1048	1106	1165	1225	1286	1348	—
Pt and (90 Pt, 10 Ir)	0	0	58	125	195	268	343	420	498	577	657
	500	737	818	899	981	1064	1147	1231	1315	1400	1485
	1000	1571	1657	1744	1831	1919	2007	2096	2185	2275	—

THERMO-E.M.F.'S AGAINST PLATINUM IN MICRO VOLTS (10^{-6} VOLT)

One junction at $0°$ C. The current flows across the other junction from the metal with the (algebraically) smaller value to the other metal. (See Watson's "Physics" and Henning in L.B.M.)

Metal.	$-190°$	$+100°$	Metal.	$-190°$	$+100°$	Metal.	$-190°$	$+100°$
Aluminium	+ 390	+ 380	Lead	+ 210	+ 410	Tantalum	—	+330
Antimony	—	+4700	Magne-			Tin	+200	+ 410
Bismuth	+12300	−6500	sium	+ 330	+ 410	Zinc	− 120	+ 750
Cadmium	− 60	+ 900	Mercury	—	0	Brass	—	c. + 400
Cobalt	—	−1520	Nickel	+2220	−1640	Constantan*	—	− 3140
Copper	− 200	+ 740	Palla-			German sil-		
Gold	− 120	+ 730	dium	+ 790	− 560	ver†	—	c. − 1000
Iron	− 2900	c. +1600	Silver	− 140	+ 710	Manganin‡	—	+ 570

* Eureka, 60 Cu, 40 Ni.   † 60 Cu, 15 Ni, 25 Zn.   ‡ 84 Cu, 4 Ni, 12 Mn.

RADIATION AND OPTICAL THERMOMETRY

Most radiation thermometers depend upon either (1) the Stefan-Boltzmann law, $E = K(\theta^4 - \theta_0^4)$, where E is the total energy (all wave-lengths) radiated per sec. by a black body at absolute temp. θ to surroundings at absolute temp. θ_0, and K is a const. ($K = 5\cdot7 \times 10^{-12}$ watts per cm.^{2} per $1°$—see p. 65); or (2) Wien's equation connecting the temperature with the intensity of some particular wave-length of light

emitted (p. 65). The Wien equation is, Intensity $I = c_1\lambda^{-5}e^{-\frac{c_2}{\lambda T}}$, where λ is the wave-length, T is the "black body" temp. on the absolute scale, c_1 and c_2 are constants, and e is the base of the Napierian logarithms. Both equations give results which agree very accurately with the gas scale over the calibrated range $0°$ to $1200°$ C. Up to about $1500°$ radiation thermometers are, in practice, almost always graduated empirically, usually against a thermo-couple.

The "black body" temperature of a radiating substance is the temperature at which an ideal black body would emit radiation of the same intensity as that from the substance, the radiation considered being of some particular wave-length. A perfectly black body absorbs all the radiation which falls upon it ; it is destitute of reflecting power. Coal, carbon, metals which when heated tarnish with a black oxide, enclosed furnaces and muffles at a uniform temperature, all conform very nearly to this definition. When a pyrometer is sighted upon a body which is not "black," the temperature recorded—the "black body" temperature—will be lower than the true temperature to an extent which increases with the reflecting power of the body, e.g. if platinum and carbon have equal "black body" temperatures, their actual temperatures may differ by $180°$ or so at $1500°$.

TEMPERATURE AND COLOUR OF FIRE

Appearance	Red—just visible.	Dull Red.	Cherry Red.	Orange.	White.	Dazzling White.
Temperature	c. 500° C.	c. 700°	c. 900°	c. 1100°	c. 1300°	c. 1500°

For **standard temperatures** for thermometer calibration, see p. 50.

MELTING AND BOILING POINTS OF THE ELEMENTS

For an account of temperature measurements, see p. 46. For melting and boiling points of chemical compounds, see p. 109 ; of fats and waxes, see p. 50.

Element.	Melting Point.	Observer.	Boiling Point at 760 mms.	Observer.
Aluminium . .	657° C.	Holborn and Day, 1900	1800° C.	Greenwood, 1909
Antimony . .	630	„ „	1440	Greenwood, 1909
Argon . . .	− 188	Ramsay and Travers, 1901	− 186	——
Arsenic . . .	volatilizes	——	(sublimes) 450	——
Barium . . .	850	Guntz, 1903	——	——
Beryllium . .	c. 1430	Just and Mayer, 1909	——	——
Bismuth . . .	269	Callendar, 1899	1420	Greenwood, 1909
Boron . . .	2000 to 2500	Weintraub, 1909	(sublimes) 3500 (?)	——
Bromine . .	− 7·3	van der Plaats, 1886	63	van der Plaats, 1886
Cadmium . .	321	Holborn and Day, 1900	778	D. Berthelot, 1902
Cæsium . . .	26·4	Eckardt and Graefe, 1900	670	Ruff & Johannsen, 1906
Calcium . . .	780	Ruff and Plato, 1903	——	——
Carbon . . .	4000 (?)	(Calculated) McCrae, 1906	——	——
Cerium . . .	623	Muthmann & Weiss, 1904	——	——
Chlorine . . .	− 102	Olszewski	− 33·6	Regnault, 1863
Chromium . .	1520	Bureau of Standards	2200	Greenwood, 1909
Cobalt . . .	1480	Bureau of Standards	——	——
Copper . . .	{ 1084 * / 1083	Holborn and Day, 1900 } Day and Sosman, 1910 }	2310	Greenwood, 1909
Erbium . . .	——	——	——	——
Fluorine . .	− 223	Moissan and Dewar, 1903	− 187	Moissan & Dewar, 1903
Gallium . .	30·2	L. de Boisbaudran, 1876	——	——
Germanium .	960	Biltz, 1911	——	——
Gold	{ 1063 / 1062 †	Holborn and Day, 1901 } Day and Sosman, 1910 }	2530 (?)	——
Helium . . .	below − 272	Onnes, 1911	− 268·8	Onnes, 1911
Hydrogen . .	− 259	Travers, 1902	− 252·7	Travers, 1902
Indium . . .	155	Thiel, 1904	1000 (?)	——
Iodine . . .	113	Lean & Whatmough, 1898	184·4	Drugmann & Ramsay, '00
Iridium . . .	2290	Mendenhall & Ingersoll, '07	2550 (?)	——
Iron	1530	Bureau of Standards	2450	Greenwood, 1909
Krypton . . .	− 169	Ramsay, 1903	− 151·7	Ramsay, 1903
Lanthanum .	810	Muthmann & Weiss, 1904	——	——
Lead	327	Holborn and Day, 1900	1525	Greenwood, 1909
Lithium . . .	186	Kahlbaum, 1900	> 1400	Ruff & Johannsen, 1906
Magnesium .	633	Heycock and Neville, 1895	1120	Greenwood, 1909
Manganese .	1260	Bureau of Standards	1900	Greenwood, 1909
Mercury . .	− 38·80	Chappuis, 1900	356·7	Callendar, 1899
Molybdenum .	2450	Pirani & Meyer, 1912	3200 (?)	——
Neodymium .	840	Muthmann & Weiss, 1904	——	——
Neon	——	——	− 239	Dewar, 1901
Nickel . . .	1452 †	Day and Sosman, 1910	2330 (?)	——
Niobium . .	1950	von Bolton, 1907	——	——
Nitrogen . .	− 210·5	Fischer and Alt, 1903	− 195·7	Fischer & Alt, 1903

* In reducing atmosphere ; 1062° in air.   † Const. vol. N. thermometer.

MELTING AND BOILING POINTS OF THE ELEMENTS (contd.)

Element.	Melting Point.	Observer.	Boiling Point at 760 mms.	Observer.
Osmium . . .	2700° C.	—	—	—
Oxygen . . .	−219	Dewar, 1911	−182°·9 C.	Travers, 1902
Palladium *—	1549 ‡	Day and Sosman, 1910	2540	
thermo-jn. (a)	1535	Holborn & Henning, 1905	—	
optical therm.	1549		—	
„	1545	Nernst & Wartenberg, 1906	—	
„	1582	Holborn & Valentiner, 1907	—	
thermo-jn. (a)	1530	Waidner & Burgess, 1907	—	
„ (b)	1543	„ „	—	
optical therm.	1546		—	
Phosphorus .	44·1$_{760}$	Hulett, 1899	287	Schrötter, 1848
Platinum *—				
thermo-jn. (a)	1710	Harker, 1905	2450 (?)	
„ (a)	1710	Holborn & Henning, 1905	—	
optical therm.	1729		—	
„	1750	Nernst & Wartenberg, 1906	—	
„	1789	Holborn & Valentiner, 1907	—	
thermo-jn. (a)	1706	Waidner & Burgess, 1907	—	
„ (b)	1731	„ „	—	
optical therm.	1770	„ „ 1909	—	
Potassium . .	62·5	Holt and Sims, 1894	758	Ruff & Johannsen, 1905
Praseodymium	940	Muthmann and Weiss, 1904	—	
Radium . . .	700	Curie and Debierne, 1910	—	
Rhodium . .	1907	Mendenhall & Ingersoll, '07	2500 (?)	
Rubidium . .	38·5	Erdmann and Köthner, 1896	696	Ruff & Johannsen, 1905
Ruthenium . .	1900 (?)	—	2520 (?)	
Samarium . .	1350	—		
Selenium . .	217	Saunders, 1900	690	Berthelot, 1902
Silicon . . .	1200 (?)		3500 (?)	
Silver . . .	{ 962 † 960 ‡	Holborn and Day, 1900 } Day and Sosman, 1910 }	1955	Greenwood, 1909
Sodium . . .	97·0	Kurnakow & Puschin, 1902	{ 877 742	Ruff & Johannsen, 1905 Permann, 1889
Strontium . .	900			
Sulphur . . .	{ 115 rhombic 119 monoclinic	—	{ 444·55 (c.p. air) 444·7 (c.v. N) 444·53 (c.p. N)	} Eumorfopoulos, 1908 (corrected, 1909) } Chappuis & Harker, 1902 } Callendar, 1899
Tantalum . .	2910	Burgess, 1907	—	
Tellurium . .	450	Matthey, 1901	1390	Deville and Troost, 1880
Thallium . .	301	Kurnakow & Puschin, 1901	1280 (?)	Wartenberg, 1907
Thorium . .	1690	Wartenberg, 1909	—	
Tin	232	Heycock & Neville, 1895	2270	Greenwood, 1909
Titanium . .	1800	—		
Tungsten . .	3500	Gen. Elect. Co. Lab.	3700 (?)	—
Vanadium . .	1720			
Xenon . . .	−140	Ramsay, 1903	−109	Ramsay, 1903
Zinc . . .	418 ‡	Day and Sosman, 1910	918	Berthelot, 1902
Zirconium . .	c. 2300			

* See section on thermo-electric thermometers, p. 46, for meaning of (a) and (b).
† In reducing atmosphere ; 995° in air.　‡ Const. vol. N. thermometer.

E

STANDARD TEMPERATURES

Melting and boiling points of elements will be found on p. 48; of chemical compounds, on p. 109.

B.P. = boiling point at 760 mm.; M.P. = melting point; T.P. = transition point.

Substance.		Temp.	Substance.		Temp.
		°C.			°C.
Hydrogen	B.P.	−253	Zinc *	M.P.	419·4
Oxygen	B.P.	−183	Sulphur *	B.P.	444·7
Carbon dioxide	B.P.	− 78·2	Aluminium	M.P.	657
Mercury	M.P.	− 38·8	NaCl (Harker)	M.P.	801
Water	M.P.	0	K_2SO_4	M.P.	1070
$Na_2SO_4 . 10H_2O$	T.P.	32·383	Palladium (p. 49)	M.P.	1550
Water	B.P.	100	Platinum (p. 49)	M.P.	1750
Naphthalene *	B.P.	218·0	Tin (Greenwood)	B.P.	2270
Tin *	M.P.	231·9	Arc † (W. & B.)‡	—	3700 abs.
Benzophenone *	B.P.	306·0	Arc † (Harker, '08) ‡	—	3620 abs.
Cadmium *	M.P.	321·0	Sun † (p. 66)	—	5800 abs.

* Const. vol. N. scale, Waidner & Burgess, 1911 ; W. & B., Waidner & Burgess, 1904.
† Black body temperature.   ‡ Positive crater.

EFFECT OF PRESSURE ON BOILING POINTS

$\delta p/\delta t$ is given as mm. Hg per degree C. for pressures not very far removed from 760 mm.

The boiling point in absolute degrees C. of a substance under 760 mm. $= t + c(760 − p,(t + 273)$, where c is a constant for the substance, and t is the B.P. in degrees C. at the pressure p mm. The constant c is the same for chemically similar substances.

(See Young, " Fractional Distillation.")

Substance.	$\delta p/\delta t$	c	Substance.	$\delta p/\delta t$	c	Substance.	$\delta p/\delta t$	c
		×10⁻⁶			×10⁻⁶			×10⁻⁶
Hydrogen	200	—	CCl_4	23	123	Benzene	23·5	121
Oxygen	77	146	Pentane, n	25·8	125	Toluene	21·7	120
Carbon dioxide	55	—	Alcohol, methyl	29·6	100	Aniline	19·6	112
Water	27·2	99	„ ethyl	30·3	94	Naphthalene	17·1	119
Mercury	13·6	118	„ amyl	25	98	Benzophenone	15·8	109
Sulphur *	11·0	114	Ether, ethyl	26·9	121	Acetone	26·4	115

* $t_p = t_{760} + ·0904(p − 760) − ·0_452 (p − 760)^2$, Harker & Sexton, 1908.

MELTING, FREEZING, AND BOILING POINTS OF FATS AND WAXES

At 760 mm. pressure.   (See Lewkowitsch's treatise.)

Substance.	M.P.	F.P.	Substance.	M.P.	F.P.	Substance.	M.P.	B.P.
	°C.	°C.		°C.	°C.		°C.	°C.
Butter	28–33	20–23	Beeswax	61–64	60–63	Paraffin wax,		
Lard	36–40	27–30	Spermaceti	42–49	42–47	Soft	38–52	350–390
Tallow, beef	40–45	27–35	Stearin	71·6	70	Hard	52–56	390–430
„ mutton	44–45	36–41	Naphthalene	80·0	—	Olive oil	—	c. 300

THERMAL CONDUCTIVITIES

The thermal conductivity, k, is given below as the number of (gram) calories conducted per sq. cm. per sec. across a slab of the substance 1 cm. thick, having a temp.-gradient of 1° C. per cm., *i.e.* calorie cm.$^{-1}$ sec.$^{-1}$ temp.$^{-1}$. (See Callendar, "Conduction of Heat," *Encyc. Brit.*, and Winkelmann's "Handbuch der Physik," III., 1906.)

METALS AND ALLOYS

k for most pure metals decreases with rise of temperature; the reverse appears to be true for alloys. If κ be the electrical conductivity and θ the absolute temp., then $k/(\kappa\theta)$ is very approximately a constant for pure metals. (See J. J. Thomson, "Corpuscular Theory of Matter," and Lees, *Phil. Trans.*, 1908.) The electrical conductivity of the same specimen of many of the substances below will be found on p. 81.

Substance.	Temp. °C.	Cond.k.	Observer.	Substance.	Temp. °C.	Cond.k.	Observer.
Metals—				Mercury ..	0	·0148	}H. F.
Aluminium * .	−160	·514	}Lees,	„ ..	50	·0189	}Weber,'79
„ .	18	·504	}P.T., '08	„ ..	50	·0177	A., 1864
„ .	18	·480	}J. & D.,	„ ..	17	·0197	R. W., '02
„ .	100	·492	}1900	Nickel . . .	−160	·129	Lees, '08
Antimony . .	0	·044	}Lorenz,	„ {97%}	18	·142	}J. & D.,
„ ..	100	·040	}1881	„ { Ni}	100	·138	}1900
Bismuth . .	−186	·025	M., 1907	Palladium . .	18	·168	}J. & D.,
„	18	·0194	}J. & D.,	„	100	·182	}1900
„ ..	100	·0161	}1900	Platinum . .	18	·166	}J. & D.,
Cadmium,pure	−160	·239	Lees, '08	„ ..	100	·173	}1900
„ .	18	·222	}J. & D.,	Silver, pure .	−160	·998	}Lees,
„ .	100	·216	}1900	„	18	·974	}1908
Copper, pure .	−160	1·079	Lees, '08	„	18	1·006	}J. & D.,
„ .	18	·918	}J. & D.,	„	100	·992	}1900
„ .	100	·908	}1900	Tin, pure . .	−160	·192	Lees, '08
Gold . . .	18	·700	}J. & D.,	„	18	·155	}J. & D.,
„ . .	100	·703	}1900	„	100	·145	}1900
Iron, pure . .	18	·161	}J. & D.,	Tungsten . .	18	·35	Coolidge
„ „ . .	100	·151	}1900	Zinc, pure . .	−160	·278	Lees, '08
„ wrought	−160	·152	Lees, '08	„	18	·265	}J. & D.,
„ „ †	18	·144	}J. & D.,	„ ..	100	·262	}1900
„ „ †	100	·143	}1900				
„ cast ‡	54	·114	}Callendar	**Alloys—**			
„ „ ‡	102	·111					
„ „ §	30	·149	Hall	Brass ‖ . .	−160	·181	}Lees,
„ steel {1%}	−160	·113	}Lees,	„	17	·260	}1908
„ „ {C.}	18	·115	}1908	Constantan }	18	·054	}J. & D.,
„ „ „	18	·108	}J. & D.,	(Eureka)¶}	100	·064	}1900
„ „ „	100	·107	}1900	German silver	0	·070	}Lorenz,
Lead, pure .	−160	·092	Lees, '08	„ „	100	·089	} 1881
„ „ .	18	·083	}J. & D.,	Manganin ** .	−160	·035	Lees, '08
„ „ .	100	·082	}1900	„ .	18	·053	}J. & D.,
Magnesium .	0 to	}·376 {	}Lorenz,	„	100	·063	}1900
„	100		1881	Platinoid . .	18	·060	Lees, '08

* 99% Al.   † ·1% C, ·2% Si, ·1% Mn.   ‡ 2% C, ·3% Si, 1% Mn.
§ 3·5% C, 1·4% Si, ·5% Mn.   ‖ 70 Cu, 30 Zn.   ¶ 60 Cu, 40 Ni.
** 84 Cu, 4 Ni, 12 Mn.

A., Ångström; J. & D., Jaeger & Diesselhorst; M., Macchia; R. W., R. Weber; *P.T., Phil. Trans.*

MISCELLANEOUS SUBSTANCES

The values below are mostly at ordinary temperatures. They must be regarded as rough average values in the case of indifferent conductors. Nearly all liquids have very approximately the same conductivity, which in most cases appears to increase with temperature.

Substance.	k	Substance.	k	Substance.	k	Substance.	k
	$\times 10^{-3}$		$\times 10^{-3}$		$\times 10^{-3}$		$\times 10^{-4}$
Glass—		Cotton wool .	·04	Quartz, $\perp$ axis	16, L.	**Liquids—**	
Crown; window	2·5, L.	Cork . . .	·13, L.	Rubber, Para	·45, L.	Alcohol, 25° .	4·3, L.
Flint	2, L.	Earth's crust †	4	Sand . . .	·13	Aniline, 12° .	4·1
Jena	1–2, L.	Ebonite . .	·42, L.	Sawdust . .	·12	Glycerine, 25°	6·8, L.
Soda	1·3–1·8	Felt . . .	·09	Silicate cotton	·19	Paraffin oil,17°	3·5
		Flannel . .	·23, L.	Silk . . .	·22, L.	Turpentine,13°	3
Woods (dry)—		Gas carbon .	10	Slate . . .	4·7, L.	Vaseline, 25°	4·4. L.
Mahogany . .	·5, L.	Graphite . .	12				

Substance.	k	Substance.	k	Substance.	Temp.	Cond. k.	Obs.
Oak, teak . .	·6	Ice	5	Water . . .	**17°**	·00131	R.W.'03
Pine, walnut .	·4, L.	Marble, white	7·1, L.	,, . . .	**20**	·00143	M. & C.
Miscellaneous		Mica * . . .	1·8, L.	,, . . .	**4**	·00138	H. F.
Asbestos paper .	·6	Paper . . .	·3, L.	,, . . .	**23·6**	·00152	} Weber
Cardboard . .	·5	Paraffin wax .	·6, L.	,, . . .	**11**	·00147	} Lees,
Cement . . .	·7, L.	Porcelain. .	2·5, L.	,, . . .	**25**	·00136	} 1898
Cotton . . .	·55, L.	Quartz, $\parallel$ axis	30, L.				

* Perp. to cleavage plane. † Average for igneous and sedimentary rocks ; see Brit. Ass. Reports. L., Lees, 1892 & 1898 ; M. & C., Milner & Chattock, 1898 ; R. W., R. Weber.

GASES

In the case of a gas the thermal conductivity $k = 1·603\eta c_v$, where η is the viscosity, and c_v the specific heat at constant volume. Stefan, and Kundt and Warburg have found, in agreement with this formula, that k for air, hydrogen, etc., is constant between the pressures 76 cm. and ·1 cm. k increases with the temperature. (See Meyer's "Kinetic Theory of Gases.")

Gas.	Temp.	Cond. k.	Gas.	Temp.	Cond. k.	Gas.	Temp.	Cond. k.	Gas.	Temp.	Cond. k.
	C.	$\times 10^{-5}$		C.	$\times 10^{-5}$		C.	$\times 10^{-5}$		C.	$\times 10^{-5}$
H_2	−150°	11·7, E.	Air	0°	5·22 *	CO	7°	5·10, W.	N_2O	0°	3·50, W.
,,	0	31·8, E.	O_2	7	5·63, W.	CO_2	0	3·07, W.	,,	100	5·06, W.
,,	0	31·9, G.	A	0	3·89, S.	,,	0	3·27, Sc.	NO	8	4·60, W.
,,	100	36·9, G.	CH_4	8	6·47, W.	,,	100	5·06, W.	Hg	203	1·85, Sc.
He	0	33·9, S.	C_2H_4	0	3·95, W.	NH_3	0	4·58, W.			
N_2	7	5·24,W.	CO	0	4·99, W.	,,	100	7·09, W.			

* Mean of five observers. E., Eckerlein, 1900 ; G., Graetz, 1885 ; S., Schwarze, 1903 ; Sc., Schleiermacher, 1889 ; W., Winkelmann, 1875.

COEFFICIENTS OF LINEAR EXPANSION OF SOLIDS

To represent accurately over any considerable range the variation of length (l) with temperature (t) requires for almost all solid substances a parabolic or cubic equation in t. But if the temperature interval is not large, a linear equation $l_t = l_0(1 + at)$ may be employed ; and this gives a definition of the mean coefficient of linear expansion (a) over that temperature range. The coefficient of **cubical expansion** $= 3a$.

There is little point in tabulating coefficients of higher-powered terms of t, since for a given specimen it is as a rule impossible without measurement to assume with any accuracy anything more definite than the average value of even the first power coefficient (a). Except in a few cases the linear coefficient as defined above increases with the temperature. The values of a subjoined are per degree C., and except when some temperature is specified, for a range round and about 20° C. Some substances expand irregularly, and extrapolation of a may therefore be dangerous. Interpolation of a from the constituent metals must be employed with caution in the case of alloys. (See Winkelmann's "Handbuch der Physik," iii. 1906.)

COEFFICIENTS OF LINEAR EXPANSION OF SOLIDS (contd.)

Element.	a.	Obs.	Element.	a.	Obs.	Element.	a.	Obs.
	$\times 10^{-6}$			$\times 10^{-6}$			$\times 10^{-6}$	
Aluminium	25·5	V. '93	Copper . .	16·7	V. '93	Palladium .	11·7	S. '03
Antimony .	12	F. '69	Gold . . .	13·9	V. '93	Platinum .	8·9	B. '88
Bismuth .	15·7	V. '93	Iridium . .	6·5	B. '88	Potassium .	83	H. '82
C. (diamond)	1·2	F. '69	Iron (cast) .	10·2	D. '02	Selenium, 40°	36·8	F. '69
„ (gas carbon) .	5·4	F. '69	„ (wrought)	11·9	H.D. '00	Silver . .	18·8	V. '93
„ (graphite)	7·9	F. '69	Steel, 10·5 to	11·6	N.P.L.	Sulphur . .	c. 70	----
Cadmium .	28·8	M. '66	Lead . . .	27·6	M. '66	Thallium, 40°	30·2	F. '69
Cobalt . .	12·3	T. '99	Magnesium .	25·4	V. '93	Tin . . .	21·4	M. '66
			Nickel . .	12·8	T. '99	Zinc, 25·8 to	26·3	N.P.L.

Substance.	a.	Obs.	Substance.	a.	Obs.
Alloys—	$\times 10^{-6}$		**Miscellaneous** (contd.)	$\times 10^{-6}$	
Aluminium bronze . . .	17·0	N.P.L.	Glass, flint, 45 SiO_2,		
Brass (ordy.) c. 66 Cu, 34 Zn	18·9	N.P.L.	8 K_2O, 46 PbO	7·8	Sc.
Bronze, 32 Cu, 2 Zn, 5 Sn§	17·7	B. '88	„ Jena, 16''' (see p. 74)	7·8	} T.S.S.
Constantan (Eureka), 60 Cu, 40 Ni	17·0	N.P.L.	„ „ 59''' (see p. 74)	5·7	} '96
			„ Verre dur (see p. 74)	7·2	C. '07
German silver, 60 Cu, 15 Ni, 25 Zn, 50°	18·4	Pf. '72	Granite	8·3	
Gunmetal (Admiralty). .	18·1	N.P.L.	Gutta-percha	198	Ru. '82
Magnalium, 86 Al, 13 Mg	24	St. '01	Ice, −10° to 0°	50·7	Vn. '02
Nickel steel,* 10 % Ni .	13·0	N.P.L.	Iceland spar, ∥ axis . .	25·1	B. '88
„ „ 20 % „ .	19·5	N.P.L.	„ „ ⊥ axis . .	−5·6	B. '88
„ „ 30 % „ .	12·0	N.P.L.	Marble, white Carrara, 15°, 1·4 to	3·5	N.P.L.
„ „ 36 % „ (Invar †)	0·9	N.P.L.	„ black	4·4	
„ „ 40 % „ .	6·0	N.P.L.	Masonry . . . 4 to	7	----
„ „ 50 % „ .	9·7	N.P.L.	Paraffin wax, 0°–40° . .	c. 110	----
„ „ 80 % „ .	12·5	N.P.L.	Porcelain, Berlin . . .	2·8	S. '03
Phosphor bronze, 97·6 Cu, 2 Sn, ·2 P	16·8	B. '88	„ „ 0°–100°	3·1	H.G. '01
			„ Bayeux . . .	3·4	Bd. '00
Platinum-iridium, 90 Pt, 10 Ir ‡	8·7	B. '88	„ „ 0°	2·5	T. '02
			Portland stone	c. 3	----
Platinum - silver, 33 Pt, 67 Ag	15	----	Quartz (crystal), ∥ axis .	7·5	B. '88
			„ „ ⊥ axis .	13·7	B. '88
Solder, 2 Pb, 1 Sn, 50° .	25	Sm.	Silica (fused), −80° to 0° .	·22	S. '07
Speculum metal, 68 Cu, 32 Sn	19·3	Sm.	„ „ 0° to 30 °.	·42	C. '03
			„ „ 0° to 100°	·50	S. '07
Type metal, c. 135° . . .	19	Dl.	„ „ 0° to 1000°	·54	R. '10
Miscellaneous—			Sandstone . . . 7 to	12	
Brick (Egyptian) . . .	9·5	N.P.L.	Slate . . . . . 6 to	10	----
Cement and concrete, 10 to	14	----	**Woods** (1) along grain—		
Ebonite . . . . 64 to	77	----	Beech ; mahogany . . .	c. 3	Vl. '68
Fluor spar, CaF_2 . . .	19	F. '68	Oak ; pine	c. 5	Vl. '68
Glass, soft, 68 SiO_2, 14 Na_2O, 7 CaO	8·5	Sc.	(2) across grain—		
			Beech	60	Vl. '68
„ hard, 64 SiO_2, 20 K_2O, 11 CaO	9·7	Sc.	Mahogany	40	Vl. '68
			Pine	34	Vl. '68

* See Guillaume's "Les Applications des Aciers au Nickel," 1904. † Invar is obtainable in three qualities, with a range of coefficients of (−·3 to + 2·5) × 10^{-6} at ordinary temperatures. ‡ Used for international prototype metre (see p. 3). § Used for Imperial Standard Yard (see p. 4). B. Benoît; Bd. Bedford; C. Chappuis; D. Dittenberger; Dl. Daniell; F. Fizeau; H. Hagen; H.D. Holborn and Day; H.G. Holborn and Grüneisen; M. Matthiessen; N.P.L. National Physical Laboratory; Pf. Pfaff; R. Randall; Ru. Russner; S. Scheel; Sc. Schott; Sm. Smeaton; St. Stadthagen; T. Tutton; T.S.S. Thiesen, Scheel, and Sell; V. Voigt; Vl. Villari; Vn. Vincent.

COEFFICIENTS OF CUBICAL EXPANSION OF GASES

The volume coefficient, a, at constant pressure is defined by $v_t = v_0(1 + at)$; the pressure coefficient, β, at constant volume is defined by $p_t = p_0(1 + \beta t)$, where v_t and p_t are the volume and pressure respectively corresponding to $t°$, the initial volume and pressure (v_0, f_0) being measured at $0°$ C. The values of both a and β depend on the initial pressure of the gas. If a gas obeys Boyle's law exactly, $a = \beta$.

Comparison of rarefied gas, H_2 and absolute temperature scales.— By graphically or otherwise extrapolating a and β to zero pressure, they become equal (as we should expect, for rarefied gases should behave as ideal gases and obey Boyle's law), and we may write $a = \beta = \gamma$. For example, Berthelot finds from Chappuis' data—

For H_2, mean $\gamma = \cdot00366207 = 1/273\cdot07$ (see p. 44)

N_2,   „ $\gamma = \cdot00366182 = 1/273\cdot09$ (see p. 44)

Kelvin's absolute temperature scale agrees with the ideal gas scale, and therefore with the rarefied gas scale. Now, as will be seen below, β for $H_2 = \gamma$ very nearly, and thus the constant-volume hydrogen scale of temperature may justifiably be taken as closely approximating to the thermodynamic scale (see also p. 44).

(See Young's "Stoichiometry"; and Berthelot and Chappuis, *Trav. et Mém. du Bur. Intl.*, 1907.)

Gas.	Temp.	p_0.	a	Obs.	Gas.	Temp.	p_0.	β	Obs.
			AT CONSTANT PRESSURE.					AT CONSTANT VOLUME.	
	C.	cm. Hg.				C.	cm. Hg.		
Air	$0°$–$100°$	$100\cdot1$	$\cdot0036728$	C., 1903	**Air**	—		$\cdot58$	$\cdot0037666$	M., 1892
„	0–100	76	3671	R., 1847	„	—		$1\cdot32$	37172	„
H₂	0–100	100	36600	C., 1903	„	—		$10\cdot0$	36630	„
„	0–100	76	3661	R., 1847	„	—		17–24	36513	R., 1847
„	0–100	76	36609	R. M.	„	—		76	36650	„
N₂	0–100	100	367313	C., 1903	„	$0°$–$100°$	$100\cdot1$	36744	C., 1903
„	0–100	139	367750	C., 1903	„		200	3690	R., 1847
„	—	200 atm.	434	A., 1890	„		2000	3887	„
„	—	1000 „	218	A., 1890	„	0–1067	23	36643	J. P.
O₂	—	100 „	486	A., 1890	**H₂**	0–100	52	36626	T. J., '02
CO	—	76	3669	R., 1847	„	0–100	70	366255	„
CO₂	0–20	$51\cdot8$	37128	C., 1903	„	0–100	100	366256	C., 1903
„	0–40	„	37100	„	„	0–100	109	36627	O, 1908
„	0–100	„	37073	„	**N₂**	0–100	53	36683	C., 1903
„	0–20	$99\cdot8$	37602	„	„	0–100	79	36718	„
„	0–40	„	37536	„	**O₂**	0–100	100	367440	„
„	0–100	„	37410	„	„	0–1067	66	36738	M N., '03
„	0–20	$137\cdot7$	37972	„	„	0–1067	18–23	36652	J. P.
„	0–40	„	37906	„	**He**	0–100	52	36627	T. J., '02
„	0–100	„	37703	„	„	0–100	70	366255	„
„	0–100	76	37282	R. M.	„	0–100	100	36616	O, 1908
N₂O	—	76	3719	R., 1847	**A**	—	$51\cdot7$	3668	K R., 96
NH₃	0–50	76/15°	3854	P.D., '06	**CO**	0–100	76	3667	R., 1847
SO₂	—	76	3903	R., 1847	„	0–1067	23	36648	J. P.
					CO₂	0–100	$51\cdot8$	36981	C., 1903
					„	0–20	$99\cdot8$	37335	„
					„	0–100	$99\cdot8$	37262	„
					„	0–1067	24	36756	J. P.
					N₂O	—	76	3676	R., 1847
					SO₂	—	76	3845	R., 1847

A., Amagat; C., Chappuis; J. P., Jacquerod & Perrot; K. R., Kuenen & Randall; M., Melander; M. N., Makower & Noble; O., Onnes; P. D., Perman & Davies; R., Regnault; R. M., Richards & Marks; T. J., Travers & Jacquerod.

COEFFICIENTS OF CUBICAL EXPANSION OF LIQUIDS

As with solids (see p. 52), if the temperature interval is not large, a linear equation $v_t = v_0(1 + at)$ may be employed to show the relation between the volume (v) of a liquid and its temperature (t). The mean coefficient (a) thus defined increases in general with the temperature. The values of a subjoined are per ° C., and for a range round 18° C. unless otherwise specified.

Liquid.	Temp. range.	Mean Coefficient from 0° C. to t° C.	Observer.
Water (see p. 22 and below)	H scale. 17 to 40	$\cdot 0_3 13019/(t) - \cdot 0_4 65769 + \cdot 0_5 86797t - \cdot 0_7 7336t^2$	Chappuis, '97
	17 to 100	Density $= 1 - \dfrac{(t - 3.982)^2}{466,700} \cdot \dfrac{t + 273}{t + 67} \cdot \dfrac{350 - t}{365 - t}$	Thiesen, '03
Mercury (see p. 22)	24 to 299	$\cdot 00018179 + \cdot 0_9 175t + \cdot 0_{10} 351t^2$	Regnault, '47 (Broch)
	0 to 100	$\cdot 00018169 - \cdot 0_8 2817t + \cdot 0_9 115t^2$	Chappuis, '07
	−10 to 300	$\cdot 000180555 + \cdot 0_7 12444t + \cdot 0_{10} 254t^2$	{ Callendar & { Moss, 1911
	0 to 180	$\cdot 000181385 + \cdot 0_6 9770t + \cdot 0_{10} 18318t^2$	Donaldson, '12

Liquid.	a	Liquid.	a	Liquid.	a	Liquid.	a
	$\times 10^{-5}$		$\times 10^{-5}$		$\times 10^{-5}$		$\times 10^{-5}$
Acetic acid .	107	Ether, ethyl .	163	Pentane . .	159	Water, 60–80	58·7
Alcohol, me. .	122	Ethyl bromide	137	Toluene . .	109		
,, ethyl	110	Glycerine . .	53	Turpentine .	94	**Solutions—**	
,, amyl	93	Mercury (see	above	Xylol (m) . .	101	CaCl₂, 5·8% .	25·0
Aniline . .	·85	Methyl iodide	121	Water, 5°–10°	5·3	,, 40·9%	45·8
Benzene . .	124	Oil, olive . .	70	,, 10–20	15·0	NaCl, 26% .	43·6
CS₂ . . .	121	,, paraffin .	90	,, 20–40	30·2	H₂SO₄, 100%	57
Chloroform .	126	,, ,, 20°–199°	110	,, 40–60	45·8		

MECHANICAL EQUIVALENT OF HEAT

Joule's equivalent, J, is here given as the number of ergs equivalent to a calorie, *i.e.* the heat required to raise 1 gram of water through 1° C. at some specified temperature. The **15° calorie** is about 1 part in 1000 greater than the **20° calorie**. (See p. 56.)
See Griffith's "Thermal Measurement of Energy," 1901.

Observer.	Calorie.	Ergs.	Observer.	Calorie.	Ergs.
	N. scale	$\times 10^7$		N. scale	$\times 10^7$
Joule, 1843	20° C.	4·169	Bousfield, *Phil. Trans.*, 1911	20° C	4·175
Rowland, 1878 . . .	20°	4·180	Crémieu & Rispail, 1908	0°	4·185
Griffiths, 1893	20°	4·184	Reynolds & Moorby, 1897	Mean	4·184
Schuster and Gannon, 1894	20°	4·181	Barnes, 1909 (deduced)	Mean	4·185
Callendar and Barnes, 1899	20°	4·180			

SPECIFIC HEATS

SPECIFIC HEAT OF WATER

Callendar and Barnes (*Phil. Trans.*, 1902) used an electrical method of determining the temperature variation of the specific heat of water. The specific heats below are reduced by Callendar ("Ency. Brit.," Art. "Calorimetry") from their results; they are relative to the specific heat at 20° C. on the C.P. nitrogen scale. The **20° calorie** (see pp. 5 and 55) is adopted as 4·180 joules = 4·180 × 10⁷ ergs, being the mean of the results of Rowland (1879) and of Reynolds and Moorby (reduced), each of whom used a mechanical method of determining "J." Thus the values of J below do not rest on the values attributed to the electrical standards employed. The specific heat of water is a minimum at 37·5° C., according to Callendar and Barnes.

The 15° calorie (according to Barnes, *Proc. Roy. Soc.*, 1909) = 4·184 joules, assuming the e.m.f. of the Clark cell at 15° C. = 1·4330 international volts.

The mean calorie ($= \frac{1}{100}$ of heat required to raise 1 gram of water from 0° to 100° C.) = 4·185 joules (Barnes, 1909); = 4·184 joules (Reynolds and Moorby, 1897, corrected by Smith).

Temp.	Specific heat.	Joules.	Temp.	Specific heat.	Joules.	Temp.	Specific heat.	Joules.
− 5° C.	1·0158	4·246	45° C.	·9983	4·173	95° C.	1·0063	4·206
0	1·0094	4·219	50	·9987	4·175	100	1·0074	4·211
5	1·0054	4·202	55	·9992	4·177	120	1·0121	4·231
10	1·0027	4·191	60	1·0000	4·180	140	1·0176	4·254
15	1·0011	4·184	65	1·0008	4·183	160	1·0238	4·280
20	1·0000	4·180	70	1·0016	4·187	180	1·0308	4·309
25	·9992	4·177	75	1·0024	4·190	200	1·0384	4·341
30	·9987	4·175	80	1·0033	4·194	220	1·0467	4·376
35	·9983	4·173	85	1·0043	4·198			
40	·9982	4·173	90	1·0053	4·202			

SPECIFIC HEAT OF MERCURY

In terms of the gram calorie at 15°·5 on the const. vol. H. scale. (Barnes and Cooke, *Phys. Rev.*, 15, 1902.) Mercury has a minimum specific heat at 140° C. (Barnes, *Brit. Ass. Rep.*, 1909.)

Temp.	0° C.	20°	40°	60°	80°	100°	200°
Specific heat	·0335	·0333	·0331	·0329	·0328	(·0327)	(·032)

SPECIFIC HEATS OF THE ELEMENTS

For gases, see p. 58. (See Waterman, *Phys. Rev.*, 1896.)

Substance.	Temperature.	Sp. heat.	Observer.	Substance.	Temperature.	Sp. heat.	Observer.
Aluminium .	−182° to 15°	·168	Tilden, 1903	Bromine, liqd.	13° to 45°	·107	Andrews, '48
	15 to 185	·219	"	Cadmium * .	−186 to −79	·050	Behn, 1900
	600	·282	Richards, '93	" pure	18 to 99	·055	Voigt, 1893
Antimony .	−186 to −79	·0462	Behn, 1900	Cæsium	0 to 26	·048	E. & G., 1900
	17 to 92	·0508	Gaede, 1902	Calcium . .	−185 to 20	·157	N. & B., 1906
Arsenic, cryst.	21 to 68	·083	B. & W., 1868		0 to 100	·149	Be., 1906
" amorph.	21 to 65	·076		Carbon—			
Barium .	−185 to 20	·068	N. & B., 1906	Gas carbon .	24 to 68	·204	B. & W., 1868
Beryllium . .	0 to 100	·425	N. & P., 1880	Charcoal .	0 to 24	·165	H.F.Weber,'75
Bismuth . .	−186	·0284	Giebe, 1903	"	0 to 224	·238	"
	22 to 100	·0304	W., 1896	Graphite .	−50	·114	"
Boron, amor.	0 to 100	·307	M. & G., 1893	" .	11	·160	"
Bromine, solid	−78 to −20	·084	Regnault, '49	" .	202	·297	"

* Contained Fe and Zn.

SPECIFIC HEATS OF THE ELEMENTS (contd.)

Substance.	Temperature.	Sp. heat.	Observer.	Substance.	Temperature.	Sp. heat.	Observer.
Carbon (contd.)				Palladium . .	−186° to 18°	·053	Behn, 1898
Graphite .	977° C.	·467	H.F.Weber,'75		18 to 100	·059	,,
Diamond .	−50	·064	,,	Phosphorus—			
,,	11	·113	,,	,, yellow	−78 to 10	·17	Regnault, 1849
,,	206	·273	,,		13 to 36	·202	Kopp, 1864
,,	985	·459	,,	,, liquid	49 to 98	·205	Person, 1847
Cerium . . .	0 to 100	·045	H., 1876	,, red	15 to 98	·17	Regnault, 1853
Chlorine, liqd.	0 to 24	·226	Knietsch	Platinum . .	−186 to 18	·0293	Behn, 1898
Chromium .	−200	·067	Adler, 1903		18 to 100	·0324	,,
(1·4% Fe & Si)	0	·104	,,		1230	·0461	Tilden, 1903
	100	·112	,,	Potassium . .	−78 to 23	·166	Schütz, 1892
	400	·133	,,	Rhodium . .	10 to 97	·058	Regnault, 1862
Cobalt . . .	−182 to 15	·082	Tilden, 1903	Ruthenium .	0 to 100	·061	Bunsen, 1870
	15 to 100	·103	,,	Selenium, cryst.	22 to 62	·084	B. & W., 1868
	15 to 630	·123	,,	,, amorph.	18 to 38	·095	,,
Copper . . .	−192 to 20	·0798	Schmitz, 1903	Silicon, cryst.	−185 to 20	·123	N. & B., 1906
	20 to 100	·0936	,,		57	·183	H.F.Weber,'75
	900	·118	Le Verrier, '92		232	·203	,,
Didymium .	0 to 100	·046	H., 1876	Silver . . .	−186 to −79	·0496	Behn, 1900
Gallium, solid	12 to 23	·079	B., 1878		15 to 100	·056	B. & S., 1895
,, liquid	12 to 119	·080	,,		427	·059	Tilden, 1903
Germanium .	0 to 100	·074	N. & P., 1887	Sodium . .	−185 to 20	·234	N. & B., 1906
Gold . . .	−185 to 20	·035	N. & B., 1906		10	·297	Bernini, 1906
	18 to 99	·0303	Voigt, 1893		128	·333	,,
Indium . .	0 to 100	·057	Bunsen, 1870	Sulphur—			
Iodine . .	9 to 98	·054	Regnault, 1840	,, rhombic	17 to 45	·163	Kopp, 1865
Iridium . .	−186 to 18	·0282	Behn, 1898	,, liquid	119 to 147	·235	Person, 1847
	18 to 100	·0323	,,	Tantalum . .	−185 to 20	·033	N. & B., 1906
Iron . . .	−192 to 20	·089	Schmitz, 1903		58	·036	v. Bolton, 1905
	20 to 100	·119	,,	Tellurium, crys.	15 to 100	·048	Fabre, 1887
	225	·137	Stücker, 1905	Thallium . .	−192 to 20	·0300	Schmitz, 1903
	0 to 1100	·153	Harker, 1905		20 to 100	·0326	,,
Lanthanum .	0 to 100	·045	H., 1876	Thorium . .	0 to 100	·028	Nilson, 1883
Lead . . .	−192 to 20	·0293	Schmitz, 1903	Tin . . .	−186 to −79	·0486	Behn, 1900
	20 to 100	·0305	,,		19 to 99	·0552	Voigt, 1893
	300	·0338	Naccari, 1888	,, molten	240	·064	Spring, 1886
Lithium . .	0 to 19	·837	Be., 1906	Titanium . .	−185 to 20	·082	N. & B., 1906
	0 to 100	1·093	,,		0 to 100	·113	N. & P., 1887
Magnesium .	−186 to −79	·189	Behn, 1900		0 to 440	·162	,,
	18 to 99	·246	Voigt, 1893	Tungsten . .	−185 to 20	·036	N. & B., 1906
	225	·281	Stücker, 1905		20 to 100	·034	Gin, 1908
Manganese .	14 to 97	·122	Regnault, 1862	Uranium . .	11 to 98	·062	Regnault, 1840
Mercury .	See preceding page.				0 to 98	·028	Blümcke, 1885
Molybdenum	−185 to 20	·063	N. & B., 1906	Vanadium . .	0 to 100	·115	Mache, 1897
	15 to 91	·072	D. & G., 1901	Zinc . . .	−192 to 20	·084	Schmitz, 1903
Nickel . . .	−186 to 18	·086	Behn, 1898		20 to 100	·093	,,
	18 to 100	·109	,,		300	·104	Naccari, 1888
Osmium . .	19 to 98	·031	Regnault, 1862	Zirconium . .	0 to 100	·066	M. & D., 1873

B., Berthelot ; Be., Bernini ; B. & S., Bartoli & Stracciati ; B. & W., Bettendorff & Wüllner ; D. & G., Defacqz & Guichard ; E. & G., Eckardt & Graefe ; H., Hillebrand ; M. & D., Mixter & Dana ; M. & G., Moissan & Gautier ; N. & B., Nordmeyer & Bernouilli ; N. & P., Nilson & Pettersson ; W., Waterman.

SPECIFIC HEATS

SPECIFIC HEATS OF GASES AND VAPOURS

The values at const. pressure are, unless otherwise stated, all at atmospheric pressure. The specifi
heats given are calories per gram of gas per degree C. at the temp. stated.

Gas.	Temp.	Sp ht.	Observer.	Gas.	Temp.	Sp. ht.	Observer.
AT CONSTANT PRESSURE (c_p)				Ammonia, NH_3 . .	23-100	·520	⎫Wiedemann
				Nitrous oxide, N_2O	26-103	·213	⎬ 1876
Air (dry)	20° C.	·2417	Swann, 1909	Nitric oxide, NO .	13-172	·232	Regnault, '6:
,, ,, . . .	100	·2430	,, ,,	N. peroxide, NO_2 .	27-67	1·625	B. & O., 188:
,, ,, . . .	20-440	·2366	H. & A., 1905	H_2S	20-206	·245	Regnault, '6:
,, ,, . . .	20-98	·2372	⎫Witkowski,	CS_2	86-190	·160	,, ,,
,, ,, . . .	-102-17	·2372	⎬ ,, 1896	Methane, CH_4 . .	—	·591	Lussana, '94
,, ,, 70 atmos.	-50	·312	,, ,,	Ethylene C_2H_4 . .	—	·404	,, ,,
Argon	20-90	·123	D., 1897	Benzene, C_6H_6 . .	34-115	·299	⎫Wiedemann
Hydrogen . . .	—	3·402	Lussana, 1894	Chloroform $CHCl_3$.	27-118	·144	⎬ 1877
,, 30 atmos.	—	3·788	,, ,,	Me. alcohol CH_4O.	101-223	·458	Regnault, '6:
Nitrogen	0	·2350	* H. & H.,'07	Et. alcohol C_2H_6O.	108-220	·453	Regnault, '6.
,, (liq.) . .	-200	·43	Alt, 1904	,, ether $(C_2H_5)_2O$.	25-111	·428	W., 1876
Oxygen	20-440	·2419	H. & A., 1905	Turpentine, $C_{10}H_{16}$	179-249	·506	Regnault, '6:
,, . . .	20-800	·2497	,, ,,				
,, (liq.) . .	-190	·347	Alt, 1904	**AT CONSTANT VOLUME (c_v)**			
Chlorine . . .	16-343	·115	Strecker, '81				
Bromine . . .	19-388	·055	,, '82	Air,† 1 atmos . .	0°	·1715	Joly, 1891
Iodine	206-377	·034	,, ,,	Hydrogen ‡ . . .	c. 50	2·402	,, ,,
Carbon monoxide .	23-99	·242	W., 1876	Carbon dioxide § .	c. 55	·1650	,, 1894
,, dioxide :	0	·2010	*H. & H., '07	Argon	0-2000	·0746	Pier, 1909
,, ,,	20	·2020	Swann, 1909	Nitrogen ‖ . . .	0	·175	,, ,,
,, ,,	100	·221	,, ,,	Water vapour . .	100	·340	,, ,,
,, ,, 30 atmos.		·2670	Lussana, '94				
Water vapour . .	100	·4652	*H. & H., '07				

B. & O., Berthelot & Ogier; D., Dittenberger; H. & A., Holborn & Austin (Reichsanstalt); W., Wiedemann.
* H. & H., Holborn ⎧**Nitrogen** (0-1400°), $c_p = ·2350 + ·000019t$ ⎫ Mean specific
and Henning ⎨**CO₂** (0-1400°), $c_p = ·2010 + ·000074 2t - ·0,18t^2$ ⎬ heats between
(Reichsanstalt). ⎩**Water vapour** (100-1400°), $c_p = ·4669 - ·0000168t + ·0,44t^2$ ⎭ 0° and $t°$ C.
† Air, $c_v = ·1715 + ·02788\rho$ where ρ is the density (gm./c.c.).   § CO_2, $c_v = ·165 + ·2125\rho + ·34\rho^2$, ρ being densit:
‡ H, c_v diminishes with increasing density and falling temp.   ‖ N, $c_v = ·175 + ·00016t$, t being the temp.

RATIO OF THE SPECIFIC HEATS FOR GASES AND VAPOURS

γ = the ratio of the specific heat at constant pressure to that at constant volume. γ is usually determined directly by some method involving an adiabatic expansion, such as the determination of the velocity of sound in the gas. From a knowledge of either (1) the pressure or (2) the temperature immediately following an adiabatic expansion (Clément and Desormes, Lummer and Pringsheim's methods respectively), γ can be deduced from $pv^\gamma = $ const. or $\theta v^{\gamma-1} = $ const. (See Capstick, "Science Progress," 1895; and Moody, *Phys. Rev.*, Ap., 1912.)

Gas.	Temp.	γ	Observer.	Gas.	Temp.	γ	Observer.
Monatomic gases				Air (dry)	0°	1·402	Koch, 1907
Helium	0° C.	1·63	B. & G., 1907	,, ,, . . .	0	1·402	F., 1908
Argon	0	1·667	Niemeyer, '02	,, ,, . . .	500	1·399	,, ,,
Neon	19	·1·642	Ramsay, 1912	,, ,, . . .	900	1·39''	Kalähne, '03
Krypton	19	1·689	,,	,, ,, . . .	-79·3	1·405	Koch, 1907
Xenon	19	1·666	K. & W., 1876	,, ,, 200 ⎫	0	1·828	,, ,,
Mercury vapour .	310	1·666	,, ,,	,, ,, atmos. ⎬	-79·3	2·333	,, ,,
Diatomic gases—				Hydrogen . . .		1·419	Hartmann,'0
Air (dry)	5-14	1·402	L. & P., 1898	,, . . .	4-16	1·408	L. & P., 1898
,, ,, . . .	0	1·401	Stevens, 1905	Nitrogen	—	1·41	Cazin, 1862
,, ,, . . .	15	1·401	Makower, '03	Oxygen	5-14	1·400	L. & P., 1898
,, ,, . . .		1·414	Hartmann, '02	Carbon monoxide .	—	1·401	Leduc, 1898
				Nitric oxide, NO .		1·394	Masson

B. & G., Behn & Geiger; F. Fürstenau; K. & W. Kundt & Warburg; L. & P., Lummer & Pringsheim.

RATIO OF THE SPECIFIC HEATS FOR GASES AND VAPOURS (contd.)

Gas.	Temp.	γ	Observer.	Gas.	Temp.	γ	Observer.
Triatomic gases				Acetylene, C_2H_2 .	—	1·26	M. & F., 1897
Ozone	—	1·29*	Jacobs, 1905	Ethylene, C_2H_4 . .	—	1·264	Capstick, '95
Water vapour . .	100° (?)	1·305	Makower, '03	Benzene, C_6H_6 . .	20°	1·40	Pagliani, '96
Carbon dioxide . .	4–11	1·300	L. & P., 1898	,, ,,	99·7	1·105	Stevens, '02
,, ,,	—	1·306	Hartmann, '05	Chloroform, {	24–42	1·110	Müller, 1883
,, ,,	500	1·26	F., 1908	$CHCl_3$ . . . {	99·8	1·150	Stephens, '02
Ammonia, NH_3 .	—	1·336	Leduc, 1898	CCl_4	—	1·130	Capstick, '95
Nitrous oxide, N_2O	—	1·324	,, ,,	Me. alcohol . .	99·7	1·256	Stevens, '02
Nitrogen} N_2O_4 .	20°	1·172	Natanson, '85	,, bromide . .	—	1·274	Capstick, '93
peroxide} NO_2 .	150	1·31	,, ,,	,, chloride . .	19–30	1·279	,, ,,
H_2S	—	1·340	Capstick, '95	,, iodide . .	—	1·286	,, ,,
CS_2	—	1·239	,, ,,	Et. alcohol . .	53	1·133	Jaeger, 1889
Sulphur dioxide. {	16–34	1·26	Müller, 1883	,, ,,	99·8	1·134	Stevens, '02
	500	1·2	F., 1908	,, bromide . .	—	1·188	Capstick, '93
Polyatomic gases				,, chloride . .	22·7	1·187	,, ,,
Methane, CH_4 . .	—	1·313	Capstick, '93	,, ether . . .	12–20	1·024	Low, 1894
Ethane, C_2H_6 . .	—	1·22	{Daniel &	,, ,, . . .	99·7	1·112	Stevens, '02
Propane, C_3H_8 . .	—	1·130	{ Pierron, '99	Acetic acid . .	136·5	1·147	,, ,,

* Extrapolated ; F., Fürstenau ; L. & P., Lummer & Pringsheim ; M. & F., Maneuvrier and Fournier.

SPECIFIC HEATS OF VARIOUS BODIES

In most cases, the specific heats given must only be regarded as rough average values.

Substance.	Temp.	Sp. ht.	Substance.	Temp.	Sp. ht.	Substance.	Temp.	Sp. ht.
Alloys—	°C.		Ether, ethyl .	18°	·56	Glass, Jena 16''' †	18°	·19
Brass, red . .	0	·090	Glycerine . .	18–50	·58	,, Jena 59''' †	18	·19
,, yellow .	0	·088	Oil, olive . .	7	·47	Granite . . .	20–100	{·19 to ·20
Eureka	18	·098	,, paraffin .	20–60	{·51 to ·54			
(Constantan)			Sea-water . .	17	·94	Ice	−21 to −1	}·502
German silver .	0–100	·095	Toluene . . .	18	·40	Indiarubber .	15–100	{·27 to ·48
			Turpentine . .	18	·42			
Liquids—						Marble, white .	18	{·21 to ·22
Alcohol, amyl .	18	·55	**Miscel-**			Paraffin wax .	0–20	·69
,, ethyl .	0	·547	**laneous—**			Porcelain ‡ . .	15–1000	·255
,, ,,	40	·648	Asbestos . .	20–100	·20	Quartz, SiO_2 .	0	·174.
,, methyl	12	·601	Basalt . . .	20–100	{·20 to ·24		350	·279
Aniline * . . .	15	·514				Rock salt, $NaCl$	18	·21
Benzene . . .	10	·340	Ebonite . . .	20–100	·33	Sand	20–100	·19
,, . . .	40	·423	Fluorspar, CaF_2	30	·21	Silica (fused) § .	15–200	·2co
Brine,	−20	·69	Glass, crown .	10–50	·16	,, ,, .	15–800	·248
density = 1·2 {	0	·71	,, flint . .	10–50	·12			
(Harker) {	15	·72						

* Griffiths, *Phil. Mag.*, 1893. † See p. 74. ‡ Harker, 1905. § Greenwood, 1911.

LATENT HEAT OF FUSION

The number of gram calories required to convert 1 gram of substance from solid into liquid without change of temperature.

ICE

Temp.	Lt. ht.	Observer, etc.
	cals.	
−6·5° C.	76·03	Pettersson, 1881.
0	79·59	Regnault, 1843, corrected.
0	80·02	Bunsen, 1870, with ice calorimeter.
0	79·77	Smith, *Phys. Rev.*, 1903 (in terms of 15° calorie = 4·184 joules, taking Clark cell = 1·433 volts at 15° C.).

VARIOUS SUBSTANCES

Substance.	Temp.	Lt. ht.	Substance.	Temp.	Lt. ht.	Substance.	Temp.	Lt. ht.
	°C.	cals.		°C.	cals.		°C.	cals.
Elements—			Platinum .	1750	27	$NaNO_3$ . .	311	63
Aluminium .	657	77	Potassium .	62	16	KNO_3 . . .	339	47
Bismuth . .	269	13	Silver . . .	960	22	H_2SO_4 . . .	10·3	24
Cadmium . .	321	14	Sulphur . .	115	9	Acetic acid .	4	44
Copper . . .	—	43	Tin	232	14	Benzene . . .	5·4	30
Lead	327	5	Zinc	418	28	Glycerine . .	13	42
Mercury . .	—	3	**Compounds—**			Naphthalene .	80	35
Palladium .	1550	36	NH_3	−75	108	Xylene . . .	16	39
Phosphorus .	44	5						

LATENT HEAT OF VAPORISATION

Latent heats are given as the number of gram calories required to convert 1 gram of substance from liquid into vapour without change of temperature. The latent heat of vaporisation vanishes at the critical temperature.

Trouton's Rule.—The latent heat of vaporisation of 1 gramme molecule of a liquid divided by the corresponding boiling point (on the absolute scale) is a constant (C). C = 21 for substances of which both liquid and vapour are unassociated. If the liquid is associated, C > 21 (*e.g.* water, C = 26); if the vapour is associated, C < 21 (*e.g.* acetic acid, C = 15). [See Nernst's "Theoretical Chemistry."]

STEAM

Regnault's equation connecting latent heat and temperature takes no account of the temperature variation of the specific heat of water (see p. 56). The equation gives values which are too large at low temperatures. The equations of Griffiths, Henning, and Smith have been reduced and are here expressed in terms of the **15° calorie** = 4·184 joules. Griffiths' and Smith's results rest further on an attributed value of 1·433 volts for the e.m.f. of the Clark cell at 15° C.

See also next page. [The critical temp. of water is about 365° C.]

Observer.	Temp. range of expts.	Latent heat L_t at t° C.
Regnault, 1847 .	63°–194° C.	$L_t = 606·5 - ·695t$
Griffiths, 1895 .	30° and 40°	$L_t = 598·0 - ·605t$
Henning, *Ann. d. Phys.*, 1906,	30°–100°	$\left\{ \begin{array}{l} L_t = 599·4 - ·60t, \text{ to } ·3\,\% \\ \text{or } L_t = 94·3\,(365 - t)^{·3125}, \text{ to } ·1\,\% \end{array} \right.$
1909	100°–180°	$L_t = 538·97 - ·6428(t - 100) - ·0_3834(t - 100)^2$
Smith, *Phys. Rev.*, 1907 .	14°–40°	$L_t = 597·2 - ·580t$

LATENT HEAT OF STEAM (contd.)

In terms of 15° calorie	Regnault, 1847.	Griffiths, 1895.	Joly, 1895.	Callendar, *	Dieterici, 1905.	Henning, 1906.	Smith, 1911.	Richards & Matthews, 1911.
L_0	606†	598†	—	595†	596·0‡	599†	—	—
L_{100}	537	537·5†	540§	540	538·9‖	539·4	540·5	538·0

* From sp. ht. of steam experiments and total heat formula.   † Extrapolated.
‡ Reduced to mean calories (4·185 joules) ; Clark cell = 1·433 volts.
§ By comparing L_{100} (by steam calorimeter) with the mean specific heat of water between 12° and 100°.
Callendar and Barnes' specific heat has been used (p. 56).   ‖ Carlton-Sutton, 1917.

LATENT HEATS OF VAPORISATION OF VARIOUS SUBSTANCES

The values below are for pure substances, and are due to Young, *Proc. Roy. Dublin Soc.*, 1910. The precise calorie employed is not stated.

Temp.	SnCl₄	CCl₄	Pentane (n)	Methyl	Ethyl	Propyl	Ethyl ether.	Methyl	Ethyl	Propyl	Acetic acid.	Benzene.
				Alcohol.				Acetate.				
C.	cals.	cals.	cals.	cals.	cals.	cals.	cals.	cals.	cals.	cals.	cals.	cals.
0°	—	—	—	289·2	220·9	—	92·52	—	—	—	—	—
20	—	—	—	284·5	220·6	—	87·54	—	—	—	84·05	—
40	—	—	84·31	277·8	218·7	—	82·83	—	—	—	87·02	—
60	—	46·00	80·07	269·4	213·4	—	78·44	98·59	—	—	89·69	—
80	—	46·00	75·33	259·0	206·4	173·0	73·50	94·07	85·78	79·80	91·59	95·45
100	31·76	44·15	69·94	246·0	197·1	164·0	68·42	88·39	82·15	76·33	92·32	91·41
120	30·54	42·08	64·48	232·0	184·2	153·0	62·24	82·87	77·53	71·84	94·38	86·58
140	29·12	39·92	56·58	216·1	171·1	142·4	55·93	76·83	72·24	67·66	91·83	82·82
160	27·69	37·95	47·42	198·3	156·9	129·0	46·07	69·96	65·91	62·80	89·63	78·94
180	26·29	35·40	35·01	177·2	139·2	116·3	31·87	61·00	59·87	57·23	87·71	74·62
200	24·57	32·61	24·68*	151·8	116·6	102·2	19·38‡	50·56	52·71	50·78	85·55	68·81
220	22·82	29·45	—	112·5	88·2	85·3	—	34·87	42·63	42·40	82·02	62·24
240	20·86	25·56	—	84·51†	40·3	63·4	—	20·99§	27·17	30·70	78·18	54·11
260	18·50	20·07	—	—	—	33·5	—	—	12·03‖	11·73¶	72·26	43·82
280	15·60	10·43	—	—	—	—	—	—	—	—	63·48	27·43
Crit. temp.	318°·7	283°·1	197°·2	240°	243°·1	263°·7	193°·8	233°·7	250°·1	276°·2	321°·6	288°·5

* At 190°.   † At 230°.   ‡ At 190°.   § At 230°.   ‖ At 249°.   ¶ At 275° C.

Substance.	Temp	Lt. ht.	Substance.	Temp	Lt. ht.	Substance.	Temp	Lt. ht.
	C.	cals.		C.	cals.		C.	cals.
Mercury	358°	68	Liquid N₂O	−20°	67	Chloroform	61°	58
Sulphur	316	362	„ NH₃	—	341	Et. bromide	38	60
Phosphorus	287	130	„ CO₂	0	57	„ propionate	100	79
Liquid H₂	—	123	„ „	22	32	„ iodide	71	47
„ O₂	−188	58	„ SO₂	−10	96	„ formate	50	98
„ N₂	—	50	„ CS₂	46	85	Am. alcohol	131	120
„ air	c. 50		Me. formate	32·5	110·5	Aniline	—	104
„ Cl	−22	67	„ iodide	42	46	Toluene	111	84
Bromine	58	46	Chloroform	0	67	Turpentine	159	70
Iodine	174	24						

THERMOCHEMISTRY

In thermochemistry the conservation of energy is assumed in accordance with experiment, and consequently (1) if a cycle of chemical change takes place so that the final state of the reacting substances is identical with the initial, then as much heat is absorbed as is given out, *i.e.* the total heat of the reaction is zero ; (2) the heat of reaction only depends on the initial and final states of the reacting substances, and not on the intermediate stages. The results below are affected by, but have not been corrected for, any changes in the accepted values of the atomic weights since the experiments were carried out.

MOLECULAR HEAT OF FORMATION

The **molecular heat of formation** (H.F.) is the heat liberated when the molecular weight in grams of a compound is formed from its elements. When the state of aggregation of an element or compound is not given, it is the state in which it occurs at room temperature and pressure. A minus sign before an H.F. means that heat is absorbed in the building up of the compound.

Unit—the gram calorie (at 15° to 20° C.) per gm. molecule of compound. Aq = solution in a large amount of water. The reactions are at constant pressure.

Example.—H.F. of $CuSO_4 = 183,000$; of $CuSO_4 . Aq = 198,800$. ∴ the heat of solution of $CuSO_4 = 198,800 - 183,000 = 15,800$ cals. per gram mol.

(T., Thomsen, "Thermochemistry," trans. by Miss K. A. Burke ; B., Berthelot, *Ann. d. Chim. et d. Phys.*, 1878 ; T.B., mean of both these observers' values. For organic compounds, see p. 64.

INORGANIC COMPOUNDS

Compound.	Mol. H.F. in calories.	Compound.	Mol. H.F. in calories.	Compound.	Mol. H.F. in calories.
Non-Metals	× 10³		× 10³		× 10³
HCl gas . .	22·0, T.	CO_2 from amorph. C	97·3, B.T.	$NH_4Cl . Aq$.	72·4
HCl . Aq . .	39·3, T.	CO_2 from diamond	94·3, B.	$(NH_4)_2SO_4$	283, T.B.
HBr gas . .	8·4, T.			$(NH_4)_2SO_4 . Aq$	280·6
HBr . Aq . .	28·6, T.	B_2O_3 ; amp. B.	273, B.	$NH_4OH . Aq$.	90, B.
HI gas . . .	−6·1, T.B.	SiO_2Aq ; crys.	180, B.	BaO	126, T.
HI . Aq . .	+13·2, T.B.	As_2O_3 . . [Si	155, T.	$Ba(OH)_2$ . .	217, T.
HF „	+38·5	As_2O_5 . . .	219, T.	$BaCl_2$ . . .	197, T.
H_2O liq. . .	68·4, T.	CCl_4 from diamond	76, B.	$BaCl_2Aq$ . .	199·1, T.
„ „ . .	69·0, B.			Bi_2O_3 . . .	20
„ gas . .	58·1, B.	$SbCl_3$ solid .	91·4, T.	$BiCl_3$ . . .	91, T.
$H_2O_2 . Aq$ . .	47·0	$SbCl_5$ liq. . .	105, T.	$Cd(OH)_2$ . .	66, T.
H_2S from rhombic S. .	2·7, T.	CS_2 from diamond & rhombic S. .	−19, B.	$Cd + O + H_2O$	
NH_3 . . .	12			$CdCl_2$ . . .	93, T.
AsH_3 . . .	−36·7	C_2N_2 gas from diam. .	−74, B.	$CdSO_4$ . . .	222, T.
SbH_3 . . .	−87, B.			$CdSO_4 . 8/3H_2O$ on sol. in Aq	+2·66, T.
SiH_4 . . .	25	H_2SO_4 liq. . .	193, T.	$CdSO_4 . Aq$.	232·7, T.
SO_2 from rhombic S. .	70	$H_2SO_4 . Aq$ from rhombic S. . . .	210, T.	Cs_2O . . .	100
SO_3 liq. from rhombic S. .	103			CaO	131, T.
		HNO_3 liq. . .	41·6, B.	„ Moissan .	145
N_2O . . .	−19	$HNO_3 . Aq$ . .	49	$Ca(OH)_2$ „	229
NO	−21·6, T.	HCN gas from diam.	−30·5	CaC_2 . . .	−7·25
N_2O_3 . . .	−21·4, B.			$CaCl_2$ . . .	170, T.
$NO_2/22°$ . .	−1·7, B.	HCN liq. . .	−24·8	$CaCl_2 . Aq$ . .	187·4, T.
„ /150° . .	−7·6, B.	H_3PO_4 liq. . .	302	$CaSO_4$ . . .	318, T.
N_2O_5 liq. . .	3·6, T.			$CaCO_3$ . . .	270, T.
P_2O_5 solid . .	369	**Metals—**		$Ca(NO_3)_2$ . .	202, B.
$P_2O_5 . Aq$ . .	405	Al_2O_3 . . .	380, B.	CoO	64
CO from amorph. C. .	29, T.	$AlCl_3$ . . .	161	$CoCl_2$ . . .	76·5. T.
		$Al_2(SO_4)_3 . Aq$	880	$CoSO_4 . 7H_2O$	234, T.
CO from diamond .	26·1, B.	NH_4Cl . . .	76·3, T.B.	$Co(NO_3)_2 . 6H_2O$	119, T.
				CuO	37·2, T.
				$CuCl_2$ . . .	51·6

INORGANIC COMPOUNDS (contd.)

Compound.	Mol. H.F. in calories.	Compound.	Mol. H.F. in calories.	Compound.	Mol. H.F. in calories.
	$\times 10^3$		$\times 10^3$		$\times 10^3$
Metals (contd.)		$MgCl_2$ . . .	151, T.	$AgCl$ . . .	29·2, T.B.
$CuSO_4$ . . .	183. T.	$MgSO_4$ . . .	302, T.	Na_2O . . .	91 to 100
$CuSO_4$. Aq	198·8, T.	$MgSO_4$. Aq .	322	$NaHO$ . . .	102·3, T.B.
$CuSO_4$. $5H_2O$ } on sol. in Aq./	− 2·75	MnO . . .	91	$NaHO$. Aq .	112·2, T.B.
$AuBr_3$ . . .	8·8, T.	$MnCl_2$ . . .	112	$NaCl$ . . .	97·8, T.B.
$AuCl_3$ . . .	23, T.	Hg_2O . . .	24·9, T.	$NaNO_3$ . . .	111, T.B.
FeO . . . }	64·6	HgO . . .	21·1	Na_2SO_4 . . .	328·3, T.B.
$Fe_2O_3/400°$. }	196	Hg_2SO_4 . .	175	Na_2CO_3 . .	272, T.B.
Le Chatelier .		$HgCl$ . . .	31·3	SrO . . .	130, T.B.
$FeSO_4$. $7H_2O$.	240	$HgCl_2$ . . .	53·2	$Sr(OH)_2$ . .	217, B.
$FeSO_4$. Aq .	236	NiO . . .	59·7	$SrCl_2$ . . .	185, T.B.
$FeCl_3$ . . .	96, T.	$NiCl_2$ . . .	74·5, T.	$SrCl_2$. Aq .	196, T.
PbO . . .	50·3, T.	$NiSO_4$. Aq .	229, T.	Tl_2O . . .	42·2, T.
PbO_2 . . .	62·4	$PtCl_4$ . . .	59·4	$TlCl$ . . .	48·6, T.
$PbCl_2$ . . .	83, T.	K_2O . . .	97	Tl_2SO_4 . .	221, T.
$PbSO_4$ . . .	216, T.	KHO . . .	104, B.T.	SnO . . .	70
$Pb(NO_3)_2$ . .	105·5	KHO . Aq . .	117, B.T.	$SnCl_2$ . . .	81, T.
$Pb(NO_3)_2$. Aq	97·9	KCl . . .	106, B.T.	$SnCl_4$ . . .	128
Li_2O . . .	140	KCl . Aq . .	101·6, T.	ZnO . . .	85·4, T.
$LiOH$ . . .	111	KNO_3 . . .	119, B.T.	$ZnCl_2$ . . .	97·3, T.B.
$LiCl$ . . .	94, T.	K_2SO_4 . . .	344, T.B.	$Zn(NO_3)_2$. Aq	132
$LiCl$. Aq . .	102·4	Ag_2O . . .	5·9, T.	$ZnSO_4$ . . .	230·3, T.B.
Li_2SO_4 . . .	334, T.	"	7, B.	$ZnSO_4$. Aq .	248·7
$LiNO_3$ . . .	112, T.	$AgNO_3$ . . .	28·7, T.B	$ZnSO_4$. $7H_2O$ } on sol. in Aq/	− 4·26
MgO . . .	143, B.	$AgNO_3$. Aq .	23·3, T.		

MOLECULAR HEAT OF NEUTRALISATION

Unit—the gram calorie (at 15° to 20°) per gram molecule of base. Thus $KOH . Aq + HCl . Aq = KCl . Aq + H_2O + 13,750$ calories. Thomsen (= T.) observed at 18° to 20° C., and the final dilution was 3600 gms. (7200 for Na salts) per gm. mol. of base. Berthelot (= B.) used at least 2000 gms. of H_2O per 17 gms. of hydroxylion, − HO.

Base.	HCl	HF	HNO_2	HCN	$\frac{1}{2}H_2SO_4$	$\frac{1}{2}H_2CO_3$	$1H_3PO_4$	1 Oxalic.
	$\times 10^3$	$\times 10^3$	$\times 10^3$	$\times 10^3$	$\times 10^3$	$\times 10^3$	$\times 10^3$	$\times 10^3$
1NaOH .	13·74,T.; 13·7, B.	16·3,T.	13·7,T.; 13·5, B.	2·8	15·64, T.	10·1, T.; 10·2, B.	14·8, T.	13·8,T.
2NaOH .	—	—	—	—	31·38‡, T.	20·2 §, T.	27·1*, T.	28·3,T.
1LiOH .	13·85, T.	16·4 †	—	2·93	15·64, T.		—	—
1KOH .	13·7, T.; 13·6, B.	16·1	13·8, T.	2·8,T.	15·7, T.B.	10·1, B.	—	13·8,B.
1NH₄OH .	12·3, T.; 12·4, B.	15·2	12·3, T.	1·3, B.	14·3, T.B.	8·4, T. ; 5·3, B.	13·5, B.	12·7
$\frac{1}{2}Ca(OH)_2$	14·0, B.	18·4 †	13·9, B.	3·2	15·6, T.	9·3,† T. ; 9·8,† B.	—	—
$\frac{1}{2}Sr(OH)_2$.	13·8, T.	17·8 †	13·9, B.	3·15	15·4, T.	10·4,†T.B.	—	—
$\frac{1}{2}Ba(OH)_2$	13·9, B.	16·1	14·1,T.; 13·9, B.	3·15	18·4, B.T.	11·0,†T.B.	—	...
$\frac{1}{2}Mg(OH)_2$	13·8, B.	15·2	13·8, T.	1·5	15·3, B.T	8·95,† B.	—	—
$\frac{1}{2}Cu(OH)_2$	7·5, T.	10·1	7·6		9·2		—	—

* 3NaOH gives $34·0 \times 10^3$, T.　　† Base in solid state.　　‡ $1H_2SO_4$.　　§ $1H_2CO_3$.

HEATS OF COMBUSTION AND FORMATION OF CARBON COMPOUNDS, COAL, ETC.

Molecular heats of formation (H.F.) of organic compounds are deduced from their heats of combustion (H.C.), by subtracting the latter from the heat generated on burning the carbon and hydrogen contained in the compound. Experimental errors in the H.C. thus become magnified in the H.F. Heats of combustion determined by Thomsen are for the vapour of the compound at 18° C.; for the liquid the H.C. and H.F. would be greater by the latent heat of evaporation. Thomsen assumes H.F. of CO_2 from amorphous C as = 96,960 cal.; of water as 68,360 cal. per gm. molecule. · For H.F. of inorganic compounds, see p. 62.

The H.C. and H.F. of carbon compounds is an **additive** property (see Thomsen's "Thermochemistry"). Berthelot's bomb calorimeter has been of considerable importance in the modern experimental side of the subject.

Unit—the gram calorie (at 15° to 20°) per gram molecule.

Example.—16 gms. of methane, CH_4, give out 212,000 gram calories of heat when burnt at **constant pressure,** to water and CO_2 at 18° C.

(T., Thomsen, "Thermochemistry;" B., Berthelot.)

Compound.	H.C.	H.F.	Compound.	H.C.	H.F.
	$\times 10^3$	$\times 10^3$		$\times 10^3$	$\times 10^3$
Methane, CH_4	212, T. 213, B.	21·7	Me. acetate, $C_3H_6O_2$	399, T.	96·7
			Carb. bisulphide, CS_2	265, T.	− 26
Ethane, C_2H_6	370, T. 372, B.	28·6	Methylamine, CH_5N	258, T.	9·5
			Dimethylamine, C_2H_7N	420, T.	12·7
Propane, C_3H_8	529, T.	35·1	Aniline, C_6H_7N	838, T.	−17·4
Acetylene, C_2H_2	310, T. 314	−47·8	Pyridine, C_5H_5N	675, T.	−19·4
			Sugar, $C_{12}H_{22}O_{11}$	1364	
Ethylene, C_2H_4	333, T.	−2·7	Illuminating gas per cub. metre	5600 to 6500	
Benzene, C_6H_6	799, T.	−12·5			
Naphthalene, $C_{10}H_8$	1239		Coal (anthracite)	7·6 to 8·4	per gm.
Toluene, C_7H_8	956, T.	−3·5			
Me. alcohol, CH_4O	182, T.	51·4	Coal (brown)	4·7	,, ,,
Me. chloride, CH_3Cl	177, T.	22·6	Coke	6·9	,, ,,
Chloroform, $CHCl_3$	107, T.	2·1	Paraffin oil	9·8	,, ,,
Et. alcohol, C_2H_6O	340, T.	58·5	Wood	3·9 to 4·4	,, ,,
Et. ether, $C_4H_{10}O$	660, T.	70			
Et. chloride, C_2H_5Cl	334, T.	30·7	**Albumens—**		
Acetic aldehyde, C_2H_4O	282, T.	48·7	Casein	5·86	,, ,,
Formic acid, CH_2O_2	69·4, T.	95·9	Flesh	5·66	,, ,,
Acetic acid, $C_2H_4O_2$	225, T.	105·3	White of egg	5·67	,, ,,
Propionic acid, $C_3H_6O_2$	387, T.	109·4	Yolk of egg	8·12	,, ,,
Me. formate, $C_2H_4O_2$	241, T.	89·4	Hæmoglobin	5·9	,, ,,

MOLECULAR HEAT OF DILUTION

The heat set free or absorbed on diluting a gram molecule of liquid with water is the molecular heat of dilution: thus on diluting HCl to (HCl, **300** H_2O), 17,300 calories per 36·5 grams of HCl are set free; diluting $2NaCl$, $nH_2O (n = 20)$ to ($2NaCl$, **100**H_2O) absorbs 1060 cal. per 2×58.65 gm. of NaCl. **Unit**—the gram calorie (at 15° to 20°) per gram molecule. (See Thomsen, "Thermochemistry.")

HCl n = 0		HNO₃ n = 0		H₂SO₄ n = 0		NaHO n = 3		NH₃,* 		2NaCl n = 20		2NaNO₃ n = 12		Na₂SO₄ n = 50		ZnCl₂ n = 5		Zn(NO₃)₂ n = 10	
H_2O	$\times 10^3$	H_2O	$\times 10^3$	H_2O	$\times 10^3$	H_2O	$\times 10^3$	n H_2O	$\times 10^3$	H_2O	$\times 10^3$	H_2O	$\times 10^3$	H_2O	$\times 10^3$	H_2O	$\times 10^3$	H_2O	$\times 10^3$
1	5·37	1	3·28	1	6·38	5	2·13	1	1·26	100	−1·06	50	−2·26	100	−·665	10	1·85	15	·91
2	11·36	5	6·6	5	13·1	7	2·9	3	·385	200	−1·31	100	−3·29	200	−1·13	20	3·15	20	1·15
5	14·96	10	7·32	49	16·7	9	3·1	5·8	·21	400	−1·41	200	−3·86	400	−1·38	50	5·32	50	1·20
50	17·1	20	7·46	199	17·1	25	3·26	9·5	·02	—	—	400	−4·19	800	−1·48	100	6·81	100	1·11
300	17·3	320	7·49	1600	17·9	200	2·94	110	·00	—	—	—	—	—	—	400	8·02	200	1·07

* Heat developed on diluting $NH_3.nH_2O$ to $NH_3.200H_2O$ (Berthelot).

ENERGY AND WAVE-LENGTH OF FULL RADIATION

The radiation from a full or black body radiator depends both in quality and quantity upon the temperature. The total energy radiated (of all wave-lengths), from unit area in unit time, is given by *Stefan's law*, $E = K\theta^4$, where K is Stefan's constant and θ is the absolute temperature (see Optical Pyrometry, p. 47, and below).

The dependence of the quality on the temperature is expressed by *Wien's displacement law*, $\lambda_m\theta = $ const., where λ_m is the length of the particular waves which have maximum emissive power. Thus the emissive power E_m of the waves of length λ_m, varies as the 5th power of the temperature (absolute) : $E_m\theta^{-5} = $ const.

The emissive power of some particular wave-length λ is expressed accurately by

$$E_\lambda = C\lambda^{-5}/(e^{a/\lambda\theta} - 1) \quad \ldots \ldots \quad Planck's\ formula$$

where $C = \cdot353$ erg.-cm.^{2} sec.$^{-1}$, $a = 1\cdot431$ cm.-deg., and e is the base of Napierian logs.

At low temperatures or for short wave-lengths ($\lambda\theta < 3$ cm.-deg.) Planck's formula becomes (to $\cdot8$ % at least)—

$$E_\lambda = C\lambda^{-5}e^{-a/\lambda\theta} \quad \ldots \quad Wien's\ formula\ (see\ p.\ 47)$$

For long waves and high temperatures ($\lambda\theta > 730$ cm. deg.), we have (to 1 % at least)—

$$E_\lambda = C\lambda^{-4}\theta e^{-a/a} \quad \ldots \ldots \quad Rayleigh's\ formula$$

(See Preston's "Heat," 2nd edit. ; Kayser's "Spectroscopie," II. ; Lorentz's "Theory of Electrons," 1910.)

WIEN'S DISPLACEMENT LAW	STEFAN'S LAW
$\lambda_m\theta = $ const. $= A$. (See above). λ is measured in cms.	Total radiation from a full radiator $= K\theta^4$ (see above). K is in erg cm.$^{-2}$ sec.$^{-1}$ deg.$^{-4}$. $K = 5\cdot72 \times 10^{-5}$ (Millikan, 1917).

A	Observer.	K	Observer.
$\cdot2940$	Lummer and Pringsheim, 1899	$5\cdot45 \times 10^{-5}$	Kurlbaum, 1912
$\cdot2888$	Paschen and Wanner, *B. B.*, 1899	$5\cdot18$,,	Lummer and Pringsheim, *A. d. P.*, 1901
$\cdot2902$	Wanner, 1900		
$\cdot2940$	Paschen, *A. d. P.*, 1901	$5\cdot67$,,	Shakespeare, 1911
$\cdot2890$	Rubens and Kurlbaum, *A. d. P.*, 1901	$5\cdot89$,,	Keene, 1912

A. d. P., Ann. der Phys. ; B. B., Berlin Ber. ; C. R., Compt. Rend.

SOLAR CONSTANT AND TEMPERATURE OF SUN

The solar constant S is the energy received from the sun by the earth (at its mean distance) per sq. cm. in unit time, corrected for the loss by absorption in the earth's atmosphere.

The determination of the absorption loss is difficult ; it is best derived from simultaneous observations at high and low stations.

Langley and Abbot ("Smithsonian Reports," 1903 *et seq.*) give the following relation between atmospheric absorption and wave-length :—

Wave-length (Å.U. $= 10^{-8}$ cm.)	4000	6000	8000	10,000	12,000
Fraction transmitted	$\cdot49$	$\cdot74$	$\cdot85$	$\cdot89$	$\cdot91$

If R is the energy radiated in unit time from a sq. cm. of the sun's surface, then

$$R = \left\{\frac{earth's\ solar\ distance}{sun's\ radius}\right\}^2 \times S = \left\{\frac{9\cdot28 \times 10^7}{4\cdot33 \times 10^5}\right\}^2 \times S = 46,000S$$

Assuming the sun to be a full or black body radiator, its "effective" absolute temperature θ may be deduced either from (1) Stefan's law, $R = K(\theta^4 - T^4)$, where K is Stefan's constant (see above) and T is the earth's absolute temperature, or (2) Wien's displacement law, $\theta\lambda_m = $ const. (see above).

Langley and Abbot (ref. above) find the distribution of the energy of solar radiation among the different wave-lengths (λ) to be as follows :—

Wave-length (Å.U.) . . .	4000	4500	5000	5500	6000	7000	8000	10,000	12,000	14,500	21,000
Relative energy, E	$15\cdot2$	$18\cdot4$	19	16	14	11	$8\cdot8$	$5\cdot4$	$3\cdot2$	2	$\cdot6$

λ for $E_{max.} = 4900 \times 10^{-8}$ cm. Taking Wien's displacement law to be $\theta\lambda_{max.} = \cdot29$, and assuming the sun to be a full radiator, its temperature $\theta = 5920°$ absolute.

F

SOLAR CONSTANT AND TEMPERATURE OF THE SUN (*contd.*)

The values of S below are expressed in both (1) calories per min. per cm.^{2}, and (2) watts per cm.^{2} (1 calorie per sec. = 4·18 watts). The sun's mean temp. θ is in degrees C. absolute. Abbot and Fowle find the solar constant varies by about 8 %. (See Poynting and Thomson's "Heat;" Chree, *Nature*, **82**, 2090; Report (1910) of the International Union for Solar Research; and "Smithsonian Reports.")

Solar Const.		Sun's Temp.	Account.	Observer.
cals. min.$^{-1}$ cm.$^{-2}$	watts cm.$^{-2}$			
		Abs.		
—	—	5770°	Comparison with const. temp. Atmos. absorp. taken as 29 %	Wilson, 1902
—	—	5920	Using Wien's displacement law (above)	Langley & Abbot, '03
2·25	·154	7060	Gorner Grat, Switzerland	Scheiner, 1908
—	—	5610	Natl. Phys. Lab., England. Atmos. absorp. taken as 29 %	Harker & Blackie, '08
2·38	·166	5630 }	Mt. Blanc. Comparison with const. temp. Atmos. absorp., 9 % with zenith sun	{ Féry & Millochau { Féry, 1909
—	—	5360 }		
—	—	5630	Mt. Blanc. Atmos. absorp., 3·4 %	Millochau, 1909
2·1	·146	5970†	Washington (sea-level) and Mt. Wilson (6000 ft.)	Abbot & Fowle, '09
2·1	·146	5970†	Review of previous work	Bellia, 1910
1·925*	·134	5840†	Mt. Wilson (6000 ft.) and Mt. Whitney (14,500 ft.)	Abbot, 1910

* Mean value for period 1904–9 (*Nature*, 1911).
† Calculated from S, taking Stefan's const. as 5·3 × 10^{-12} watts cm.$^{-2}$ sec.$^{-1}$ deg.$^{-4}$.

THE CRYOSCOPIC CONSTANT

The cryoscopic constant, K, would be the depression of the freezing-point of a solvent when the molecular weight in grams of any substance (which does not dissociate or associate) is dissolved in 100 grams of the solvent, supposing the laws for dilute solutions held for such a concentration (Raoult, 1882). Van't Hoff (1887) showed that $K = R\theta^2/(100L)$, where R = the gas constant (see p. 5), θ the absolute freezing-point of the solvent, L its latent heat of fusion in ergs. **Example.**—For 1 gram-molecule of solute in 100 gms. of water—

$$K = 8·315 \times 10^7 \times (273·1)^2/(79·67 \times 4·184 \times 10^9) = 18·60$$

(See Whetham's "Theory of Solution," p. 149.)

Solvent.	M. pt.	Lat. ht. (cals.)	K		Solvent.	M. pt.	Lat. ht. (cals.)	K	
			Calcd.	Obsd.				Calcd.	Obsd.
Water	0° C.	79·6	18·6 {	18·58, G. 18·52 *	Benzene	5° C.	29·1, P.W.	53·3	49, R.
$H_2SO_4.H_2O$	8·4	31·7, B.	50	48, L.	"	5·5	30·1, F.	51·6	51·2, P.
$SbCl_3$	73·2	13·4, T.	174	184, T.	Formic acid	8	57·4, Pe.	27·5	28, R.
Acetic acid	17	43·7, Pe.	38·5	39, R.	Phenol	40	24·9, P.W.	78·6	72·7, E.
Aniline	− 6	—	—	58·7, A.R.	p. Xylol	16	39·3, C.	42·5	43, P.M.

* Mean of six observers; A.R., Ampola and Rimatori, 1897; B., Berthelot; C., Colson; E., Eykman, 1889; F., Fischer; G., Griffiths (who used 0·0005 to 0·02 normal sugar solutions); L., Lespieau, 1894; P., Paternò, 1889; Pe., Pettersson; P M., Paternò and Montemartini, 1894; P.W., Pettersson and Widman; R., Raoult; T., Tolloczko, 1899.

VELOCITY OF SOUND

The velocity of sound (longitudinal waves) in a body, $V = \sqrt{E/\rho}$, E being the elasticity, and ρ the density. In gases and liquids E is the adiabatic volume elasticity; in isotropic solid rods or pipes E is Young's Modulus. For gases $V = \sqrt{\gamma P/\rho}$, P being the pressure, and γ the ratio of the specific heat of the gas at constant pressure to that at constant volume. For values of γ, see p. 58.

For moderate temperature variations, the velocity of sound in gases is given by $V_t = V_0(1 + \frac{1}{2}at) = V_0 + 61t$ in cms. per sec. for dry air ($a = \cdot 00367$).

The velocity of sound decreases with decreasing intensity down to the normal value. In gases in tubes the velocity increases with the diameter up to a limiting value for free space. The values below are for free space. Barton's "Sound" and Poynting and Thomson's "Sound" may be consulted. [1 foot = 30·48 cms.]

Substance.	Temp.	Velocity.	Observer.
Gases—		cms./sec.	
Air (dry)	0° C	$(3\cdot3133) \times 10^4$	Calcd. ($\gamma = 1\cdot401$)
,, ...	0	3·3136 ,,	Violle, 1900
,, ...	0	3·3132 ,,	Stevens, 1900
,, ...	0	3·3129 ,,	Hebb, 1905
,, ...	0	3·3192 * ,,	Thiesen, 1908 ‡
,, ...	− 45·6	3·056 ,,	Greely, 1890
,, ...	−182 4	1·815 ,,	Cook, 1906
,, ...	100	3·865 ,,	Stevens, 1900
,, ...	500	5·53 ,,	,,
,, ...	1000	7·0 ,,	,,
,, (Krakatoa wave)	—	3·21 ,,	1883
,,Sound-waves from sparks	0	3·50–4·45 ,, †	Töpler, 1908
Hydrogen	0	12·86 ,,	Zoch, 1866
Oxygen	0	3·172 ,,	Dulong, 1829
,,	−184·7	1·737 ,,	Cook, 1906
Nitrous oxide, N_2O .	0	2·60 ,,	Wullner, 1878
Ammonia, NH_3 . .	0	4·16 ,,	,,
Carbon monoxide .	0	3·371 ,,	,,
Carbon dioxide . .	10–24	2·573 ,,	Low, 1894
Coal-gas	0	4·9–5·15 ,,	—
Sulphur dioxide . .	0	2·09 ,,	Masson, 1857
Water-vapour . .	0	4·0 ,,	,,
,, (satd.)	110	4·13 ,,	Treitz, 1903
Liquids—			
Water	8·1	$14\cdot35 \times 10^4$	Colladon & Sturm, 1827
,, ...	4	13·99 ,,	Martini, 1888
,, ...	25	14·57 ,,	,,
,, (sea) Explosion waves	18	17·3–20·1 ,, †	Threlfall & Adair, 1889
Alcohol (abs.), C_2H_6O	8·4	12·6 ,,	Martini, 1888
Ether, $(C_2H_5)_2O$.	0	11·4 ,,	,,
Turpentine, $C_{10}H_{16}$.	3·5	13·7 ,,	,,

* Free from CO_2. † The range of speeds is given by varying intensities. ‡ Reichsanstalt.

The values for metals are due to Wertheim, 1849; Masson, 1857; and Gerossa, 1888.

Solid.	Velocity cms./sec.	Solid.	Velocity cms./sec.	Solid.	Velocity cms./sec.
Aluminium . .	$51\cdot0 \times 10^4$	Lead . . .	$12\cdot3 \times 10^4$	Brass . . .	$c.\ 36\cdot5 \times 10^4$
Cadmium . .	23·1 ,,	Nickel . . .	49·7 ,,	Deal (along	49–50 ,,
Cobalt . . .	47·2 ,,	Platinum . .	26·8 ,,	grain)	
Copper . . .	39·7 ,,	Silver . . .	26·4 ,,	Fir ,,	42–53 ,,
Gold	20·8 ,,	Tin	24·9 ,,	Mahogany ,,	41–46 ,,
Iron (wrought)	49–51 ,,	Zinc	36·8 ,,	Oak · ,,	40–44 ,,
,, (cast) . .	$c.\ 43$,,	Glass (soda) .	50–53 ,,	Pine ,,	$c.\ 33$,,
Steel	47–52 ,,	,, (flint) .	$c.\ 40$,,	Indiarubber .	·5–·7 ,,

VELOCITY (IN AIR) AND PRESSURE Koch (1907).			SENSITIVENESS OF EAR TO PITCH Rayleigh (1907).	

VELOCITY (IN AIR) AND PRESSURE. Koch (1907).

Press. in atmos.	Relative Velocity of Sound.		Frequency.	Condensation for same audibility.
	0° C.	−79·3° C.		
1	1·000	·842	512	1
·25	1·008	·831	256	1·6
50	1·022	·830	128	3·2
100	1·064	·885	85	6·4
150	1·132	1·047		
200	1·220	1·239		

ORGAN PIPES
End Correction.

For a pipe with a flange at the open end, the antinode is situated ·82 (radius of pipe) beyond end. With no flange, the end-correction is ·57 (radius). (See Lamb's "Sound." [1910.]

Wave-length.
L = length of pipe.

Closed pipe . . $4L, \dfrac{4L}{3}, \dfrac{4L}{5}$, etc.

Open pipe . . . $2L, \dfrac{2L}{2}, \dfrac{2L}{3}$, etc.

TRANSVERSE VIBRATIONS OF RODS

L, length ; K, radius of gyration of cross-section ; E, Young's Modulus ; ρ, density.

	No. of Nodes.	Distance of Nodes from one end.	Frequency $\propto \dfrac{K}{L^2}\sqrt{\dfrac{E}{\rho}}$
Both ends free	2	·224 L ; ·776L	1
	3	·132L ; ·5L ; ·868L	2·76
	4	{ ·094L ; ·356L } { ·644L ; ·906L }	5·40
One end fixed	0	—	1
	1		6·27
	2	·226L	17·5
	3	·132L ; ·5L	34·4
		·094L ; ·356L ; ·644L	

Temp. correction of Frequency (n) of a Tuning-fork.
(M'Leod and Clarke, 1880, and König)
$$n_t = n_0(1 - ·00011t)$$

The pressure exerted by Sound waves has been measured directly up to ·24 dyne/cm². (Altberg, 1903)

THE EAR

Shortest time perceivable by ear (Hill, 1908)	·007 sec.
Amplitude of faintest audible sound (Rayleigh, 1877)	10·4 × 10⁻⁸cm.
Ditto (Shaw, 1904)	1·4 × 10⁻⁸ cm.
Pressure variation to which normal ear can respond (Abraham, 1907)	c.4 × 10⁻⁷mm. mercury.
Lower limit of audition in vibns./sec.	About 30.
Upper limit of audition in vibns./sec.	24,000 to 41,000.
Extreme range of ear	c. 11 octaves.
Musically available	c. 7 ,,
Highest pitch in piano	3520
Highest pitch in orchestra (piccolo d′)	4752
Lowest pitch in largest organs (64-foot pipe)	8

FREQUENCY RATIOS OF MUSICAL SCALE

	C Doh	D Ray	E Me	F Fah	G Soh	A Lah	B Te	c Doh
Natural scale	1 24 1·000	9/8 27 1·125	5/4 30 1·250	4/3 32 1·333	3/2 36 1·500	5/3 40 1·667	15/8 45 1·875	2 48 2·000
Equally tempered scale	1·000	1·122	1·260	1·335	1·498	1·682	1·888	2·000
Standard forks (König) (marked c′ = 512 and so on)	c′ 256	d′ 288	e′ 320	f′ 341·3	g′ 384	a′ 426·7	b′ 480	c″ 512

The French Standard, "Diapason Normal" of 1859 (which adopts a fork having c″ = 522 at 20° C.) is coming into general adoption for organs and pianos in England, the Continent, and America, as the result of a makers' conference in 1899. Other scales in vogue are Concert Pitch (c″ = 546), Society of Arts (c″ = 528), Tonic Sol-fa (c″ = 507), Philharmonic (c″ = 540). (The "middle" c of the piano is c′.)

VELOCITY OF LIGHT IN VACUO

Mean value *in vacuo* = $2 \cdot 9986 \times 10^{10}$ cm./sec. = 186,326 miles/sec.   For values of v, the ratio between the E.M. and E.S. units, see below.

cm./sec.	Method.	Observer.	cm./sec.	Method.	Observer.
$\times 10^{10}$			$\times 10^{10}$		
3·07	⎱ Eclipse of one of	Römer, 1676	2·999	Rotating mirror	Michelson, 1879
2·998	⎰ Jupiter's moons	,, corrected	3·014	Toothed wheel	Young & Forbes, '81
3·153	Toothed wheel	Fizeau, 1849	2·9985	Rotating mirror	Michelson, 1882
2·986	Rotating mirror	Foucault, 1862	2·9986	,, ,,	Newcomb, 1882
3·004	Toothed wheel	Cornu, 1878	2·9986	Toothed wheel	Perrotin, 1900

VELOCITY OF LIGHT IN LIQUIDS

Liquid.	Vel. in vacuo / Vel. in liquid	Refractive index for Na D line.	Method.	Observer.
Water . .	1·330	1·333/20°	Rotating mirror	Michelson, 1883
CS₂ . . .	1·758	1·627/20°	,, ,,	,, ,,

VELOCITY OF HERTZIAN WAVES

(See Blondlot and Gutton, *Rep. Cong. Phys.*, Paris, 1900.)

cm./sec.	Observer.	cm./sec.	Observer.	cm./sec.	Observer.
$\times 10^{10}$		$\times 10^{10}$		$\times 10^{10}$	
2·989	Blondlot	3·003	Trowbridge	2·989	Saunders
2·991	McClean		and Duane	**2·991**	**Mean**

RATIO OF ELECTROMAGNETIC TO ELECTROSTATIC UNIT OF CHARGE

This ratio "v" is a pure number, and is numerically equal to $\sqrt{\mu k}$, *i.e.* on Maxwell's theory, to the velocity of electric disturbances, such as light and Hertzian waves, through a medium whose magnetic permeability is μ and specific inductive capacity k.   (See pp. 7 and 84.)   For the velocity of light, see above.

Most observers have used a "capacity method" of determining v.   (See Gray, "Absolute Measurements; and Rosa, *Bull. Bureau of Standards*, 1907.)

v	Observer.	v	Observer.	v	Observer.
$\times 10^{10}$		$\times 10^{10}$		$\times 10^{10}$	
2·963	J. J. Thomson, 1883	2·997	Thomson and Searle, 1890	3·001	Hurmuzescu, '96
2·982	Rowland, 1889	3·009	Pellat, 1891	2·997	Perot and Fabry
3·000	Rosa, 1889	2·993	Abraham, 1892	2·997	Rosa & Dorsey, 1907

PHOTOMETRIC STANDARDS

The Geneva Congress of 1896 proposed a set of units for measuring (1) luminous intensity, (2) flux (the "lumen"), (3) illumination (the "lux"), (4) brightness, and (5) quantity of light (see *Electrician*, July 14, 1911). The British unit of intensity is the "candle." The **mean spherical candlepower** of a light is the mean of the intensities measured in all directions from the light. The **mean horizontal candlepower** is the mean of all the intensities in a horizontal plane through the lamp.

The **British "candle"** is a spermaceti candle, $\frac{7}{8}$ inch in diameter (6 to the lb.) which burns at the rate of 120 grains per hour. This is, however, found to be an unsatisfactory standard, and in modern photometry the British unit is taken as being one-tenth part of the light given out by the Harcourt 10 candlepower Pentane lamp, burning at a pressure of 760 mms. mercury in an atmosphere containing 8 parts in 1000 by volume of water-vapour as measured by a ventilated hygrometer. The candlepower of this lamp

$$= 10 + \cdot 066(8 - w) - \cdot 008(760 - H)$$

where w is the number of parts in 1000 (by vol.) of water-vapour in air at a barometric pressure of H mms. of mercury.

The **United States "candle"** prior to April 1, 1909, was 1·6% greater than the British.

The **French unit** is the Bougie decimale, which is the 20th part of the light given out by a sq. cm. of platinum at its solidifying point. This is a difficult unit to reproduce, and the Carcel lamp burning colza oil is used in practice. The Carcel unit is taken (with some uncertainty) as 4% less than the Bougie decimale.

The **German unit** is the light given out by the Hefner lamp (which burns amyl acetate), burning at a pressure of 760 mms. mercury in an atmosphere containing 8·8 parts in 1000 (by vol.) of water-vapour as measured by a ventilated hygrometer.

The National Physical Laboratory, the Bureau of Standards of America, and the Laboratoire Central d'Electricité of Paris have come to an agreement which involves the reduction of the old value of the American candle by 1·6%. They agree in future to employ as a common unit the proposed **International candle** = 1 British Pentane candle = 1 American candle = 1 French Bougie decimale = 10/9 German Hefner unit = ·104 Carcel unit (see Paterson, *Phil. Mag.*, 1909).

EFFICIENCIES OF VARIOUS LIGHTS

It has become customary to express efficiencies (or rather inefficiencies) in watts per candle. The value of a luminous efficiency cannot be properly appreciated without a knowledge of the distribution of the intensity. Estimates of the proportion of light energy to the total energy vary widely. S. P. Thompson ("Manufacture of Light") quotes from 1 part in 7000 for a gas flame to 1% for the most efficient lights.

The usual accepted "efficiencies" are given below in watts per mean spherical candlepower. They must only be regarded as approximate (see Solomon, "Electric Lamps," 1908).

Light.	Efficiency.	Light.	Efficiency.
Bat's-wing gas flame	c. 100	Tantalum lamps	1·7–2·1
Paraffin lamps	c. 50	Tungsten (osram, etc.) lamps .	1·3
Welsbach mantle, etc.	c. 15	Open arc lamps	1·1–1·4
High-pressure gas	c. 8	Enclosed arc lamps	2·3
Carbon filament lamps	3·5–4·5	Yellow flame arc lamps . . .	·4
Metallized carbon filament lamps	2·8	Mercury vapour lamps	·3–·4
Nernst lamps	2·1–2·4		

In high-grade standard photometry the Lummer Brodhun photometer head is usually employed. A unit of light may be maintained and reproduced with an accuracy of the order of $\frac{1}{10}$%, by means of sets of properly seasoned glow lamps.

The candlepower of a carbon glow lamp varies as the 6th power (approx.) of the voltage; of a metallic filament lamp, as the 3·6th power.

A candle is visible at about a mile on a clear dark night. The energy in the luminous radiation from a standard candle is about 5×10^5 ergs/sec. (Rayleigh, "Collected Papers"), whence the energy falling on 1 sq. cm. at a distance of 1 metre would be 4 ergs per sec. Angström (1902) gets values about double these.

GASEOUS REFRACTIVE INDICES AND DISPERSIONS

Dispersion.—Cauchy's equation is $\mu - 1 = A(1 + B/\lambda^2)$, where μ is the refractive index for the wave-length λ; A and B are constants. B is the coefficient of dispersion. The **refractivity** $(\mu - 1) = A$, when $\lambda = \infty$. The values of A and B are for wave-lengths measured in cms. The refractive indices are mostly for the sodium D line ($\lambda = 5893 \times 10^{-8}$ cm.). The values of μ are reduced to a standard density at $0°$ and 760 mms. by assuming that $(\mu - 1)/\rho$ is a constant for each gas, ρ being the density. Cauchy's formula is in general inadequate over large dispersions. (See Cuthbertson, *Science Progress*, 1908 ; and *Proc. & Trans. Roy. Soc.* for 1905 *et seq.*)

Gas or Vapour.	Refractive Index μ for Na D line.	Cauchy's Constants. A.	B.	Observer.
Air . . .	1·0002918	$28·71 \times 10^{-5}$	$5·67 \times 10^{-11}$	Scheel (Reichsanstalt), 1907
Hydrogen . .	1·0001384	13·58 ,,	7·52 ,,	,,
Helium . .	1·0000350	3·48 ,,	2·3 ,,	Burton; Cuthbertson & Metcalfe,1907
Neon . .	1·0000671	6·66 ,,	2·4 ,,	C. & M. Cuthbertson, 1909
Argon . .	1·0002837	27·92 ,,	5·6 ,,	Burton, 1907
Krypton .	1·0004273	41·89 ,,	6·97 ,,	C. & M. Cuthbertson, 1908
Xenon . .	1·000702	68·23 ,,	10·14 ,,	,, ,,
Fluorine .	1·000195	—	—	Cuthbertson & Prideaux, 1906
Chlorine .	1·000768	—	—	Mascart, 1878
Bromine .	1·001125	—	—	,, ,,
Iodine . .	1·00192 †	—	—	Hurion, 1877
Oxygen .	1·000272	26·63 ,,	5·07 ,,	Rentschler, 1908
Sulphur . .	1·001111	104·6 ,,	21·2 ,,	Cuthbertson & Metcalfe, 1908
Selenium .	1·001565	—	—	,, ,,
Tellurium .	1·002495	—	—	,, ,,
Nitrogen .	1·000297	29·06 ,,	7·7 ,,	Scheel (Reichsanstalt), 1907
Phosphorus .	1·001212	116·2 ,,	15·3 ,,	Cuthbertson & Metcalfe, 1908
Arsenic . .	1·001552	—	—	,, ,,
Zinc . . .	1·002050	—	—	,, ,,
Cadmium .	1·002675	—	—	,, ,,
Mercury .	1·000933	87·8 ,,	22·65 ,,	,, ,,

Gas or Vapour.	Refractive Index μ for Na D line.	Observer.	Gas or Vapour.	Refractive Index μ for Na D line.	Observer.
Water-vapour . .	1·000257	Mascart, '78	Tellurium tetra-chloride . . .	1·002600	P. & M.
,,	1·000250	Lorenz, '74	Phosph. hydrogen	1·000786 *	Dulong, '26
Ammonia . . .	1·000377	Mascart, '78	Phosphorus tri-chloride . . .	1·001730	Mascart, '78
,, . .	1·000373	Lorenz, '74	Methane, CH_4 .	1·000441	,, ,,
Nitrous oxide . .	1·000515	Mascart, '78	Pentane, C_5H_{12} .	1·001701	,, ,,
Nitric oxide .	1·000297	,, ,,	Acetylene, C_2H_2 .	1·000606	,, ,,
Hydrochloric acid	1·000444	,, ,,	Ethylene, C_2H_4 .	1·000719	,, ,,
Hydrobromic acid	1·000570	,, ,,	,,	1·000674	Prytz, '80
Hydriodic acid .	1·000906	Hurion, '77	Benzene, C_6H_6 .	1·001812	Mascart, '78
Carbon monoxide	1·000334	Mascart, '78	,,	1·001765	Prytz, '91
,, dioxide	1·0004498	Perreau, '96	Methyl fluoride .	1·000449	Cuthbertson
,, bisulphide	1·001476	Mascart, '78	,, chloride .	1·000865	Mascart, '78
Sulph. hydrogen	1·000641 *	Dulong, '26	,, alcohol .	1·000552	Prytz, '80
,, ,,	1·000619	Mascart, '78	,,	1·000619	Mascart, '78
Sulphur dioxide .	1·000660	Walker, '03	Chloroform, $CHCl_3$	1·001455	,, ,,
,, trioxide .	1·000737	C. & M., '08	Carbon tetra-chloride . . .	1·001768	,, ,,
,, hexafluoride	1·000783	,,			
Selenium ,,	1·000895	,,			
Tellurium ,,	1·000991	,,			

* White light.   † Violet light. $\mu = 1·00205$ for red light.   Iodine shows anomalous dispersion.
C. & M., Cuthbertson & Metcalfe ; P. & M., Prideaux & Metcalfe.

REFRACTIVE INDICES

Refractive indices, μ, (against air) at 15° C. for various wave-lengths.

The **temperature coefficient** given below is the change of refractive index per 1° C. rise of temperature for the case of the sodium D line.

The refractive indices are due chiefly to Gifford (*Proc. Roy. Soc.*, 1902, 1904, 1910); Rubens and Paschen (for the infra-red) and Martens (1902). The two Jena glasses are selected as typical. Other glasses are dealt with on p. 74.

Wave-length in A.U. (10⁻⁸ cm.).	Calcspar, 18° ord. ray.	ext. ray.	Jena glass. Crown*	flint.†	Fluorite, CaF₂ 18°.	Quartz, 18° ord. ray.	ext. ray.	Fused silica.	Rock salt, 18°.	Sylvin, KCl 18°.	Water at 20°.
Infra-red.	1·	1·	1·	1·	1·	1·	1·	1·	1·	1·	1·
223,000	----	----	----	----	----	----	----	----	3403	3712‡	----
94,290	----	----	----	----	3161	----	----	----	4983	4587	----
42,000	----	----	----	----	4078	4569	----	----	5213	4720	----
21,720	6210	4746	4946	6153	4230	·5180	5261	----	5262	4750	----
12,560	6388	4782	5042	6268	4275	5316	5402	----	5297	4778	3210
Visible.											
Li, (r) 6708	6537	4813	5140	6434	4323	5415	5505	4561	5400	4866	3308
H, (C) 6563	6514	4846	5145	6414	4325	5419	5509	4564	5407	4872	3311
Cd, (r) 6438	6550	4847	5149	6453	4327	5423	5514	4568	5412	4877	3314
Na, (D) 5893	6584	4864	5170	6499	4339	5443	5534	4585	5443	4904	3330
Hg, (g) 5461	6616	4879	5191	6546	4350	5462	5553	4602	5475	4931	3345
Cd, (g) 5086	6653	4895	5213	6598	4362	5482	5575	4619	5509	4961	3360
H, (F) 4861	6678	4907	5230	6637	4371	5497	5590	4632	5534	4983	3371
Cd, (b) 4800	6686	4911	5235	6648	4369	5501	5594	4636	5541	4990	3374
Hg, (v) 4047	6813	4969	5318	6852	4415	5572	5667	4697	5665	5097	3428.
Ultra-violet.											
Sn 3034	7196	5136	5552	----	4534	5770	5872	4869	6085	5440	3581
Cd 2144	8459	5600	----	----	4846	6305	6427	5339	7322	6618	4032
Al 1852	----	----	----	----	5099	6759	6901	5743	8933	8270	----
Temp. coefficient (D)	$+\cdot0_55$	$+\cdot0_414$	$-\cdot0_51$	$+\cdot0_53$	$-\cdot0_41$	$-\cdot0_55$	$-\cdot0_56$	$-\cdot0_53$	$-\cdot0_44$	$-\cdot0_44$	$-\cdot0_48$

* Light barium crown. † Dense silicate flint. ‡ $\mu = 1\cdot3692$ for $\lambda = 225,000$.

REFRACTIVE INDICES

Refractive indices μ_D (against air) at 15° C. for sodium D line ($\lambda = 5893 \times 10^{-8}$ cm.).

Substance.	μ_D	Substance.	μ_D	Substance.	μ_D
Solids.		Alcohol, ethyl . .	1·362	Monobrom benzene	1·563
Alum (potash) . .	1·456	„ amyl . .	1·41	„ „ naphthalene.	1·660
Cyanin . . .	1·71	Aniline	1·590	Nitrobenzene . .	1·553
Diamond . . .	2·417	Benzene	1·504	Oil, cedar	1·516
Glass (see above and p. 74)		Bromoform . . .	1·591	„ cloves . . .	1·532
Ice	1·31	Canada balsam . .	1·53	„ cinnamon . .	1·601
Mica . . 1·56 to	1·60	Carb. bisulphide .	1·632	„ olive . . .	1·46
Ruby	1·76	„ tetrachloride	1·464	„ paraffin . . .	1·44
Sugar	1·56	Chloroform . . .	1·449	Sulphuric acid . .	1·43
Topaz	1·63	Ether, ethyl . . .	1·354	Turpentine . . .	1·47
Liquids.		Ethylene dibromide	1·540	Water (see above).	1·333
Alcohol, methyl .	1·33	Glycerine	1·47		
		Methylene iodide .	1·744		

DISPERSIVE POWERS

The dispersive power (ω) given below $= (\mu_C - \mu_F)/(\mu_D - 1)$, where μ_C, μ_D, μ_F are the refractive indices corresponding to the red (C) H line (6563), the yellow Na (D) line (5893), and the green-blue (F) hydrogen line (4862).

Substance.	ω	Substance.	ω	Substance.	ω
Solids.		Quartz, ord. . .	·0143	**Liquids.**	
Calcite, ord. . .	·0204	„ ext. . .	·0146	Carb. bisulphide .	·0545
„ ext. . .	·0125	Fused silica . . .	·0147	Alcohol	·0171
Fluorite	·0105	Rock salt . . .	·0233	Turpentine . . .	·0206
Glass (see p. 74)		Sylvin	·0226	Water	·0180

SILVERING SOLUTION

Due to the late Dr. Common. Other recipes will be found in Baly's "Spectroscopy" (Longmans) and Woollatt's "Laboratory Arts" (Longmans).

Make up 10 % solutions of (1) pure nitrate of silver, $AgNO_3$; (2) pure caustic potash, KOH; (3) loaf sugar; and (4) ammonia (90 % water, 10 % ammonia of sp. gr. ·880). To the sugar soln. add ½ % of pure nitric acid and 10 % of alcohol. The sugar soln. is very much improved by keeping. Make up also a 1 % soln. of $AgNO_3$. Distilled water must be used for all the solns.

For silvering say a 12-in. mirror, take 400 c.c. of the $AgNO_3$ soln. and add strong ammonia until the brown precipitate first formed is nearly dissolved, then use the 10 % ammonia until the soln. is just clear. Add 200 c.c. of the KOH soln. A brown precipitate is again formed, which must be dissolved in ammonia exactly as before, the ammonia being added until the liquid is just clear. Now add the 1 % soln. of $AgNO_3$ until the liquid becomes a light brown colour about equal in density of colour to sherry. This colour is important, and can only be properly obtained by the use of the weak soln. Dilute the liquids to 1500 c.c. with distilled water.

The mirror should be thoroughly cleaned with acid and placed in a dish of distilled water.

All being ready, add 200 c.c. of the sugar soln. to 500 c.c. of water; add the mixture to the silver-potash soln., mix thoroughly, and pour them into a clean empty dish. Then lift the mirror out of its dish of distilled water and place it face downwards in this soln., taking care to exclude all air-bubbles.

The liquid will turn light brown, dark brown, and finally black. In four or five minutes, often sooner, a thin film of silver will commence to form on the mirror, and this will thicken until in about 20 minutes the whole liquid has acquired a yellowish-brown colour, with a thin film of metallic silver floating on the surface. Half an hour is the usual time taken in silvering, but this is shortened by using warmer liquids. About 18° C. is the best temperature.

Lift the mirror out, thoroughly wash with distilled water, and stand on its edge for say 12 hours in an inclined position until it is dry. The slight yellowish "bloom" can then be polished off by rubbing softly with a pad of chamois leather and cotton-wool. The subsequent polishing is done with a little dry well-washed rouge on the leather pad. The film should be opaque and brilliant, and with careful handling will be very little changed with long use.

Porcelain, glass, or earthenware dishes should be used.

If a very thick film is required, two silvering baths can be used, the article being left in the first bath for 15 minutes, then lifted out, rinsed with distilled water and at once immersed in the second bath, which should be ready in another dish. The film should not be allowed to dry during the operation of changing baths.

NOTE.—The silver-potash solution will not keep beyond a couple of hours. Any excess of this solution unused should have the silver precipitated at once with HCl. If the silver-potash is kept, say for 10 or 12 hours, a black powder collects on the surface. This powder, which is probably some form of fulminate of silver, is explosive, and may shatter the vessel.

GLASS

The **raw materials** for the manufacture of glass are (1) silica—usually as sand or felspar ; (2) salts of the alkali metals—Na_2SO_4, Na_2CO_3, or K_2CO_3 ; (3) salts of bases other than alkalies—red lead, limestone or chalk, $BaCO_3$ or $BaSO_4$, $MgCO_3$, ZnO, MnO_2, Al_2O_3, As_2O_3, etc. In general, glasses rich in silica and lime are hard, while glasses in which alkali, lead, or barium preponderate are soft. Hardness is, of course, also largely dependent on annealing. Ordinary " soft " (*i.e.* easily fusible) German glass is a soda-lime glass rather rich in alkali ; "hard" (refractory) glass is a potash-lime glass rather rich in lime. Jena combustion tubing is a borosilicate containing some magnesia.

Thermometry Glasses.—Glasses which contain **both** soda and potash to any extent give a large temporary zero depression (see p. 45). Data concerning *Verre dur* (71% SiO_2, 12% Na_2O, ½% K_2O, 14% CaO, 2% Al_2O_3 and MgO), *Jena 16'''* (67% SiO_2, 14% Na_2O, 7% CaO, 12% ZnO, Al_2O_3 and B_2O_3), *Jena 59'''* (72% SiO_2, 12% B_2O_3, 11% Na_2O, 5% Al_2O_3), *Kew glass* (44% SiO_2, 34% PbO, 12% K_2O, 2% Na_2O, 2% CaO, MgO, etc.), will be found on p. 45.

Optical Glasses.—In building up achromatic lens systems a knowledge of the dispersive power (ω) of each glass employed is essential. This is defined as the ratio of the difference of the deviations (*i.e.* the dispersion) for any two colours to the deviation of some mean intermediate colour. ω thus depends on the colours selected ; for visual work they are usually the red (C) line of hydrogen (wave-length $\lambda_C = 6563 \times 10^{-8}$ cm.), the yellow sodium (D) line ($\lambda_D = 5893$), and the green-blue (F) hydrogen line ($\lambda_F = 4862$). If μ_C, μ_D, μ_F are the corresponding refractive indices, $\omega = (\mu_C - \mu_F)/(\mu_D - 1)$ for the brightest part of the visible spectrum.

Flint glass—a term which survives from times when ground flints were extensively employed in making the best glass—now always implies a dense glass which contains lead and has a high refractive index and dispersive power.

Crown glass, originally designating only lime-silicate glasses, is now applied generally to glasses having a low dispersive power.

Jena Optical Glasses.—For ordinary flints and crowns ω and μ are roughly proportional, and this was true for all commercially available glasses prior to the advances initiated in 1881 by Abbé and Schott at Jena. They succeeded (*e.g.* by the addition of barium) in producing glasses which do not obey any such proportionality ; *e.g.* the very valuable barium crown glasses (below) combine the high refractive index of a flint glass with the low dispersive power of a crown. Such glasses have brought about the excellent achromatism and flatness of field which now obtain in photographic lenses and large telescopic objectives. The introduction of boron into a glass lengthens the blue end of the spectrum relatively to the red ; the addition of phosphorus, fluorine, potassium, or sodium has the opposite effect : such control over the dispersion has made the modern microscope possible.

Some typical examples of Jena glasses are subjoined. For a complete list, see the catalogue of Schott and Genossen, Jena. The simple phosphate and borate glasses have been withdrawn on account of their lack of durability. The borosilicate crowns are among the most durable and chemically resistant of all glasses. The U.V. glasses are markedly transparent to ultra-violet light as far as about $\lambda = 2880$.

See p. 72, and Zschimmer's "History of the Jena Glass Works," Hovestadt's "Jena Glass," and Rosenhain's "Glass Manufacture," 1908 (with bibliography).

(After Zschimmer, *Zeit. Inst.,* 1908.)

Glass.	μ_D	$\omega_{(C,D,F)}$	Dens.	Glass.	μ_D	$\omega_{(C,D,F)}$	Dens.
			grms. c.c.				grms. c.c.
Crowns—	1·4782	·0152	2·23	**Flints** (*contd.*)—			
(Silicate) crown .	1·5127	·0175	—	U.V. flint 3492 . .	1·5329	·0131	—
	1·5215	·0168	2·50	Telescope (Sb) flint	1·5286	·0194	2·50
U.V. crown 3199 .	1·5035	·0155	—	Borosilicate flint .	1·5503	·0203	2·81
Borosilicate crown	1·4944	·0151	2·33		1·5753	·0218	2·90
	1·5141	·0156	2·47		1·5489	·0187	—
Barium crown .	1·5726	·0174	3·21		1·5825	·0216	—
	1·6120	·0180	—	Barium flint . .	1·5848	·0189	—
Heavy barium crown	1·6130	·0178	3·60		1·6235	·0256	3·67
Flints—					1·6570	·0276	3·95
	1·5794	·0244	3·25		1·7174	·0340	4·49
(Silicate) flint .	1·6138	·0271	3·58	Heavy flint . .	1·7782	·0378	4·99
	1·6489	·0296	3·87		1·9044	·0461	5·92
					1·9625	·0508	—

SPECTROSCOPY

It is now agreed that the use of the diffraction-grating in fundamental work must be limited to interpolation between standard wave-lengths obtained by other means. The accepted standard lines are three in the spectrum of cadmium. Their wave-lengths (λ) obtained by interference methods, and measured (by direct comparison with the standard metre at Paris) in dry air at 15° C. (H-scale) and 760 mms. mercury pressure, are given below in tenth-metres ($= 10^{-8}$ cm. $= 1$ Angström unit). (See Michelson's "Light Waves and their Uses.") [$\mu = 10^{-4}$ cm.; $\mu\mu = 10^{-7}$ cm.]

Observer.	λ Cd red.	λ Cd green.	λ Cd blue.
Michelson and Benoit, 1894	6438·4700	5085·8218	4799·9085
Benoit, Fabry, and Perot, 1907 . . .	6438·4702	—	—

The following values (all in tenth-metres) are of course only approximate :—

Hertzian Waves.	Infra-red.	Red.	Orange.	Yellow.	Green.	Blue.	Violet.	Ultra-violet.
$10^{14} - 4 \times 10^7$	$3\cdot1 \times 10^6$ §7700	6470	5880	5500	4920	4550	3600	600 ‖

STANDARD LINES—IRON ARC SPECTRUM

Obtained by an interference method, and based on Benoit, Fabry, and Perot's value for the wave-length of the red line of cadmium. The wave-lengths below are given in tenth-metres (10^{-8} cm.), measured in dry air at 15° (H-scale) and 760 mms. mercury. (Buisson and Fabry, *Compt. Rend.*, 1907 and 1909.)

2373·737	2987·293	3724·379	4352·741	4878·226	5405·780	5952·739
2413·310	3030·152	3753·615	4375·935	4903·324	5434·530	6003·039
2435·159 *	3075·725	3805·346	4427·314	4919·006	5455·616	6027·059
2506·904 *	3125·661	3843·261	4456·554	4966·104	5497·521	6065·493
2528·516 *	3175·447	3865·526	4494·572	5001·880	5506·783	6137·700
2562·541	3225·790	3906·481	4531·155	5012·072	5535·418	6191·569
2588·016	3271·003	3935·818	4547·854	5049·827	5569·632	6230·732
2628·296	3323·739	3977·745	4592·658	5083·343	5586·770	6265·147
2679·065	3370·789	4021·872	4602·944	5110·415	5615·658	6318·029
2714·419	3399·337	4076·641	4647·437	5127·364	5658·835	6335·343
2739·550	3445·155	4118·552	4678·855	5167·492	5709·396	6393·612
2778·225	3485·344	4134·685	4707·287	5192·362	5760·843 ‡	6430·859
2813·290	3513·820	4147·677	4736·785	5232·958	5763·013	6494·994
2851·800	3556·879	4191·441	4754·046 †	5266·568	5805·211 ‡	
2874·176	3606·681	4233·615	4789·657	5302·316	5857·760 ‡	* Si.
2912·157	3640·391	4282·407	4823·521 †	5324·196	5892·882 ‡	† Mn.
2941·347	3677·628	4315·089	4859·756	5371·498	5934·683	‡ Ni.

CHIEF ABSORPTION (FRAUNHOFER) LINES IN SOLAR SPECTRUM

Rowland's wave-lengths corrected approximately by the use of Fabry and Perot's results, measured in tenth-metres (10^{-8} cm.) in air at 20° and 760 mms. Owing to atmospheric absorption, the sun's spectrum extends only to about wave-length 3000.

Line.	Subst.	Rel. Intens.	Line.	Subst.	Rel. Intens.	Line.	Subst.	Rel. Intens.
3047·5	Fe	20	L 3820·4	Fe-C	25	(H_γ)4340·4	H	20
3057·3	Ti-Fe	20	3825·8	Fe	20	F 4861·37	H (β)	30
3059·0	Fe	20	3838·2	Mg-C	25	b_2 5172·7	Mg	20
O {3440·6	Fe	20	3859·8	Fe-C	20	b_1 5178·22	Mg	30
{3441·0	Fe	15	K 3933·6	Ca	1000	E 5269·56	Fe	8
3524·5	Ni	20	3961·5	Al	20	($D_3$5875·62)†	He	—
N 3581·2	Fe	30	H 3968·4	Ca	700	D_2 5889·97	Na	30
3608·8	Fe	20	4045·8	Fe	30	D_1 5895·93	Na	20
3618·7	Fe	20	4063·6	Fe	20	C 6562·8	H (α)	40
M 3719·9	Fe	40	(H_δ)4101·8	H	40	B 6867·3	‡	6
3734·8	Fe	40	4226·7	Ca	20	A 7661 *	‡	—
3737·1	Fe	30	G 4307·9	Fe	6	Z 8228 *	—	

* Langley, 1900.   † Emission line in chromosphere alone.
‡ Oxygen in earth's atmos.   § Wood, 1911.   ‖ X and γ rays 8·4 to 0·07.

EMISSION SPECTRA

EMISSION SPECTRA OF SOLIDS

For a fuller treatment of wave-lengths see Watts' "Index of Spectra" and appendices, Kayser's "Handbuch der Spectroscopie," Hagenbach and Konen's "Atlas of Emission Spectra," 1905. For recent work consult the *Astrophysical Journal*. The wave-lengths below are measured in tenth-metres (10^{-8} cm.) in air at 15° C. and 760 mms. The visible spectrum colours are indicated—*r, o, y, g, b, v.* The brightest lines are emphasized and the approximate boundary of the ultra-violet region is indicated thus . . . . . .

ALUMINIUM (arc).

3083
3093
. . . . .
3944 *v*
3962 *v*
4663 *b*
5057 *g*
5696 *y*
5723 *y*

BARIUM (BaCl$_2$ in flame).

Full of bands, some diffuse, and some resolvable.

3501
. . . . .
3910 *v*
3994 *v*
4131 *v*
4554 *b*
4934 *g*
m 5536 *gy*
5778 *y*
5854 *y*
6142 *o*
6497 *r*

BORON (Boric acid in flame).

Diffuse maxima at

4500 *b*
4700 *b*
4900 *b*
5200 *g*
5450 *g*
5800 *y*
6000 *o*

CADMIUM (arc).

3261
3404
3466
3611
. . . . .
3982 *v*

CADMIUM (contd.)

4413 *b*
4678 *b*
4799·908 *b*
5085·822 *g*
5338 *g*
5379 *g*
6438·470 *r*

CÆSIUM (CsCl in flame).

3611·8
3617
3877
3889
. . . . .
4555 *b*
4593 *b*
5664 *y*
5845 *y*
6011 *o*
6213 *o*
6724 *r*
6974 *r*

CALCIUM (CaCl$_2$ in flame).

Bands predominate; line at

4227

(Flame arc).

3362
3644
. . . . .
(K) 3934 *v*
(H) 3968 *v*
4227 *v*
4303 *b*
4426 *b*
4435 *b*
4455 *b*
4586 *b*
4878 *b*
5270 *g*
5350 *g*
5589 *y*
5595 *y*
5858 *y*

CALCIUM (contd.)

6122 *o*
6162 *o*
6440 *o*
6463 *o*
6500 *r*

COPPER (arc in vacuo).

Fabry and Perot, 1902.

3248
3274
. . . . .
4023 *v*
4063 *v*
5105·543 *g*
5153·251 *g*
5218·202 *g*
5700 *y*
5782·090 *y*
5782·159 *y*

INDIUM (In(OH)$_2$ in flame).

4102 *v*
4511 *v*

IRON (see p. 75).

LITHIUM (LiCl in flame).

4132 *v*
4602 *b*
6104 *o*
6707·846 *r* [1]

[1] Fabry and Perot, 1902.

MAGNESIUM (arc).

3091
3093
3097
3330
3332
3337
3830

MAGNESIUM (contd.)

3832
3838
5168 *g*
(b$_2$) 5173 *g*
5184 *g*
5529 *y*

MERCURY (Mercury lamp).

Stiles, *Astro. Journ.*, 1909.

3126
3131
3650
. . . . .
4046·8 *v*
4078·1 *v*
4358·343 *v* [2]
4916·4 *bg*
4959·7 *g*
5460·742 *g* [2]
5769·598 *y* [2]
5790·659 *y* [2]
6152 *o*
6232·0 *o*

[2] Fabry and Perot, 1902, and Rayleigh, 1906.

POTASSIUM (KCl in flame).

3446
3447
. . . . .
4044 *v*
4047 *v*
5802 *y*
7668 *r*
7702 *r*

RADIUM (RaBr$_2$ in flame).

Runge and Precht, 1903.

3650
3815
. . . . .
4341 *v*

RADIUM (contd.)

4683 *v*
4826 *b*
5210 *g*
5360 *g*
5655 *y*
5685 *y*
6210 *o* [3]
6216 *o* [3]
6228 *o* [3]
6247 *o* [3]
6250 *o* [3]
6260 *o* [3]
6269 *o* [3]
6285 *o* [3]
6329 *o* [3]
6349 *o*
(6530 *r* [3]
to
6700 *r* [3]
6653 *r*

[3] Bands.

RUBIDIUM (RbCl in flame).

3349
3351
3587
3592
. . . . .
4202 *v*
4216 *v*
5618 *y*
5724 *y*
6207 *o*
6298·7

SILVER (arc in vacuo).

3281
3383
. . . . .
4055 *v*
4212 *v*
4669 *b*
5209·081 *g* [4]
5465·489 *g* [4]
5472 *g*
5623 *g*

[4] Fabry and Perot, 1902.

SODIUM (NaCl in flame).

Fabry and Perot, 1902 ; Rayleigh, '06.

(D$_2$) 5889·965 *o*
(D$_1$) 5895·932 *o*

STRONTIUM (SrCl$_2$ in flame).

Band spectr'm with lines at

4607·5 *b*
6387 *o*

THALLIUM (Tl or TlCl$_2$ in flame).

5350·7 *g*

TIN (spark).

3009
3034
3175
3262
3283
3331
3596
3746
. . . . .
4525 *v*
5563 *y*
5589 *y*
5799 *y*
6453 *o*

ZINC (arc in vacuo).

3036
3072
3345
. . . . .
4680·138 *b* [5]
4722·164 *b* [5]
4810·535 *b* [5]
4912 *b*
4925 *gb*
6103 *o*
6362·345 *o* [5]

[5] Fabry and Perot, 1902.

EMISSION SPECTRA OF GASES

The gases are all in vacuum tubes (2–4 mms. press.); only the brightest lines are given. The visible spectrum colours are indicated—*r, o, y, g, b, v.*
See the general remarks on last page.

ARGON, Red spectrum (small current density).	CARBON MONOXIDE or DIOXIDE (of common occurrence in many vacuum-tube spectra). Numerous bands shaded towards violet edges at	HYDROGEN Elementary spectrum.	NEON (*contd.*)	NITROGEN (*contd.*)
4159 *v*		3750	5853 *y*	5804 *y*
4192 *v*		3771	5882 *o*	5854 *y*
4198 *v*		3798	5945 *o*	5906 *o*
4201 *v*		3836	5976 *o*	5959 *o*
4259 *b*		3889	6030 *o*	6013 *o*
4300 *b*			6075 *o*	6069 *o*
4334 *b*	3590 (CN)	. . . . .	6096 *o*	With large current densities, N gives a line spectrum.
4511 *b*	3884 (CN)	3970 *v*	6129 *o*	
4703 *b*		**4102** (δ) *v*	6143 *o*	
5452 *g*		**4340** (γ) *b*	6164 *o*	
5607 *y*	4123 *v*	(F) **4861** (β) *gb*	6182 *o*	
5912 *o*	4216 (CN) *v*	(C) **6563** (α) *r*	6217 *o*	OXYGEN Elementary line spectrum.
6031 *o*	4393 *b*	For very short wave-lengths (1030–1675) see Lyman, *Astro. Journ.*, 1906.	6267 *o*	
6059 *o*	4511 *b*		6305 *o*	3919
	4735 (C) *b*		6383 *o*	3973
	4835 *b*		6402 *o*	. . . . .
	5165 (C) *g*	Secondary spectrum (see Watson, *Proc. Roy. Soc.*, 1909).	6507 *r*	4070 *v*
	5198 *g*			4072 *v*
	5610 *y*			4076 *v*
	6079 *o*		NITROGEN Band spectrum from positive column. Many bands all made up of fine lines. From 3000 to 4574 the edges occur at intervals of about 60 A.U. Other bands have edges at	4415 *b*
		KRYPTON AND XENON *Brit. Ass. Rep.*, 1905.		5208 *g* Diffuse maxima at
Blue spectrum (large current density).	HELIUM Rayleigh, 1908.			5335 *g*
				5440 *g*
3583	3188	NEON Baly, *Phil. Trans.*, 1903. Very rich in red rays.		6110 *o*
. . . . .	. . . . .			6170 *o*
4072 *v*	3889 *v*		4648 *b*	There are three other oxygen spectra: continuous, band, and series spectra.
4104 *v*	4026 *v*		4666 *b*	
4228 *v*	**4471·482** *b*	3448	4723 *b*	
4331 *b*	4713·144 *b*	3473	4813 *b*	
4348 *b*	4921·930 *gb*	3521	5340 *g*	RADIUM EMANATION Royds, *Phil. Mag.*, 1909.
4426 *b*	5015·680 *g*	3594	5614 *y*	
4430 *b*	(D₃) **5875·625** *y*	. . . . .	5755 *y*	
4431 *b*	6678·150 *r*	5765 *y*		
4610 *b*	7065·200 *r*			
4806 *b*				

ABSORPTION SPECTRA

For wave-lengths of the Fraunhofer lines in the sun's spectrum, see p. 75.
Among the enormous literature on absorption spectra, reference may be made to Kayser's "Handbuch der Spectroscopie," Baly's "Spectroscopy," Vogel's "Praktische Spectralanalyse," the writings of Prof. Hartley, Jones and Anderson's "Absorption Spectra of Solutions," 1909, Smiles' "Chemical Constitution and Physical Properties," and the British Association Reports of 1901 *et seq.*
Convenient substances which show good absorption spectra are—neodymium and praseodymium salts and didymium glass (which yield some extremely narrow absorption lines), iodine vapour, nitrogen peroxide, chlorine, chlorophyll, blood, and potassium permanganate solution.

OPTICAL ROTATIONS OF PURE LIQUIDS AND SOLUTIONS

A_t = the rotation in degrees (for light of some given wave-length) of the plane of polarization by a liquid when at the temperature $t°$ C.

l_t = the length of the column of liquid in **decimetres** (*i.e.* 10 cms.).

p = the number of grams of active substance in 100 **grams** of **solution.**

$q = (100 - p)$ = the percentage (by weight) of inactive solvent in the solution.

ρ_t = the density in grams per c.c. of the liquid or solution at $t°$.

$c_t = p\rho_t$ = the concentration expressed as grams of active substance per 100 **c.cs. of solution** at $t°$.

$[\alpha]_t$ = the **specific rotation** (at $t°$) = $\dfrac{\text{rotation per decimetre of sol.}}{\text{grams of active substance per c.c of sol.}}$

For a pure liquid $[\alpha]_t = \dfrac{A_t}{l_t\rho_t}$.

For an active substance in solution $[\alpha] = \dfrac{A_t}{l_t} \Big/ \Big(\dfrac{p}{p+q}\rho_t\Big) = \dfrac{100A_t}{l_t p\rho_t} = \dfrac{100A_t}{l_t c_t}$, since $(p + q) = 100$.

The rotation depends on the wave-length of the light used; it increases as the wave-length (λ) diminishes ($\alpha \propto \frac{1}{\lambda^2}$ approx.). α also varies with the nature of the inactive solvent and with the concentration of the solution.

The rotation is called positive or right-handed (dextro, *d*) if the plane of polarization appears to be rotated in an anti-clockwise direction when looking through the liquid **away** from the source of light. The contrary rotation is called lævo (*l*). The **molecular rotation** is the specific rotation multiplied by the molecular weight.

$[\alpha]_{20}^D$ indicates that the specific rotation is measured at 20° C. using sodium (D) light.

(See Landolt's "Optical Rotations of Organic Substances and their Practical Application.")

Optically Active Substance.	Solvent.	Conditions.	Specific Rotation $[\alpha]_t$
Cane Sugar or Candy (*d*), $C_{12}H_{22}O_{11}$ (Landolt, 1888; Pellat, 1901)	water	c = 4 to 28	$[\alpha]_{20}^D = +66.67 - .0095c$
		t = 14° to 30° C.	$[\alpha]_t^D = [\alpha]_{20}^D \{1 - .00037(t - 20)\}$
Invert Sugar(*l*),[*] $C_6H_{12}O_6$ = 1 mol. of dextrose + 1 mol. of levulose (Gubbe, 1885)	water	c = 9 to 35	$[\alpha]_{20}^D = -19°.7 - .036c$
		t = 3° to 30° C.	$[\alpha]_t^D = [\alpha]_{20}^D + .304(t - 20) + .00165(t - 20)^2$
Dextrose (*d* — glucose), $C_6H_{12}O_6$ (Parcus and Tollens, 1890; Tollens, 1884)	water	c = 9·1	$[\alpha]_{20}^D = +105°.2$ after 5·5 mins. (α modification) $= +52°.5$ after 6 hrs. (β modification)
	water	p = 1 to 18	$[\alpha]_{20}^D = +52°.5 + .025p$
l — **Glucose**, $C_6H_{12}O_6$ (Fischer, 1890)	water	p = 4	$[\alpha]_{20}^D = -94°.4$ after 7 mins. $= -51°.4$ after 7 hrs.
Levulose (*l*) (fruit sugar), $C_6H_{12}O_6$ (Parcus and Tollens, 1890; Ost, 1891)	water	c = 10	$[\alpha]_{20}^D = -104°$ after 6 mins. $= -92°$ after 33 mins.
	water	p = 2 to 31	$[\alpha]_{20}^D = -91°.9 - .11p$

[*] The molecular weight of cane-sugar is 342; which, after conversion to invert sugar, becomes 360. Hence the new concentration of the invert sugar solution is $\frac{360}{342}c$, where c is the number of grams of cane-sugar in 100 c.cs. of the original solution.

OPTICAL ROTATIONS

Optically Active Substance.	Solvent.	Conditions.	Specific Rotation $[a]_t$
Galactose (d), $C_6H_{12}O_6$ (Meissl, 1880)	water	$p = 4$ to 36 $t = 10°$ to $30°$ C.	$[a]_t^D = +83°·9 + ·078p$ $\quad - ·21t$
Ordy. **Tartaric acid** (d), $H_2C_4H_4O_6$	water	—	$[a]_{20}^D = +15·06 - ·131c$
Potassium tartrate (d), $K_2C_4H_4O_6$ (Thomsen, 1886)	water	$c = 8$ to 50	$[a]_{20}^D = +27·14 + ·0792c$ $\quad - ·00094c^2$
Rochelle salt (d), $KNaC_4H_4O_6$	water	—	$[a]_{20}^D = +29·73 - ·0078c$
l − **Turpentine**, $C_{10}H_{16}$ (Gernez, 1864 ; Landolt, 1877)	pure liquid	—	$[a]_{20}^D = -37°$
	vapour	at $761·7$ mms.	$[a]_{168}^D = -35°·5$ for mean yellow
	alcohol $(\rho_{20} = ·796)$	$q = 0$ to 90	$[a]_{20}^D = -37° - ·00482q$ $\quad - ·00013q^2$
	benzene	$q = 0$ to 91	$[a]_{21}^D = -37° - ·0265q$
	paraffin oil	Within wide limits $[a]$ **increases** with the percentage of paraffin.	
Quinine sulphate (l), $C_{20}H_{24}N_2O_2.H_2SO_4$ (Oudemans, 1876)	water	c about $1·6$ % of alkaloid (calculated)	Salt $[a]_{17}^D = -214°$ Alkaloid $[a]_{17}^D = -278°$
Nicotine (l), $C_{10}H_{14}N_2$ (Landolt, 1877 ; Hein, 1898)	pure	$t = 10°$ to $30°$C.	$[a]_{20}^D = -162°$
	benzene	$p = 8$ to 100	$[a]_{20}^D = -164°$
	water	$p = 1$ to 16	$[a]_{20}^D = -77°$
Ethyl malate (l), $(C_2H_5)_2C_4H_4O_5$ (Purdie & Williamson, '96)	pure liquid	—	$[a]_{11}^D = -10°·3$ to $-12°·4$
Camphor (d), $C_{10}H_{16}O$ (Landolt, 1877 ; Rimbach, 1892)	alcohol	$q = 45$ to 91	$[a]_{20}^D = +54°·4 - ·135q$
	benzene	$q = 47$ to 90	$[a]_{28}^D = +56° - ·166q$

OPTICAL ROTATION AND WAVE-LENGTH

Wave-length (λ) in 10^{-8} cm.	Specific Rotation at 20° C. $[a]_{20}^\lambda$				QUARTZ AT 20° C.	
	Cane-sugar or Candy in H_2O.	Turpentine (pure liq.).	Tartaric acid in H_2O $(p = 41\%)$.	Nicotine (pure liq.).	Wave-length (λ) in 10^{-8} cm.	Rotation for 1 mm. thickness.
H (C) 6563 (r)	52°·9	−29°·5	7°·75	−126°	**Li** 6708 (r)	16°·4
Na (D) 5893 (o)	66·5	−37	8·86	−162	**H** (C) 6563 (r)	17·3
Tl 5351 (g)	81·8	−45	9·65	−207·5	**Na**(D) 5893 (o)	21·72*
H (F) 4861 (g)	100·3	−54·5	9·37	−253·5	**Tl** 5351 (g)	26·53
					H (F) 4861 (g)	32·7
					H (δ) 4102 (b)	47·48

* For quartz at temperature $t°$, rotation $= 21°·72 \{1 + 0·000147(t - 20)\}$ for D line.

MAGNETIC ROTATION OF POLARIZED LIGHT

This effect was discovered by Faraday in 1845. The rotation per cm. per unit magnetic field—**Verdet's constant,** $r = a/(Hl)$, where a is the rotation in minutes for the substance in a magnetic field of H gauss, and l is the length of light-path parallel to the lines of force. r varies with the temperature and is roughly inversely proportional to the square of the wave-length of the light used. Films of Fe, Ni, and Co are exceptions to this rule.

If the light is travelling with the lines of force (*i.e.* from N. to S.), then the direction of rotation is positive, if the plane of polarization is rotated clockwise, to an observer looking in the direction in which the light is moving. If the light is reflected back on its path, the rotation is increased.

The **Molecular rotation** $r_m = rM/d$, where M is the molecular weight of the substance, and d is its density. r_m is an additive property in organic compounds (Perkin, *Journ. Chem. Soc.,* 1884).

The rotations below are for the sodium D line ($\lambda = 5893 \times 10^{-8}$ cm.).

(For Voigt's theory of magneto-rotation, see Schusters, "Optics," 1909. See also Becquerel's papers in *Compt. Rend.,* etc.)

Substance.	Temp.	Rotation r in mins. of arc.	Substance.	Temp.	Rotation relative to Water.
Water	0°C.	+·01311,R.W.	Ethyl alcohol . .	16·8	·8637, P.
,, ...	20	+·01312,R.W.	n. propyl alcohol . .	15·6	·9139, P.
Carbon bisulphide .	0	+·04347,R.W.	Amyl(iso) alcohol . .	19·9	·9888, P.
,, ,,	18	+·04200, Ra.	Ethyl bromide . . .	19·7	1·395, P.
Quartz, ⊥ axis . .	20	+·01368,* Bo.	,, chloride . .	5·0	1·035, P.
,, ,, . .	20	+·01664, Bo.	,, iodide . .	18·1	2·251, P.
,, ,, . .	20	+·1587,† Bo.	Formic acid . . .	20·8	·7990, P.
Jena {phosphate crown	18	+·0161, D.B.	Acetic ,, . . .	21·0	·7976, P.
glass {heaviest flint .	18	+·0888, D.B.	Propionic acid . . .	20·3	·8369, P.
FeCl₃ dens. = 1·693	15	−·2026, B.	Benzene	15	2·062, B.
,, ,, 1·023	15	+·0122, B.			

* $\lambda = 6439$. † $\lambda = 2194$. B., Becquerel; Bo., Borel, 1903; D.B., Du Bois, 1894; P., Perkin; Ra., Rayleigh, 1884; R.W., Rodger and Watson, 1896.

METALLIC REFLECTION OF LIGHT

(The percentage of normally incident light reflected from different surfaces.)

The column of figures (below) in the case of **speculum metal** (7 Cu, 3 Sn) reads 30% (for λ = 2510); 51%, 56%, 64%, 67%, 71%, 89%, 94% (for λ = 140,000).

Wave-length λ in A.U. (10⁻⁸ cm.).		Cu.	Au.	Ni.	Pt.	Ag.	Steel.	Magna- lium.*	Glass mirror.	
									Ag back.	Hg back.
Ultra- violet	{ 2,510	26%	39%	38%	34%	34%	33%	67%	—	—
	{ 3,570	27	28	49	43	74	45	81	—	—
Visible	{ 4,200	33	29	57	52	87	52	83	86%†	73%†
	{ 5,500	48	74	63	61	93	55	83	88	71
	{ 7,000	83	92	69	69	95	58	83	90	73
Infra-red	{ 10,000	90	95	72	73	97	63	84		
	{ 40,000	97	97	91	91	98	88	89	* 69 Al, 31 Mg.	
	{ 140,000	98	98	97	96	99	96	92	† λ = 4500.	

DIOPTER

In applied optics the "power" of a lens or mirror is expressed in diopters. The number of diopters equals the reciprocal of the focal length expressed in metres.

ELECTRICAL RESISTIVITIES

Electrical specific resistances or resistivities in ohm-cms. **Conductivities** (in reciprocal ohms) are the reciprocals of resistivities. For a table of reciprocals, see p. 136.

METALS AND ALLOYS

The resistivity depends to some extent on the state of the metal. In general, cold drawing increases, while annealing diminishes the resistance. The winding of a wire into a coil increases its resistance.

For pure metals, the resistance is roughly proportional to the absolute temperature, and would apparently vanish not far from the absolute zero. This rule does not hold even approximately for alloys.

For wire resistances, see p. 83; for temperature coefficients, next page. The thermal conductivities of the same samples of many of the substances below will be found on p. 51.

Substance.	Temp.	Sp. Re.	Observer.	Substance.	Temp.	Sp. Re.	Observer.
Metals	°C.	×10⁻⁶		**Metals** (contd.)	°C.	×10⁻⁶	
Aluminium*	−160	0·81	Lees,	Platinum	−203	2·4	D.&F., '96
,,	18	2·94	P. T., '08	,,	18	11·0	J. & D.,
,,	18	3·21	J. & D.,	,,	100	14·0	1900
,,	100	4·13	1900	Potassium	0	6·64	B., '04
Antimony	15	40·5	Berget, '90	Rhodium	18	6·0	——
Bismuth	18	119·0	J. & D.	Silver, 99·9 %	−160	0·56	Lees,
,,	100	160·3	1900	,,	18	1·66	1908
Cadmium, drawn	−160	2·72	Lees, '08	,,	18	1·63	J. & D.,
,,	18	7·54	J. & D.,	,,	100	2·13	1900
,,	100	9·82	1900	Sodium	0	4·74	B., 1904
Copper, drawn	−160	0·49	Lees, '08	Strontium	20	25	M., 1857
,,	18	1·78	J. & D.,	Tantalum	18	14·6	
,,	100	2·36	1900	Tellurium	20	21·	M., 1858
,, annealed	18	1·59	Mean	Thallium, pure	0	17·6	D.&F., '96
Calcium	20	10·5	M.&C., '05	Thorium	15	40·1	Bo., '09
Cobalt	20	9·71	R., 1901	Tin, drawn	−160	3·5	Lees, '08
Gold	−183	0·68	D.&F., '96	,,	18	11·3	J. & D.,
,,	18	2·42	J. & D.,	,,	100	15·3	1900
,,	100	3·11	1900	Tungsten	25	5·0	Fink, '10
Iridium	18	5·3	——	Zinc, pure	−160	2·2	Lees, '08
Iron	18	9-15	Mean	,,	18	6·1	J. & D.,
,, {·1% C.}	18	12·0	J. & D.,	,,	100	7·9	1900
,,	100	16·8	1900				
,, wrought	−160	5·4	Lees, '08	**Alloys**			
,, ,, †	18	13·9	J. & D.,	Brass	−160	4·1	Lees,
,, ,,	100	18·8	1900	,, ‡	17	6·6	1908
,, steel {·1% C.}	18	19·9	J. & D.,	,, ‡	18	6-9	Mean
,,	100	25·6	1900	Constantan	18	49·0	J. & D.,
Lead, drawn	−160	7·43	Lees, '08	(Eureka)§	100	49·1	1900
,,	18	20·8	J. & D.,	German silver ‖	18	16-40	Mean
,,	100	27·7	1900	,, ,,	0	26·6	Lorenz,
Lithium	0	8·4	B., '04	,,	100	27·6	1881
Magnesium	0	4·35	D. & F.	Manganin ¶	−160	43·13	Lees,
Mercury	0	94·07	See	,,	18	44·50	1908
,,	20	95·76	pp. 6, 82.	,,	18	42·05	J. & D.,
Molybdenum	25	4·1	Fink, '10	,,	100	42·11	1900
Nickel	−160	5·9	Lees, '08	Phosphor-bronze	18	5-10	Mean
,, {97% Ni.}	18	11·8	J. & D.,	Platinoid ‖	−160	32·5	Lees,
,,	100	15·7	1900	,,	18	34·4	1908
Osmium	20	9·5	Blair, '05	90 Pt, 10 Rh	0	21·1	D.&F., '96
Palladium	18	10·7	J. & D.,	67 Pt, 33 Ag	0	24·2	——
,,	100	13·8	1900				

* 99 % Al.   † ·1% C, ·2 % Si, ·1 % Mn.   ‡ 70 Cu, 30 Zn.
§ 60 Cu, 40 Ni.   ‖ 62 Cu, 15 Ni, 22 Zn.   ¶ 84 Cu, 4 Ni, 12 Mn.
B., Bernini; Bo., Bolton; D. & F., Dewar & Fleming; J. & D., Jaeger and Diesselhorst; M., Matthiessen; M. & C., Moissan & Chavanne; R., Reichardt; P. T., *Phil. Trans.*

ELECTRICAL RESISTIVITIES (*contd.*)

NON-METALS AND INSULATORS

The resistivities are in ohm-cms. at room temperatures unless otherwise stated. The values for insulators naturally vary widely, and the figures below are merely typical and are probably, in many cases, nothing more than the resistances of the surfaces. For a discussion of some electrical insulators, see Kaye, *Proc. Phy. Soc. Lond.*, 1911.

Substance.	Sp. Re.	Substance.	Sp. Re.	Substance.	Sp. Re.
Gas carbon . .	{ ·004 to ·007	Sulphur, 70° . .	$4 . 10^{15}$	Guttapercha . .	$2 . 10^9$
Graphite . . .	·003	Ebonite	$2 . 10^{15}$	Mica	$9 . 10^{15}$
C. lamp filament	·004	Glass, soda-lime *	$5 . 10^{11}$	Paraffin wax . .	$3 . 10^{18}$
Selenium ‡ (1907)	$2 . 10^{16}$	,, Jena, combustion * }	$>2 . 10^{14}$	Porcelain, 50°. .	$2 . 10^{15}$
Silicon §	·06	,, conducting†	$5 . 10^{8}$	Quartz	$1\cdot2 . 10^{14}$
				Fused silica * . .	$>2 . 10^{14}$

* National Physical Laboratory.   † Phillips.   ‡ In dark.   § Wick, 1908.

TEMPERATURE COEFFICIENTS OF RESISTANCE

To represent accurately over any considerable range the variation of electrical resistance (R) with temperature (*t*) requires for almost all substances a parabolic or cubic equation in *t*. But if the temperature interval is not large, a linear equation $R_t = R_0(1 + \alpha t)$ may be employed ; and this gives a definition of the mean temperature coefficient (α) over that temperature range. The table of resistivities above will readily yield the associated values of α. The coefficients given below are average ones.

Substance.	Temp.	α	Substance.	Temp.	α
Metals—		$\times 10^{-4}$	**Metals** (*contd.*)—		$\times 10^{-4}$
Aluminium	18–100	38	Silver	0–100	40
Bismuth	18	42	Tantalum	0–100	33
Cadmium	18–100	40	Tin	0–100	45
Copper *	18	42·8	Tungsten (1910) . .	0–170	51
Cobalt	0–160	33	Zinc	18–100	37
Gold	0–100	40			
Iron, pure	18	62	**Alloys**—		
Steel	18	16–42	Brass	18	10‡
Lead	18	43	Constantan (Eureka) .	18	{ −·4 to +·1‡
Mercury †	0–24	9·0			
Nickel, electrolytic	0–100	62	German silver . . .	18	2·3–6
,, commercial	0–1000	27	Manganin §	20	·02–·5‡
Palladium	18–100	37	Platinoid	18	2·5
Platinum	−100–0	35	90 Pt, 10 Ir	16	15
,, ...	0–100	38	90 Pt, 10 Rh	15	17
Molybdenum (1910) .	0–170	50	Platinum-silver (coils)	16	2·4–3·3

* High conductivity annealed commercial.   † $R_t = R_0(1 + ·0_388t + ·0_31t^2)$—Smith (N. P. L.), 1904.   ‡ N. P. L.   § Most samples of manganin have a zero temp. coeff. at from 30° C. to 40° C.

STANDARD WIRE GAUGE

The sizes of wires are ordinarily expressed by an arbitrary series of numbers. There are, unfortunately, four or five independent systems of numbering, so that the wire gauge used must be specified. The following are English Legal Standard wire gauge values. (See Foster's "Electrical Engineers' Pocket Book.")

Size.	Diameter.		Size.	Diameter.		Size.	Diameter.	
S.W.G.	Mm.	Inch.	S.W.G.	Mm.	Inch.	S.W.G.	Mm.	Inch.
6	4·88	·192	20	·914	·036	34	·234	·0092
8	4·06	·160	22	·711	·028	36	·193	·0076
10	3·25	·128	24	·559	·022	38	·152	·0060
12	2·64	·104	26	·457	·018	40	·122	·0048
14	2·03	·080	28	·376	·0148	42	·102	·0040
16	1·63	·064	30	·315	·0124	44	·081	·0032
18	1·22	·048	32	·274	·0108	46	·061	·0024

WIRE RESISTANCES

Average values in ohms per metre at 15° C. The **safe currents** for copper (high conductivity annealed commercial) are calculated at the rate of about 270 amps./cm.^{2} for No. 12 wire, 430 amps./cm.^{2} for No. 22 wire, and 500 amps./cm.^{2} for smaller diameters. Larger current densities than these are allowed in the revised "Wiring Rules" of the Institution of Electrical Engineers. Eureka is practically identical with constantan.

The average **temperature coefficient** of resistance of copper is ·00428; of nickel, ·0027; of manganin, ·00001; of German silver, ·00044; of Eureka, — ·00002; of platinoid, ·00025 per degree Centigrade. The values for the alloys may vary considerably. The **composition** of manganin is 84Cu, 4Ni, 12Mn; of German silver, 60Cu, 15Ni, 25Zn; of Eureka, c. 60Cu, 40Ni. Platinoid is said to be German silver with a little tungsten. For specific resistances, see p. 81.

S.W.G.	COPPER.		MANGANIN.	GERMAN SILVER.	S.W.G.	COPPER.		MANGANIN.	GERMAN SILVER.
	Ohms per metre.	Safe current.	Ohms per metre.	Ohms per metre.		Ohms per metre.	Safe current.	Ohms per metre.	Ohms per metre.
		amps.					amp.		
12	·0032	15·0	·077	·041	30	·222	·4	5·45	2·90
14	·0054	9·8	·131	·070	32	·293	·3	7·18	3·83
16	·0083	6·8	·204	·109	34	·404	·2	9·90	5·27
18	·0148	4·2	·361	·193	36	·590	·15	14·5	7·74
20	·0260	2·6	·645	·345	38	·950	·1	23·2	12·4
22	·0435	1·7	1·07	·57	40	1·48	·06	36·3	19·4
24	·070	1·1	1·73	·92	42	2·10	·05	53·4	27·8
26	·105	·7	2·58	1·38	44	3·30	·03	81·7	43·5
28	·155	·5	3·82	2·02	46	5·90	·02	145·5	77·4

EUREKA or CONSTANTAN.						PLATINOID (Martino's).			
S.W.G.	Ohms per metre.	20° C. temp.-rise caused by	S.W.G.	Ohms per metre.	20° C. temp.-rise caused by	S.W.G.	Ohms per metre.	S.W.G.	Ohms per metre.
		amps.			amps.				
12	·086	12·2	20	·722	1·5	20	·622	28	3·69
14	·146	8·2	22	1·20	·7	22	1·03	30	5·25
16	·228	4·9	24	1·93	·3	24	1·67	32	6·81
18	·405	2·7	26	2·89	·1	26	2·50	34	9·55

FUSES

The fusing currents are for wires mounted horizontally.

	Fusing current.	1 amp.	3	5	10	20	30	40	50
Tin	S.W.G.	37	28	24	21	18	16	14	13
Copper	S.W.G.	47	41	38	33	28	25	23	22

DIELECTRIC CONSTANTS

The inductivity, dielectric constant, or specific inductive capacity k of a material may be defined as—

(1) The ratio of the capacity of a condenser with the material as dielectric to its capacity when the dielectric is a vacuum.

(2) The square of the ratio of the velocity of electromagnetic waves in a vacuum to their velocity in the material. This ratio is dependent on the wave-length, λ, of the waves; in most cases k increases with λ. Unless otherwise stated, the inductivities below are for very long waves ($\lambda = \infty$) and at room temperatures.

If μ is the refractive index, then on Maxwell's theory of light, $k = \mu^2$, provided the frequency of the electrical oscillations is the same as that of the light vibrations. In practice we cannot find k for vibrations as rapid as those of the visible rays; the alternative is to obtain (by extrapolation) the refractive index for waves of very great wave-length, e.g. by the use of Cauchy's formula, p. 71. When such data are available Maxwell's relation is found to hold fairly exactly in the case of a number of gases and liquids, though there are many substances which provide marked exceptions.

In general, a rise of temperature diminishes the inductivity. The **temperature coefficient** a between $t°$ and $T°$ is defined by $k_T = k_t\{1 - a(T - t)\}$. In the case of water Palmer (1903) finds that a increases slightly with the frequency of oscillation.

The **Clausius-Mossotti relation** $\dfrac{k-1}{\rho(k+2)}$ = const. (ρ being the density) has been shown by Tangl (*Ann. d. Phys.*, 1908) to hold from 1 to 100 atmos. in the case of H_2, N_2, and air.

Substance.	k.	Substance.	k.	Substance.	k.
Solids—					
Calcite	7·5–7·7	Bromine . . .	3·1	Oil, paraffin . .	4·6–4·8
Ebonite	2·7–2·9	Carb. bisulphide .	2·62	Petroleum . . .	2·0–2·2
Fluorite	6·8	„ tetrachloride	2·25/18°	Toluene, a = ·001	2·3
Glass, crown . .	5–7	Chloroform, 18° .	5·2	Turpentine . . .	2·2–2·3
„ heavy crown	7–9	Ethyl acetate . .	6	Vaseline oil . .	1·9
„ flint . .	7–10	„ chloride . .	10·9	Water, λ = ∞ . .	81
„ mirror . .	6·7	„ ether, a = ·005	4·37	„ λ = 3600 cms.	3·32 *
Gypsum	6·3	Glycerine, λ = 200	39·1/15°	„ λ = 1200 „	2·79 *
Ice (−2°) . . .	93·9	Nitrobenzene . .	34/17°	„ a_{17} = ·0045 .	
Indiarubber . .	2·1–2·3	Oil, castor . . .	4·6–4·8	Xylene, m, a = ·035	2·4
Marble	8·3	„ olive . . .	3·1–3·2		
Mica	5·7–7				
Paper, dry . . .	2–2·5				
Paraffin wax . .	2–2·3	**Substance.**	**Temp.**	**k.**	**Observer.**
Pitch	1·8				
Porcelain . . .	4·4–6·8			76 cm. Hg. ; λ = ∞	
Quartz	4·5	**Gases—**			
Resin	1·8–2·6	Air	0°C.	1·000586	Klemencic, 188
Rock salt . . .	5·6	„	−20	1·000576	Tangl, 1908
Selenium (16°) .	6·1	Hydrogen . . .	0	1·000264	Boltzmann, 187
Shellac . . .	3–3·7	„	20	1·000273	Tangl, 1908
Silica, fused . .	3·5–3·6	Helium	0°	1·000074	Hockheim, 190
Spermaceti . .	c. 2·2	Nitrogen . . .	−20	1·000581	Tangl, 1908
Sulphur . . .	3·6–4·3	Nitrous oxide, N_2O	0	1·00099	Klemencic, 188
Sylvin	4·9	Carbon monoxide	0	1·000695	„ „
Vaseline . . .	2·2	„ dioxide .	0	1·000985	„ „
		„ bisulphide	15	1·0029	„ „
Liquids—		Ethylene . . .	15	1·00146	„ „
Alcohol, methyl .	35·4/13°·4	Sulphur dioxide .	14·7	1·00905	„ „
„ ethyl .	26·8/14°·7	Ammonia . . .	20	1·00718	Bädeker, 1901
„ amyl .	16·0/20°	Alcohol, methyl .	110	1·00600	„ „
Aniline, a = ·004 .	7·30	„ ethyl .	110	1·00647	„ „
Benzene, a = ·037 .	2·29/18°	Benzene . . .	110	1·00292	„ „

* Beaulard, 1908.

IONIC DISSOCIATION THEORY

On the Dissociation Theory (Arrhenius, 1887), the solute is dissociated into electrically positive cathions and negative anions. For example, KCl in water exists as KCl, K^+, Cl^-; sulphuric acid as H_2SO_4, H^+, H^-, SO_4^{++}, HSO_4^+. Probably, in many cases, these ions are attached to molecules of solvent. **The degree of dissociation** a = (number of dissociated solute molecules)/(total number of solute molecules). a is deduced from the osmotic pressure of the solution, and from its electric conductivity at different dilutions. The osmotic pressure is determined (1) directly, (2) from the raising of the boiling-point, and (3) from the depression of the freezing-point of the solvent by the presence of the solute. The equivalent conductivity (Λ) for different concentrations of any dilute solution is assumed to be proportional to the number of ions present. Λ approaches asymptotically a limiting conductivity (Λ_∞) for extreme dilutions, a state of things when, on this theory, the solute is completely dissociated. $\Lambda_m/\Lambda_\infty = a$ for the equivalent concentration m. The cathion and anion with their charges $+e$ and $-e$ (for monovalent ions) move in unit electric field in opposite directions with speeds or **mobilities** u_+ and u_-. The electrolytic current also obeys Ohm's Law, so that $X\kappa = (u_+ + u_-)ne$ (Kohlrausch, 1879), where X is the potential gradient in volts per cm., n the number of $+$ive or $-$ive ions per c.c., κ the conductivity of the solution in ohm^{-1} cm.$^{-1}$. This becomes $u_+ + u_- = 1.037 \times 10^{-5} \Lambda$ cm./sec., since $\kappa/n = \Lambda/N$, and $Ne = 96,740$ coulombs per gm. equivalent of ions.

The mobility of electrolytic ions has been directly observed by Lodge (1886), Whetham, Orme Masson, and D. B. Steele. The ratio $u_-/(u_+ + u_-) = n$ is for the negative ion, the **migration ratio** or transport number of Hittorf (1853–9). n can be determined, when complex ions are absent, from the change of concentration at the anode and cathode during electrolysis. The **mobility** of certain organic ions is approximately inversely proportional to their linear dimension a (Laby and Carse). The existence of this relation of Ohm's Law and of a relation between the viscosity (η) of the solvent and the ionic mobilities (Kohlrausch, Hosking, and Lyle) indicates that the motion of the ion through the solution may follow Stokes' Law ($v = F/6\pi\eta a$, where F is the driving force), with the numerical constant, 6π, possibly changed.

The dissociation theory postulates the conditions existing in very dilute solutions. The rôle of the medium is rather neglected (Lowry, *Science Progress*, 1908). The dissociation should be large for a solvent with a high dielectric constant, for then the attraction between the cathion and anion is small (Thomson and Nernst). This is generally true (Walden).

(Kohlrausch and Holborn, "Leitvermögen der Elektrolyten;" Whetham's "Theory of Solution.")

MIGRATION RATIOS

Hittorf's migration ratio or transport number of the anion, $n = u_-/(u_+ + u_-)$; m = equivalent concentration per litre; $t°$ = temp. of observation.

Solute.	$t°$C.	Conc. m.	Ratio n.	Solute.	$t°$C.	Conc. m.	Ratio n.	Solute.	$t°$C.	Conc. m.	Ratio n.
KCl	—	.003	.505, S.D.	AgNO₃	17°	.4to.02	.526, H.	CuSO₄	18°	{.08 to .02}	.625, M.
KBr	18°	{.03 to .01}	.504, B.	NH₄Cl	20	.05	.507, Be.				
KI	25	.05	.505, Be.	TlCl	22	.01	.516, Be.	HCl	10	{.05 to .02}	.159, N.S.
KNO₃	8	.1	.497, H.	CaCl₂	—	.005	.562, S.D.				
NaCl	18	{.03 to .009}	.604, B.	SrCl₂	21	.01	.56, Be.	HNO₃	18	.25	.17
				BaCl₂	18	.01	.55	H₂SO₄	11	.05	.17, Be.
NaNO₃	19	.05	.629, Be.	MgCl₂	21	.05	.615, Be.	KOH	—	.1	.74
LiCl	18	{.03 to .008}	.67	ZnSO₄	—	.05	.64, H.	NaOH	25	.04	.8, Be.
				CdBr₂	18	{.12 to .007}	.57	NH₃	21	.05	.56, Be.
								AgC₂H₃O₂	25	.01	.376, L.N.

B., Bogdan; Be., Bein; H., Hittorf; L.N., Löb and Nernst; M., Metelka; N.S., Noyes and Sammet; S.D., Steele and Denison.

ELECTRICAL CONDUCTIVITY OF SOLUTIONS

κ_{18} = pecific electric conductivity (in ohms⁻¹ cm.⁻¹) of the solution at 18° C.

$p.$ = mass of anhydrous solute per 100 gms. of solution.

η = the number of gm. equivalents in 1 c.c. of solution. Gm. equiv. per litre = 1000η. To find η note that $\kappa/\Lambda = \eta$.

v = volume in litres containing one gm. equivalent of solute = $1/1000\eta$.

Λ = equivalent conductivity = κ/η, = the conductivity in reciprocal ohms of 1 gm. equiv. in solution between electrodes 1 cm. apart. The chemical equiv. of, for example, "$1/2CaCl_2$" is 111/2.

Temp. coefficient = $(d\kappa/dt)/\kappa_{18}$. (See Kohlrausch and Holborn, "Das Leitvermögen der Elektrolyten" (Teubner).) K = Kohlrausch; G = Grotrian.

CONCENTRATED SOLUTIONS

1 KCl (K.G.)

p %	κ_{18}	$\Lambda = \frac{\kappa}{\eta}$	Temp. coef.
5	·0690	99·9	201
10	·1359	95·2	188
15	·2020	91·5	179
20	·2677	88·9	168
21	·2810	87·5	166

1 NaCl (K.G.)

p %	κ_{18}	$\Lambda = \frac{\kappa}{\eta}$	Temp. coef. ('0)
5	·0672	76	217
10	·1211	66·2	214
15	·1642	57·8	212
20	·1957	49·9	216
25	·2135	42·0	227
26·4	·2156	39·8	233

½ CaCl₂ (K.G.)

p %	κ_{18}	$\Lambda = \frac{\kappa}{\eta}$	Temp. coef. ('0)
5	·0643	68·6	213
10	·1141	58·3	206
15	·1505	49·2	202
20	·1728	40·6	200
25	·1781	32·1	204
30	·1658	23·9	216
35	·1366	16·1	236

½ CdCl₂ (G.)

p %	κ_{18}	$\Lambda = \frac{\kappa}{\eta}$	Temp. coef. ('0)
1	·0055	50·1	222
10	·0241	20·2	217

½ CdCl₂ (G.) (contd.)

p %	κ_{18}	$\Lambda = \frac{\kappa}{\eta}$	Temp. coef.
30	·0282	6·5	252
50	·0137	1·49	353

1 AgNO₃ (K.)

p %	κ_{18}	$\Lambda = \frac{\kappa}{\eta}$	Temp. coef. ('0)
5	·0256	83·4	218
10	·0476	74·3	217
15	·0683	67·9	215
40	·1565	45·0	205
60	·2101	31·1	209

1 (NH₄)₂SO₄ (K.)

p %	κ_{18}	$\Lambda = \frac{\kappa}{\eta}$	Temp. coef. ('0)
5	·0552	71·0	215
10	·1010	63·1	203
20	·1779	52·7	193
30	·2292	43·1	191

½ CuSO₄ (K.)

p %	κ_{18}	$\Lambda = \frac{\kappa}{\eta}$	Temp. coef. ('0)
2·5	·0109	34·0	213
5	·0189	28·7	216
10	·0320	23·1	218
17·5	·0458	17·4	236

½ CdSO₄ (G.)

p %	κ_{18}	$\Lambda = \frac{\kappa}{\eta}$	Temp. coef. ('0)
1	·0042	42·9	210
5	·0146	29·0	206
25	·0430	13·8	223
36	·0421	8·25	255

1 HCl (K.)

p %	κ_{18}	$\Lambda = \frac{\kappa}{\eta}$	Temp. coef. ('0)
5	·3948	281·0	158
10	·6302	219·1	156
20	·7615	126·2	154
30	·6620	69·8	152
40	·5152	39·1	

1 HNO₃ (K.G.)

p %	κ_{18}	$\Lambda = \frac{\kappa}{\eta}$	Temp. coef. ('0)
6·2	·312	307	147
12·4	·542	257	142
18·6	·690	211	137
24·8	·768	169	137
31	·782	133	139
49·6	·634	61	157
62	·496	36·4	157

½ H₂SO₄ (K.)

p %	κ_{18}	$\Lambda = \frac{\kappa}{\eta}$	Temp. coef. ('0)
5	·208	198	121
10	·391	180	128
15	·543	161	136
20	·653	140	145
25	·717	119	154
30	·739	99	162
35	·724	80	170
40	·680	64	178
50	·540	38	193
60	·373	20·3	213

½ H₂SO₄ (K.) (contd.)

p %	κ_{18}	$\Lambda = \frac{\kappa}{\eta}$	Temp. coef. ('0)
70	·216	9·4	256
80	·110	3·9	349
90	·107	3·22	320
100	·0157	—	031

1 KOH (K.)

p %	κ_{18}	$\Lambda = \frac{\kappa}{\eta}$	Temp. coef. ('0)
4·2	·1464	188	187
8·4	·272	169	186
12·6	·376	150	188
16·8	·456	131	193
29·4	·543	81	221
42·0	·421	39	283

1 NaOH (K.)

p %	κ_{18}	$\Lambda = \frac{\kappa}{\eta}$	Temp. coef. ('0)
2·5	·109	170	194
5	·197	149	201
10	·312	112	217
15	·346	79	249
20	·327	53	299
30	·202	20	450
40	·116	8·1	65

1 NH₃ (K.)

p %	κ_{18}	$\Lambda = \frac{\kappa}{\eta}$	Temp. coef. ('0)
·1	·00025	4·25	246
1·6	·00087	·93	238
8	·00104	·23	262
30·5	·00019	·012	

STANDARD SOLUTIONS FOR CALIBRATING CONDUCTIVITY VESSELS

κ_{18} for the **purest water** in a vacuum = ·04 × 10⁻⁶ ohms⁻¹ cm.⁻¹ (Kohlrausch and Heydweiller); κ_{18} for **conductivity water** in air is about 10⁻⁶ ohms⁻¹ cm.⁻¹; KCl 1 n = normal KCl = 74·59 gm./litre at 18° C.; NaCl sat. = saturated NaCl at temp. $t.$ of experiment. **Unit**—ohm⁻¹ cm.⁻¹. (See Kohlrausch, Holborn, and Diesselhorst.)

Solution.	0° C.	8°	12°	16°	20°	24°
NaCl, sat.	·1345	·1688	·1872	·2063	·2260	·2462
KCl, 1 n	·06541	·07954	·08689	·09441	·10207	·10984
KCl, 1/10 n	·00715	·00888	·00979	·01072	·01167	·01264
KCl, 1/50 n	·00152	·00190	·00209	·00229	·00250	·00271
KCl, 1/100 n	·00078	·00097	·00107	·001173	·001278	·001386

CONDUCTIVITY OF SOLUTIONS

EQUIVALENT ELECTRIC CONDUCTIVITY Λ OF DILUTE AQUEOUS SOLUTIONS

Extrapolated numbers are indicated by ( ). Λ for infinite dilution is given under "O." Observers: inorganic solutes, Kohlrausch; organic, Bredig, *Zeit. Phys. Chem.*, 1894.

Solute at 18° C.	Gm. equiv. per litre = 1000η.				Solute at 18° C.	Gm. equiv. per litre = 1000η.			
	0	·0001	·01	·5		·0001	·0002	·01	·5
KCl .	130·1	129·1	122	102	½ CaCl₂ .	115·2	114·5	103	74·9
KBr .	132·3	131·1	124	105	½ SrNO₃ .	111·7	111·1	99	62·7
KI .	131·0	129·8	123	106	½ BaCl₂ .	[117/·0005]		107	77·3
KF .	111·3	110·5	104	83	½ MgCl₂ .	109·4	108·9	98·1	69·5
KSCN	121·3	120·2	114	95·7	½ ZnSO₄ .	109·5	107·5	72·8	—
KNO₃ .	126·5	125·5	118	89·2	½ CdNO₃ .	[100/·005]		96	63·9
NaCl .	109·0	108·1	102	80·9	½ CuSO₄ .	109·9	107·9	71·7	—
NaF .	90·15	89·3	83·5	60·0	½ PbN₂O₆ .	120·7	119·9	103	53·2
NaNO₃	105·3	104·5	98·2	74·0					
LiCl .	98·9	98·1	99·2	70·7					
AgNO₃	115·8	115·0	108	77·5		·001	·002	·01	·5
CsCl .	133·6	132·3	125	—					
RbCl .	—	132·3	125	—	Acids.				
NH₄Cl	—	129·2	122	101	HCl . .	(377)	376	370	327
TlCl .	131·5	130·3	120	—	HNO₃ .	(375)	374	368	324
					½ H₂SO₄ .	361	351	308	205
					⅓ H₃PO₄ .	(106)	102	85	—
					Alkalies.				
					KOH .	(234)	(233)	228	197
					NaOH .	—	204·5	203·4	174
					NH₃ . .	53/·0002	38/·0005	9·6	1·35

Solute at 25° C.	Λ₁₀₂₄	Λ∞	Solute at 25° C.	Λ₁₀₂₄	Λ∞
Na formate . . .	98·1	100·4	Hydrochloride of—		
Na acetate . . .	85·7	87·5	-Propylamine . .	107·5	110·3
Na propionate . .	81·0	83·5	(CH₃)₄PCl . . .	107·4	109·8
Na butyrate . . .	77·4	79·9	(C₂H₅)₄PCl . . .	98·3	100·8
Na isobutyrate . .	77·7	80·1	(CH₃)₄AsCl . . .	105·5	108·2
Hydrochlorides of—			Hydrochlorides of—	Λ₂₅₆	
-Methylamine . .	125·1	127·8	-Aniline	100·3	106·1
-Ethylamine . .	114·3	117·0	-Methylaniline . .	99·4	105·2
-Dimethylamine .	117·5	120·3	-o-Toluidine . .	97·4	103·7
-Allylamine . . .	109·2	111·7			

EQUIVALENT ELECTRIC CONDUCTIVITY OF NON-AQUEOUS SOLUTIONS

$v = 1/m$ = volume in litres in which 1 gm. equivalent is dissolved. (See Tower, "Conductivity of Liquids," 1908.)

Solvent.	Solute.	t° C.	v	Λ	v	Λ	Solvent.	Solute.	t° C.	v	Λ	v	Λ
NH₃	KBr	−38°	5740	317·6	12410	329·7	POCl₃	N(C₂H₅)₄I	25°	750	38·5	1500	44·3
,,	AgNO₃	−15	94	188	192	110	Formic acid	KCl	25	256	58	512	61
HCN	KI	0	392	298	1024	308	,,	HCl	25	5·86	32·8	46·9	33·2
,,	S(CH₃)₃I	0	512	327	1024	332	Acetone	KI	18	1157	155	2315	163
SO₂ .	KI	0	1024	112·5	2048	134·5	,,	LiCl	18	10	49·8	13·8	99·5
,,	N(C₂H₅)₄I	0	512	157·1	1024	167·7	,,	AgNO₃	18	288	15·7	576	17·6
AsCl₃	N(C₂H₅)₄I	25	150	63·2	750	59·7							

MOBILITIES OF IONS IN LIQUIDS

The mobility of the anion $= u_- = 1·037 \times 10^{-5} \Lambda n$. ($n =$ Hittorf's number.)

Example.—For KCl, $\Lambda_\infty = 130·1$, $n = ·505$, $\therefore u_- = 1·037 \times 10^{-5} \times ·505 \times 130·1 = 6·8 \times 10^{-4}$ cm./sec. for Cl ions at 18°. Observers, Kohlrausch and Bredig; the latter's values have been multiplied by $1·1 \times 10^{-5}$ to bring them to cm./sec. **Unit**—10^{-5} cm./sec. * ½ Ca, etc. : the actual ionic velocity of the divalent ions is half the value stated here ; these values, however, fit the equations given on p. 85.

Ion.	u 18°.	Ion.	u 18°.	Ion.	u 18°.	Ion.	u 18°.	Ion.	u 25°.	Ion.	u 25°.
H .	330	NH₄	66·3	Zn* .	·48·4	F .	·48·3	HCO₂	56·3	C₂H₇H₃N	51·5
Li .	34·6	Tl .	68·4	Cu* .	·49	Cl .	67·8	CH₃CO₂	42·1	(C₂H₅)₄P .	33·7
Na .	45·2	Ca* .	53·7	Ag .	·56	Br .	70	C₂H₅CO₂	37·7	C₆H₆H₃N)	
K .	67	Sr* .	53·6	Cd* .	·49·2	I .	68·8	n.C₃H₇CO₂	33·8	aniline }	39·5
Rb .	70·5	Ba* .	57·5	Pb* .	·63·5	NO₃	64	Iso- ,,	34·0	C₆H₅HN	48·5
Cs .	70·5	Mg*	47·7	OH .	180	SO₄*	71	CH₃H₃N	53·4	(CH₃)₄As .	41·8

DIRECTLY OBSERVED MOBILITIES

Deduced from the observed movement of an ionic boundary. $m =$ equivalent concentration. **Unit**—10^{-5} cm./sec. at 18° C. (See Denison and Steel, *Phil. Trans.*, 1906.)

Ion.	m	u	Ion.	m	u	Ion.	m	u	Ion.	m	u	Ion.	m	u	Ion.	m	u
K	·5	55·3	Na	·1	31·8	Ba	·5	33	Mg	·2	16·7	Cl	·5	52·9	SO₄	·2	30·4

ELECTROMOTIVE FORCES AND RESISTANCES OF CELLS

The E.M.F.'s given are for cells on open circuit, and are only approximate ; in the case of primary batteries they refer to freshly made up cells. The internal resistances quoted are only typical ; they vary very widely in practice. With many primary cells the E.M.F. drops and the internal resistance increases as the cell ages. Nearly all modern dry cells are modified Leclanché batteries.

(See Slingo and Brooker's " Electrical Engineering.")

Cell.	Description.	E.M.F.	Resistance.
		Volts.	Ohms.
Bichromate . . .	Zn and C in 1 vol. strong H₂SO₄ and 20 vols. sat. K₂Cr₂O₇ sol.	c. 2·0	very low
Bunsen	Zn in 1 vol. H₂SO₄ and 12 vols. H₂O; C in strong HNO₃	1·8–1·9	——
Clark (see p. 8) .	Zn amalgam and Hg in sat. ZnSO₄ sol.	1·433	c. 500
Daniell	Zn in ZnSO₄ sol. or H₂SO₄ (1 to 12); Cu in sat. CuSO₄ sol.	1·07–1·08	c. 4
Grove . . .	Like Bunsen with Pt instead of C	1·8–1·9	——
Leclanché . . .	Zn and C in NH₄Cl, C, and MnO₂	c. 1·5	0·25–4
Secondary . . .	Pb and PbO₂ (etc.) in H₂SO₄ of density 1·2	2·2–1·9	negligible
Tucker	" Hygroscopic cell." Zn and C with sat. CaCl₂ sol.	1·4	——
Weston (see p. 8).	Cd amalgam and Hg in sat. CdSO₄ sol.	1·018	c. 500

MAGNETIC INDUCTION

$\mathfrak{H}$ = **magnetic force**
$\mathfrak{I}$ = **intensity of magnetization**
  = magnetic moment per cm.^{3}
  = pole strength per cm.^{2}
$\mathfrak{B}$ = **magnetic induction, or flux density**
  = $\mathfrak{H} + 4\pi\mathfrak{I}$

$\mathfrak{B}$, $\mathfrak{H}$, and $\mathfrak{I}$ are in lines per cm.^{2}, and are vector quantities.
Unit: 4π lines start from unit magnetic pole.

μ = **permeability** = $\mathfrak{B}/\mathfrak{H}$. See p. 6.
H = **susceptibility** = $\mathfrak{I}/\mathfrak{H} = (\mu - 1)/(4\pi)$. See p. 6.

Coercivity, $\mathfrak{H}_{B=0}$, is the demagnetizing force required to make $\mathfrak{B} = 0$ after saturation.
Coercive force is the demagnetizing force required to make $\mathfrak{B} = 0$ after some particular field strength.

Remanence, $\mathfrak{B}_{H=0}$, is the induction remaining when the magnetic force is removed after some particular field strength.

The work done, *i.e.* **hysteresis loss**, Q_c, in taking a cm.^{3} of magnetic material through a magnetic cycle between limits $\pm H_o = \int \mathfrak{H} d\mathfrak{I} = \frac{1}{4\pi} \int \mathfrak{H} d\mathfrak{B}$. Steinmetz's empirical formula for the hysteresis loss is $\eta \mathfrak{B}_{max}^n$, where η is a constant, and generally $n = 1\cdot6$. The magnetic properties of a material depend not only on its chemical composition, but on its previous mechanical and heat treatment ; thus only general characteristics are indicated below.

Heusler alloys (discovered by Heusler in 1903) are composed of Cu, Mn, and Al. They do not show the Kerr effect.

Good permanent **magnet steel** contains about ·5 % W and ·6 % C, is free from Mn, Cu, Ni, and Ti, and is hardened at 850° C. (Hannack, 1909). Cast iron, chilled from 1000° C., may also be used (Peirce and Campbell).

References.—Pure iron, Peirce, *Amer. Jour. Sci.*, 27 and 28, 1909 ; Terry, *Phy. Rev.*, 1909 ; iron and manganese, Burgess and Aston, *Phil. Mag.*, 1909 ; Heusler alloys, Stephenson, *Phy. Rev.*, 1910. (Ewing, " Magnetic Induction in Iron," and Kohlrausch, " Prakt. Phys.")

Material.	Permeability μ.						Coercivity.	Remanence.	$\mathfrak{H}_c$.	Hyst. loss, Q_c.
	$\mathfrak{H}=\cdot5$	$\mathfrak{H}=1$	$\mathfrak{H}=5$	$\mathfrak{H}=20$	$\mathfrak{H}=60$	$\mathfrak{H}=150$				ergs/cm.^{3}
Swedish wrought iron	2500	3710	2060	736	274	120	·08	4,000	**200**	6,700
Annealed cast steel	1450	3500	2100	747	280	123	0·97	7,100	**151**	11,700
Unannealed cast steel	490	970	1700	680	270	122	2·08	9,000	**156**	20,400
Cast iron	—	—	81	182	117	65	11·9	4,230	**155**	34,300
Magnet steel { Hardened	—	—	68/15	78	193	100	52·6	11,700	**234**	211,000
Magnet steel { Tungsten	—	—	80/10	119	204	105	27·5	9,880	**505**	116,000

Material.	$\mathfrak{H}_{max}$.	Induction, $\mathfrak{B}$, for		μ_{max}.	For $\mathfrak{H}_{max}$.		
		$\mathfrak{B}_{max}$.	$\mathfrak{H}=100$.		Coer.	Reman.	Hyst. loss.
							ergs/cm.^{3}
Mild steel **	129	18,190	17,700	8350	0·6	10,300	4,900
Steel, 2·8 % Cr, ·8 % C.					·56	6,400 ‡	
" 5·5 % W, ·6 % C.	Hardened at 770°		—		·72	7,000 ‡	280,000
" 7·7 % W, 1·9 % C.	" " 800°		—		·85	4,700 ‡	
" 4 % Mo, 1·2 % C	" " 800°		—		·85	6,700	
Iron †	50	17,100	—	1750	2·2*	c. 53 % $\mathfrak{B}_{max}$.	
Silicon iron, ·6 % Si †	50	16,000	—	1900	1·6*	c. 43 % "	
" " 4·5 % Si †	50	15,100	—	2500	1·2*	c. 39 % "	
Electrolytic iron (very pure) §	210	21,250	—	·18	10,000		
" " "	Heated to 1200° C.	16,000	—	—	2·5	12,500	
Hadfield's manganese steel ‖	—	—	—	1·3–1·5	v. small		
Nickel, annealed	100	5,137	—	296	8	3,570	—
Cobalt	140	10,000	9,500	17·4	12	3,400	—
" 96 %	114	8,237	7,800	177			19,000
Heusler alloy ¶	·92	2,735	—	115			

* H = 16.   † Otto, *Deut. Phys. Ges. Berlin*, 1910.   ‡ Bar magnet.
§ Burgess and Taylor, 1906.   ‖ 12 % Mn, 1 % C.
¶ 24 Mn, 16 Al, 60 Cu. McClennan, 1907.   ** Gumlich and Schmidt (Reichsanstalt), 1901.

MAGNETIC SUSCEPTIBILITIES OF THE ELEMENTS, ETC.

The susceptibility $H = \mathfrak{I}/\mathfrak{H} = (\mu - 1)/(4\pi)$. $H = 0$ for a vacuum. The susceptibility depends very much on the purity of the material, especially upon the absence of iron. For pure elements H appears to be independent of $\mathfrak{H}$, except possibly in the case of Mg, Sb, and Ru. H is a periodic property of the atomic weight; for example, P, As, Sb, and Bi are comparatively strongly diamagnetic.

The values below are per grm. at 18° C., except where some temperature is specified. The **gases** are per cm.^{3} at 1 atmos. [Honda (*Ann. d. Phys.*, 1910) used purest available materials and corrected H for any traces of iron; see also P. Curie, Œuvres, Paris, 1908.] + means paramagnetic; –, diamagnetic.

Elem.	H	Obs.	Elem.	H	Obs.	Elem.	H	Obs.
Solids	× 10⁻⁶		**Solids**			**Solids**		
Al .	+ ·65	L., W., H.	(*contd.*)	× 10⁻⁶		(*contd.*)	× 10⁻⁶	
Sb .	– ·95	H.	P . .	– ·9	H., B., C., Q.	V . .	+ 1·5	H.
As .	– ·31	H.	Pt. .	+ 1·32	—	Zn . .	– ·15	K., L., H.
Bi .	– 1·4	B. C. D. E. W.	K. .	+ ·4	H.	Zr . .	– ·45	H.
B . .	– ·71	H.	Rh .	+ 1·1	H., F.			
Cd .	– ·17	H.	Ru .	+ ·56	H.	**Liquids**		
Cr .	+ 3·7	H.	Se .	– ·32	H., C.	Br . .	– ·41	C., Q.
Cu .	– ·087	H.	Si .	– ·12	H.	Hg . .	– ·19	Q., M., H.
Au .	– ·15	K., H.	Ag .	– ·2	H.	N liq. .	+ ·28	F., D.
I . .	– ·36	B., C., H.	Na .	+ ·51	H.	O liq. .	+ 324	F., D.
Ir . .	+ ·15	H.	S . .	– ·5	B.,C.,L.,K.,H.	H₂O,15°	– ·837	Du B.
Fe .	See	p. 89.	Ta .	+ ·93	H.	H₂O,15°	– ·77	S.
Pb .	– ·12	H., K., L.	Te .	– ·32	E., C., H.	**Gases**		
Mg .	+ ·55	H.	Tl .	*c.* – ·3	H.	Air,16°	+ ·032	Du B.
Mn .	+ 10·6?	H.	Th .	+ 1·8	H.	A . .	– ·010	T.
Mo .	+ ·04	H.	Sn .	+ ·025	K., H.	He . .	– ·002	T.
Nb .	+ 1·3?	H.	Ti .	*c.* + 2	H.	H . .	– ·008	Q.
Os .	+ ·04	H.	W .	+ ·33	H.	N . .	+ ·024	Du B.
Pd .	+ 5·8	H., K., C., F.	U . .	*c.* + ·9	M., H.	O . .	+ ·123	Du B., Q.

B., E. Becquerel, 1855; C., Curie, 1895; D., Dewar, 1892; Du B., Du Bois; E., Ettingshausen; F., Finke; F. D., Fleming and Dewar; H., Honda; K., Königsberger, 1901; L., Lombardi, 1897; M., St. Meyer; Q., Quincke; S., Scarpa, 1905; T., Tänzler, 1907; W., Wills, 1898.

TEMPERATURE AND MAGNETIZATION

The magnetic moment (M) of a magnet diminishes as the temperature (t) rises. In $M_t = M_0(1 - at)$, a varies widely, but is of the order ·0003 to ·001. The permeability μ also depends on the temperature. There is a **critical temperature** above which μ is very small; in the case of iron it is one of the recalescence temperatures, and is the same as for carbon steels containing up to ·45 % of C.

The critical temperature of a metal is not perfectly definite, but depends to some extent on whether the metal is being heated or cooled.

Substance.	Crit. Temp.	Observer.	Substance.	Crit. Temp.	Observer.
Iron . . .	690°–870° C.	Hopkinson	Nickel, 95% .	310°	Hopkinson
" . . .	*c.* 895	Roberts-Austen	" "	377	Weiss, 1907
" . . .	855–867	Osmond	Magnetite . .	582	" "
" . . .	757	Weiss, 1907	Heusler alloys	*c.* 300	Gray, 1908
Cobalt . .	1075	Stifler, 1911	Stalloy . . .	760	Hadfield

Nickel steel (25 % Ni); **0°** to **50°** $\mu = 1·4$ to 60; **50°** to **580°** $\mu = 60$ to 0·4.

STEINMETZ'S COEFFICIENT

Values of η in Steinmetz's formula $\eta B_{max}^{1\cdot6}$ for the hysteresis loss in ergs per c.c. per cycle. B_{max} is the maximum value of the induction.

Substance.	η	Substance.	η
3½% Silicon iron (Stalloy) .	·0007	Grey cast iron	·013
Good transformer iron . .	·0011	Nickel	·012 to ·038
Dynamo cast steel	·0026	Cobalt	·012
High carbon steel, hardened	·025		

TERRESTRIAL MAGNETIC CONSTANTS

Magnetic observatories no longer remain in large cities owing to electric tram disturbances, and thus many of the places for which reliable data exist are not generally known. The general locality of the station is indicated in many cases below.

Magnetic constants obtained in most physical laboratories are usually abnormal owing to the proximity of iron in some form.

Much of the data below is derived from the Reports of Kew Observatory, and the publications of the United States Coast and Geodetic Survey.

A W declination means that the N-seeking end of the magnetic needle points west of true north ; a N inclination means that the same end of the needle points downwards.

H and V are the horizontal and vertical components of the earth's magnetic field. (See Chree, "Terrestrial Magnetism," Encyc. Brit., 11th edit., 1911.)

Place.	Latitude.	Longitude.	Year.	Declination.	Inclination.	H.	V.
	° ′	° ′		° ′	° ′	c.g.s.	c.g.s.
North magnetic pole . .	70 5 N	96 45 W	—	—	90 0 N	0	—
South magnetic pole* . .	72 25 S	154 E	1908	—	90 0 S	0	—
British Isles—							
Aberdeen (University) .	57 9 N	2 7 W	1909	16 34 W	70 39 N	·163	·464
Eskdalemuir (Dumfries)	55 19 N	3 12 W	1909	18 30 W	69 39 N	·1684	·4519
Falmouth (Cornwall). .	50 9 N	5 5 W	1909	17 48 W	66 31 N	·1880	·4327
Greenwich	51 28 N	0 0	1916	14 47 W	66 53 N	·1849	·4333
Kew	51 28 N	0 19 W	1909	16 11 W	67 0 N	·1851	·4359
Leeds (University) . .	53 48 N	1 33 W	1909	18 2 W†	68 35 N	·176	·449
St. Helier (Jersey). . .	49 12 N	2 5 W	1907	16 27 W	65 35 N	—	—
Stonyhurst (Lancs.) . .	53 51 N	2 28 W	1909	17 29 W	68 43 N	·1742	·4472
Valencia (S. W. Ireland)	51 56 N	10 15 W	1909	20 50 W	68 15 N	·1788	·4481
Africa—							
Cape Town	33 56 S	18 29 E	1885	30 15 W	56 0 S	·199	·295
Helvan (Cairo). . . .	29 52 N	31 21 E	1908	2 56 W	40 39 N	·3003	·2579
Mauritius.	20 6 S	57 33 E	1908	9 14 W	53 45 S	·2342	·3193
America—							
Agincourt (Toronto) . .	43 47 N	79 16 W	1906	5 45 W	74 36 N	·1640	·5950
Cheltenham (Washington) . .	38 44 N	76 50 W	1909	5 34 W	70 31 N	·1988	·5620
Fairhaven (Mass.) . .	41 37 N	70 54 W	1908	12 27 W	73 8 N	·1736	·5724
Goat Island (California)	37 49 N	122 22 W	1909	17 53 E	62 11 N	·2525	·4786
Greenwich (New York) .	41 0 N	73 37 W	1908	10 14 W	72 13 N	·1822	·5680
Rio de Janeiro	22 55 S	43 11 W	1906	8 55 W	13 57 S	·2477	·0616
Santiago (Chili) . . .	33 27 S	70 42 W	1906	14 19 E	30 12 S	—	—
Sitka (Alaska)	57 3 N	135 20 W	1909	30 12 E	74 37 N	·1557	·5659
Waukegan (Chicago). .	42 21 N	87 51 W	1908	2 39 W	72 46 N	·1830	·5898

* Mawson and David (with Shackleton), 1908.   † 1907.

TERRESTRIAL MAGNETIC CONSTANTS (*contd.*)

Place.	Latitude.	Longitude.	Year.	Declination.	Inclination.	H.	V.
	° ′	° ′		° ′	° ′	c.g.s.	
Asia—							
Alibag (Bombay) . . .	18 39 N	72 52 E	1908	1 2 E	23 22 N	·3686	·1592
Barrackpore (Calcutta).	22 46 N	88 22 E	1907	1 10 E	30 30 N	·3729	·2197
Hong Kong	22 18 N	114 10 E	1909	0 2 E	31 1 N	·3709	·2229
Australasia—							
Christchurch (N.Z.) . .	43 32 S	172 37 E	1903	16 18 E	67 42 S	·2266	·5526
Honolulu (Hawaii) . .	21 19 N	158 4 W	1909	9 26 E	40 54 N	·2917	·2527
Melbourne	37 50 S	144 58 E	1901	8 27 E	67 25 S	·2331	·5602
Sydney	33 52 S	151 12 E	1885	9 30 E	62 30 S	·268	·515
Europe—							
Arctic ⎰ (Norway) .	69 56 N	22 58 E	1903	0 43 W	76 21 N	·1258	·5178
Regions ⎱ (Spitzbergen).	77 41 N	14 56 E	1903	10 55 W	80 8 N	·0942	·5417
Odessa	46 24 N	30 48 E	1901	4 27 W	62 18 N	·2188	·4168
Pawlowsk (Petrograd)	59 41 N	30 29 E	1906	1 4 E	70 37 N	·1653	·4696
Potsdam	52 23 N	13 4 E	1909	9 11 W	66 20 N	·1883	·4297
Rude Skov (Copenhagen)	55 51 N	12 27 E	1908	9 43 W	68 45 N	·1741	·4476
Uccle (Brussels) . . .	50 48 N	4 21 E	1908	13 37 W	66 2 N	·1906	·4287
Val Joyeux (Paris) . .	48 49 N	2 1 E	1909	14 33 W	64 44 N	·1973	·4179

SECULAR MAGNETIC CHANGES

At the present period we are going through a remarkable secular alteration. For generations H had been steadily rising in Western Europe, but during the last few years a wave of depression has travelled across from the east. H has steadily fallen at Petrograd since about 1900, at Potsdam since about 1905, at Greenwich and Kew since 1907, while in 1909 H was still rising at Falmouth and Valencia. The easterly motion of the declination needle has also increased notably since 1900. Thus secular change data based on, say, the last five years will not serve to prospect the future.

Mean change per annum at	1908-1909.		1904-1909.			
	Decln.	H.	Decln.	Incln.	H.	V.
		c.g.s.			c.g.s.	c.g.s.
Greenwich . .	− 5·9	− 5 × 10⁻⁵	− 5·5	− 0·7	+ 1 × 10⁻⁵	− 29 × 10⁻⁵
Kew	− 6·1	− 9 ,,	− 5·4	− 1·1	+ 2 ,,	− 35 ,,
Stonyhurst . .	− 7·0	− 10 ,,	− 5·9	− 1·1	+ 6 ,,	− 25 ,,
Falmouth . . .	− 6·3	+ 4 ,,	− 4·7	− 1·4	+ 9 ,,	− 30 ,,
Valencia . . .	− 5·4	+ 7 ,,	− 5·0	− 1·2	+ 7 ,,	− 25 ,,

SECULAR CHANGES AT LONDON (GREENWICH)

Year.	Decln.	Incln.	Year.	Decln.	Incln.	H.
	° ′	° ′		° ′	° ′	c.g.s.
1580	11 17 E	72 0 N	1851	22 25 W	68 47 N	·1729
1660	0 0	73 15 N	1875	19 21 W	67 42 N	·1795
1720	13 0 W	74 40 N*	1907	16 0 W	66 56 N	·1853*
1815	24 27 W*	70 30 N	1916	14 47 W	66 53 N	·1849

* Maximum.

SPARKING POTENTIALS

The sparking voltages given below are those which will break down non-ionized air at atmospheric pressure and room temperature. The electrodes are equal smooth polished metal balls of various diameters. Russell (*Phil. Mag.* 1906) gives the dielectric strength of air at atmospheric pressures as between 38,000 and 39,000 volts per cm. for either direct or alternating potentials.

(See Kaye's "X Rays" (Longmans, 1916) for further values.)

Spark gap.	Diameter of balls in cms.				Spark gap.	Diameter of balls in cms.			
	0·5	1·0	2·0	5·0		0·5	1·0	2·0	5·0
cm.	volts. ×10³	volts. ×10³	volts. ×10³	volts. ×10³	cm.	volts. ×10³	volts. ×10³	volts. ×10³	volts. ×10³
0·1	4·8	4·8	4·7	—	0·9	19·6	25·6	28·6	30·1
0·2	8·4	8·4	8·1	—	1·0	20·2	26·7	30·8	32·7
0·3	11·3	11·4	11·4	—	1·5	22	31·6	39	46
0·4	13·8	14·4	14·5	—	2·0	23	36	47	58
0·5	15·7	17·3	17·5	18·4	3·0	24	42	57	77
0·6	17·2	19·9	20·4	21·6	4·0	25	45	64	92
0·7	18·3	22·0	23·2	24·6	5·0	26	47	69	105
0·8	19·0	24·1	26·0	27·4					

HOMOGENEOUS X-RAYS

Mass absorption coefficients, λ/ρ, measured in Al foil. λ is the absorption coefficient (see p. 107) of the homogeneous characteristic (K) X radiation from a metal; ρ is the density of aluminium foil. For a complete set of values, see Kaye's "X Rays" (Longmans, 1916).

Radiator.	Al	Cr	Fe	Ni	Co	Cu	Zn	As	Se	Ag
λ/ρ	3400	136	88·5	59·1	71·6	47·7	39·4	22·5	18·5	2·5

CATHODE DARK SPACE

The thickness (d) of the Crookes dark space is given by $d = (A/p) + B/\sqrt{i}$, where p is the pressure, i the current density, and A and B are constants for each gas. This equation is satisfied very exactly by the ordinary elementary gases, and a little less so by the gases of the helium group. Unfortunately for the use of the dark space as a pressure indicator, the current density term in the formula is almost as large as the pressure term for pressures about 1/10 mm.

The values of A and B below are for large plane aluminium electrodes. d is measured in cms., p in mms. of mercury. The unit of i is 1/10 milliampere per sq. cm. of cathode, which is about the sort of current density that obtains with an average coil discharge and a moderate-sized cathode.

(See Aston and Watson, *Proc. Roy. Soc.*, 1911.)

Gas.	Hydrogen	Nitrogen	Air	Oxygen
A	·26	·068	·065	·057
B	·43	·40	·42	·50

COEFFICIENTS OF RECOMBINATION a

a is given below in terms of $1000e$, where e is the numerical value of the ionic charge : $4\cdot7 \times 10^{-10}$ in electrostatic units. For air, $a = 3320e = 1\cdot56 \times 10^{-10}$ cm.^{3} sec^{-1}. Room temp. and pressure.

Gas.	Air.				O_2	CO_2	H_2
a	$3\cdot42$, T.; $3\cdot38$, Mc.; $3\cdot2$, L.; $3\cdot3$, H.; $3\cdot32$ *, E.				$3\cdot38$, T.	$3\cdot5$, T.	$3\cdot02$, T ; $2\cdot94$, Mc.

E., Erikson, *P.M.*, 1909; H., Hendren, *P.R.*, 1905; L., Langevin, *A.C.P.*, 1902; Mc., McClung, *P.M.*, 1902 ; T., Townsend, *P.T.*, 1899.   * 17° C., 760 mm. Hg.

a IN AIR AND PRESSURE

Press. in atmos. . . .	$\cdot2$	$\cdot5$	1	2	3	5	L., Langevin. H., Hendren.
a (relative values), L. .	5	12	27	30	26	18	

Press. in cms.	76	45	25	15	10	5	3·5	2	1
a (absolute values), H. .	3·3	2·65	2·07	1·75	1·55	1·31	1·25	1·15	1·00

a IN AIR AND TEMPERATURE
Air at constant density.   (E., Erikson ; P., Phillips, *Electrician*, 1909.)

Temp. °C. . . .	$-179°$	-68	12	64	100	155	Temp. °C. . . .	15°	100	155	176
a (in terms 1000e), E.	7·5	5·6	3·47	2·31	1·73	1·38	a (relative values), P.	1	·50	·40	·36

IONIC COEFFICIENTS OF DIFFUSION D

Rate of interdiffusion (in cm.^{3} sec^{-1}) of gaseous ions in dry air : D+ for positive, D− for negative ions. (Townsend, *Phil. Trans.*, 1899, 1900.)

Ionization	Röntgen Rays.	β and γ Rays.	Ultra-violet light.	Point discharge.
D+ at 76 cm.	·028	·032	—	·0247, ·0216
D− at 76 cm.	·043	·043	·043	·037, ·032

GASES IONIZED BY RÖNTGEN RAYS
Air, CO_2, and hydrogen at 15° C. and 760 mm.

Dry Gas.	D+	D−	Dry Gas.	D+	D−	Moist Gas.	D+	D−	Moist Gas.	D+	D−
Air O_2 {dried by $CaCl_2$}	·028 ·025	·043 ·04	CO_2 H_2 {dried by $CaCl_2$}	·023 ·123	·026 ·19	Air O_2 {sat. with H_2O}	·032 ·029	·035 ·036	CO_2 H_2 {sat. with H_2O}	·024 ·128	·025 ·142

AIR IONIZED BY β AND γ RAYS

Press. p. in cms.	77·2	55	40	30	20	Press. p. in cms.	77·2	55	40	30	20
D+ at 15° C.	·0317	·042	·0578	·078	·118	D− at 15° C.	·0429	·0542	·078	·103	1·55
pD+ „	2·45	2·31	2·31	2·34	2·36	pD− „	3·3	2·98	3·12	3·09	3·1

A.C.P., Ann. de Chim. et de Phys. ; *P.M.*, Phil. Mag. ; *P.R.*, Physical Review ; *P.T.*, Phil. Trans.

MOBILITIES OF IONS IN GASES

Velocities of ions are in cm. per sec. for unit field, or in cm.^{2} sec.$^{-1}$ volt ^{1} at temp. and press. of room. K_+ = mobility of positive ion, K_- of negative.

For **moist air** (i.e. saturated with H_2O), $K_+ = 1.37$, $K_- = 1.51$.

For **dry air** (dried by $CaCl_2$), $K_+ = 1.36$, $K_- = 1.87$. (Zeleny (air blast method), *Phil. Trans.*, 1900.) $\qquad$ * Mean = $(K_+ + K_-)/2$.

For mobilities of natural ions in air, see p. 105.

Dry Gas.	K_+	K_-	Ionization and Observer.	Dry Gas.	K_+	K_-	Ionization and Observer.
	76 cm. Hg				76 cm. Hg		
Air .	1·32	1·80	Point disch., Chattock, *P.M.*, 1899, 1901.	CO$_2$ . . .	0·76	0·81	X-rays, Zeleny, 1900.
,, .	1·54	1·78	X-rays, Wellisch, *Phil Trans.*, 1909.	,, . . .	0·86	0·90	,, Langevin, '03.
				,, . . .	0·81	0·85	,, Wellisch, '09.
,, .	1·40	1·70	,, Langevin, *A.C.P.*, 1903.	HCl . . .	1·27*		,, Rutherford.
				SO$_2$ . . .	0·44	0·41	,, Wellisch, '09.
,, .	1·39	1·78	,, Phillips,*P.R.S.*, 1906.	Cl$_2$ . . .	1·0*		,, Rutherford.
				N$_2$O . . .	0·82	0·90	,, Wellisch, '09.
,, .	1·36	1 87	,, Zeleny, *Phil. Trans.*, 1900.	NH$_3$ . .	0·74	0·80	,, ,,
				Me. acetate .	0·33	0·36	,, ,,
,, .	**1·40**	**1·78**	Mean value.	Me. bromide	0·29	0·28	,, ,,
H$_2$ . .	5·4	7·43	Point disch., Chattock.	Me. iodide .	0·21	0·22	,, ,,
,, . .	6·7	7·9	X-rays, Zeleny, 1900.	Et. alcohol .	0·34	0·27	,, ,,
He .	5·09	6·31	,, Franck and Pohl, *V.D.P.G.*, '07.	Et. acetate .	0·31	0·28	,, ,,
				Et. aldehyde	0·31	0·30	,, ,,
N$_2$ . .	1·6*	—	X-rays, Rutherford, *P.M.*, 1897.	Et. chloride .	0·33	0·31	,, ,,
				Et. ether . .	0·29	0·31	,, ,,
O$_2$ . .	1·36	1·80	,, Zeleny, 1900.	Et. formate .	0·30	0·31	,, ,,
,, .	1·3	1·85	Point disch., Chattock.	Et. iodide .	0·17	0·16	,, ,,
CO .	1·1	1·14	X-rays, Wellisch, '09.	C.Cl$_4$ . .	0·30	0·31	,, ,,
CO$_2$.	0·83	0·92	Point disch., Chattock.	Pentane . .	0·36	0·35	,, ,,
				Acetone . .	0·31	0·29	,, ,,

IONIC MOBILITY AND PRESSURE

Air ionized by Röntgen rays. (Langevin, *A.C.P.*, 1903.)

Press. cm.	7·5	20	41 5	76	143·5	Press. cm.	7·5	20	41·5	76	142
K_+	14·8	5·45	2·61	1·40	0·75	K_-	21·9	7·35	3·31	1·7	0·9

IONIC MOBILITY AND TEMPERATURE

Air at 76 cm. press. ionized by Röntgen rays. (Phillips, *P.R.S.*, 1906.)

Temp. ° C.	138°	126°	110°	100°	75°	60°	12°	−64°	−179°
K_+	2·00	1·95	1·85	1·81	1·67	1·60	1·39	0·945	0·235
K_-	2·49	2·40	2·30	2·21	2·12	2·00	1·785	1·23	0·235

IONIC MOBILITIES IN LIQUIDS AND SOLIDS

Ionized by radium rays. (Bohm-Wendt and v. Schweidler, *Phys. Zeit.*, 1909; Bialobjeski, *Compt. Rend.*, 1909.)

Substance.	$(K_+ + K_-)$	Substance.	$(K_+ + K_-)$
Petroleum ether	$3·8 \times 10^{-4}$	Ozokerite at 100° . . .	$5·1 \times 10^{-4}$
Vaseline	$5·3 \times 10^{-6}$	,, ,, 80° . . .	$35·0 \times 10^{-4}$

A.C.P., *Ann. de Chim. et de Phys.*, *P.M.*, *Phil. Mag.*; *P.R.S.*, *Proc. Roy. Soc.*; *V.D.P.G.*, *Verh. Deutsch. Phys. Gesell.*

IONIC MOBILITIES AT HIGH TEMPS

K in cm. sec.$^{-1}$ per volt cm.$^{-1}$ for coal-gas flames in most instances. The ionic mobility is independent of the acid of the salt. Gold's and Wilson's values for K agree the best with existing theory, which makes $K = Xe\lambda/mu = 17,000$ at 1800° C. (Gold). X is the electric field per cm., λ is the mean free path, and u the velocity of the corpuscle.

Salt.	Temp.	K$_+$	K$_-$	Observer.
Cs, Rb, K, Na, Li . .	Flame, c. 2000° C.	62	c. 1000	H. A Wilson, P.T., 1899
1/20 normal KCl . .	Flame	260	1400 }	Marx. Ann. der Phys.,
NaCl	,,	340	1800 }	1900
1/256 normal K salt .	Flame, c. 2000°	—	1320 }	
1/16 normal Na salt .	,, ,,	—	1280 }	Moreau, Journ. de Phys.,
Concentrated sols. of alkalies	,, ,,	80	— }	1903
Cs, Rb, K, Na, Li .	Air at 1000°	7·2	26 }	H. A. Wilson, P.T., 1899
Ba, Sr, Ca . . .	,, ,,	3·8	— }	and P.M., 1906
K, Na	Flame, c. 1800°	—	8000 }	Gold, P.R.S., 1907, ratio of potential grad. to current
K.	Flame, c. 1800°	—	13,000	Poten. grad., and gas velocity
K$_2$CO$_3$	Bunsen burner	—	9600	H. A. Wilson, P.R.S., 1909
Na	Flame, c. 2000°	—	1170	Moreau, C.R., 1909

CONDENSATION OF VAPOURS

Expansion = v_2/v_1, where v_1 is the volume of the gas before, and v_2 the volume after expansion. **Supersaturation** of the vapour (at end of cooling by expansion) necessary for condensation = S = (density of vapour when drops are formed)/(density of saturated vapour at the same temp.). (See J. J. Thomson, "Conduction of Electricity through Gases.")

CONDENSATION ON NATURAL IONS AND MOLECULES

Dust-free gas saturated with water-vapour. (C. T. R. Wilson, P.T., '97, '99, '00.)

Gas.	Rain-like Condensation.		Cloud-like Condensation.		Gas.	Rain-like Condensation.		Cloud-like Condensation.	
	v_2/v_1	S.	v_2/v_1	S.		v_2/v_1	S.	v_2/v_1	S.
Air . .	1·252	4·2	1·38	7·9	CO$_2$ . .	1·365	4·2	1·535	7·3
O$_2$ . .	1·257	4·3	1·38	7·9	Cl$_2$ . .	1·3	3·4	1·45	5·9
N$_2$ . .	1·262	4·4	1·38	7·9	H$_2$ . .	—	—	1·38	7·9

CONDENSATION IN AIR IONIZED BY RÖNTGEN AND RADIUM RAYS

(L., Laby, Phil. Trans., 1908; P., Przibram, Wien Ber., 1906.)

Vapour and Observer.	Ion.	v_2/v_1	S.	Vapour and Observer.	Ion.	v_2/v_1	S.
Water (C. T. R. Wilson)	−	1·25	4·15	n-Butyric acid, L. . .	?	1·38	15·0
Water (C. T. R. Wilson)	+	1·31	5·8	iso-Butyric acid, L. . .	?	1·36	13·3
Et. acetate, L. . . .	+	1·48	8·9	iso-Valeric acid, L. . .	?	1·22	6·0
Me. butyrate, L. . .	+	1·33	5·3	Methyl alcohol, P. . .	+	1·25	3·1
Me. iso-butyrate, L. .	?	1·35	5·2	Ethyl alcohol, P. . .	+	1·17	2·3
Propyl-acetate, L. . .	+	1·31	5·0	Propyl alcohol, P. . .	?	1·18	3·0
Et. propionate, L. . .	?	1·41	7·8	iso-Butyl alcohol, P. .	?	1·2	3·6
Formic acid, L. . .	?	1·78	25·1	iso-Amyl alcohol, P. .	+	1·22	5·5
Acetic acid, L. . . .	+	1·44	9·3	,, ,, L. .	+	1·18	4·1
Propionic acid, L. . .	?	1·34	9·4	Chloroform, P. . . .	+	1·54	3·0

A.C.P., Ann. de Chim. et de Phys.; C.R., Compt. Rend.; P.M., Phil. Mag.; P.R.S., Proc. Roy. Soc.; P.T., Phil. Trans.

NE FOR ELECTROLYTIC IONS

NE is given both in electrostatic units (E.S.U.) and electromagnetic units (E.M.U.). N is the number of molecules in a c.c. of gas at 76 cm. Hg ($g = 980.6$) and $t°$ C., and E is the charge on the monovalent ion in electrolysis.

Antecedent data.—1 coulomb deposits 1.11827 mgm. Ag. At. wt. of Ag, 107.88 ; of H, 1.008. Density of $H_2 = 8.987 \times 10^{-5}$ gm. per c.c. at 0° C.

Gas.	E.S.U.	E.M.U.	Gas.	E.S.U.	E.M.U.	Gas.	E.S.U.	E.M.U.
	$\times 10^{10}$			$\times 10^{10}$			$\times 10^{10}$	
H_2 at 0° C.	1.29015	0.4300	O_2 at 0°	1.2924	0.4308	Ideal ⎰ at 0°	1.2913	0.43044
H_2 at 15° C.	1.2230	0.4077	O_2 at 15°	1.2248	0.4083	gas ⎱ at 15°	1.2241	0.40803

Ne FOR GASEOUS IONS

N is the number of molecules per c.c. of **air** at room temp. and 76 cm. Hg ; e is the ionic charge in E.S.U., e_- for negative and e_+ for positive ions.

Ionization.	Ne_	Ne+	Observer.
X rays	1.23×10^{10}	2.41×10^{10}	Townsend, P.R.S., 1908, 1909.
Ra rays . . .	1.24×10^{10}	1.26 to 1.37×10^{10}	Haselfoot, P.R.S., 1909.

Ne CALCULATED

In E.S.U., $Ne = 3.04 \times 10^8 \times K/D = 3.04 \times 10^8 \times 1.40/0.028 = 1.52 \times 10^{10}$ for positive air ions at 76 cm. and room temp. For D and K, see pp. 94, 95.

Gas.	Ne+	Ne_	Gas.	Ne+	Ne_		Ne+	Ne_
Air .	$1.52.10^{10}$	$1.26.10^{10}$	H_2 .	$1.50.10^{10}$	$1.23.10^{10}$	Mean ⎰	$1.42.10^{10}$	$1.22.10^{10}$
O_2 .	$1.62.10^{10}$	$1.38.10^{10}$	CO_2 .	$1.07.10^{10}$	$1.02.10^{10}$	⎱	$1.32.10^{10}$	

THE IONIC CHARGE e

$e = 4.77 \times 10^{-10}$ **E.S.U.** $= 1.59 \times 10^{-20}$ **E.M.U.**, as a mean of the latest determinations. See Millikan, P.M., July, 1917.

Ionization.	Method.	e in E.S.U.	Observer.
Röntgen rays ; negative ions. ⎱	By measuring total charge on a cloud and obtaining number of ions from size of drops by Stokes' law.	$6.5.10^{-10}$	J. J. Thomson, P.M., 1898.
Ultra - violet light on metal ; negative ions ⎰		6.8 ,,	J. J. Thomson, P.M., 1899.
Röntgen rays ; negative ions.	Force (by Stokes' law) exerted by an electric field on a singly charged drop.	3.1 ,,	H. A. Wilson, P.M., 1903.
Radium rays ; negative ions.	The observer's original method.	3.4 ,,	J. J. Thomson, Proc. Camb. Phil. Soc., 1903.
Charged spray of electrolytic O_2.	Total charge on a cloud. No. of ions from weight of cloud and size of drops, using Stokes' law.	3.0 ,,	Townsend, Proc. Camb. Phil. Soc., 1897.
a particles (Ra,) assuming charge = +2e.	By counting a particles and measuring their total charge.	4.65 ,,	Rutherford & Geiger, P.R.S., 1908.
Electrolytic ions.	By counting colloid particles.	4.1 ,,	Perrin, C.R., 1908.
Charged spray of electrolytic O_2.	By H. A. Wilson's method, above.	4.7 ,,	Lattey, P.M., 1909.
a particles (Polonium) ; charge = +2e.	By counting a particles, and measuring their total charge.	4.79 ,,	Regener, Berl. Ber., 1909.
Electrolytic ions.	From Brownian movements.	4.5 ,,	Broglie, Le R., 1909.
Radium rays ; negative ions.	By H. A. Wilson's method, above	4.67 ,, 4.77 ,,	Begeman. [1909. Millikan, P.M.,'17

C.R., Comptes Rendus ; Le R., Le Radium ; P.M., Phil. Mag. ; P.R.S., Proc. Roy. Soc.

NUMBER OF MOLECULES IN A GAS

N = the number of molecules in a **gram molecule** of gas (Perrin, *Compt. Rend.*, 1908 ; Perrin and Dabrowski, *C.R.*, 1909—by observations on colloidal particles). The theoretical value is $N = NE/e = 2.894 \times 10^{14}/(4.77 \times 10^{-10}) = 6.06 \times 10^{23}$.

Method.	Gum mastic.	Gamboge.	Method.	Gum mastic.	Gamboge.
Counting by ultra microscope . .	$N = 7 . 10^{23}$	$N = 7.05 . 10^{23}$	Brownian movements	$N = 7.3 . 10^{23}$	$N = 7 . 10^{23}$

e/m FOR NEGATIVE ELECTRONS

e/m in E.M.U. gm.$^{-1}$. Velocities v in cm. sec.$^{-1}$. For some other values of *e/m* see J. J. Thomson's "Conduction of Electricity through Gases," and Wolz, *A.d.P.*, **30**, 274, 1909. The **mean** of Simon's, Becker's, Classen's, Kaufmann's, Wolz's, Bucherer's, and Bestelmeyer's values is $e/m_0 = 1.772 \times 10^7$ **E.M.U. gm.$^{-1}$**, where m_0 is the mass of the electron associated with very small velocities. For the variation of *e/m* with velocity see p. 99. (See also Schuster, *P.R.S.*, 1890.)

e/m	v	Observer.	e/m	v	Observer.
CATHODE RAYS			**LENARD RAYS**		
1.2×10^7	2.4 to $3.2 . 10^9$	J. J. Thomson, *P.M.*, 1897	$0.68 . 10^7$	3.4 to $10.7 . 10^9$	Lenard, *A.d.P.*, 1898
1.77 to 1.8 ,,	—	Kaufmann, *A.d.P.*, 1897, 1898			
1.86 ,,	—	Simon, *A.d.P.*, 1899	**INCANDESCENT OXIDES, etc.**		
1.88 ,,	$= e/m_0$				
1.87 ,,	5.7 to $7.5 . 10^9$	Seitz, *A.d.P.*, 1902	$0.87 . 10^7$	—	J. J. Thomson, *P.M.*, 1899
1.84 ,,	3.8 to 13 ,,	Starke, *V.D.P.G.*, 1903	0.56 ,,	—	Owen, *P.M.*, 1904
1.75 ,,	11.1 ,,	Becker, *A.d.P.*, 1905	1.5 ,,	0.1 to $1.0 . 10^9$	Wehnelt, *A.d.P.*, 1904
1.85 ,,	$= e/m_0$				
1.774 ,,	$1.9 . 10^9$				
1.767 ,,	$3.8 . 10^9$	Classen,*P.Z.*,1908	**SECONDARY CORPUSCULAR RAYS,** from X-rays incident on platinum		
1.771 ,,	$= e/m_0$				
			$1.773 . 10^7$	$= e/m_0$ (on Lorentz's theory)	Bestelmeyer, *A.d.P.*, 1907.
β RAYS					
$0.1 . 10^7$	—	Becquerel, *Rap. C.P.*, 1900			
1.77 ,,	$= e/m_0$	Kaufmann, *Gött. Nachr.*, 1901	**ULTRA VIOLET LIGHT ON METAL**		
1.66 ,,	$= e/m_0$ (on Lorentz's theory)	Kaufmann,*A.d.P.*, 1906	$0.76 . 10^7$	—	J. J. Thomson, *P.M.*, 1899
1.82 ,,	$= e/m_0$ (on Abraham's theory)	Kaufmann,*A.d.P.*, 1906	$1.1 . 10^7$	—	Lenard, *A.d.P.*, 1900
1.763 ,,	$= e/m_0$	Bucherer,*A.d.P.*, 1909			
,, ,,	9.5 to $20.6.10^9$				
1.767 ,,	$= e/m_0$	Wolz,*A.d.P.*,1909	**ZEEMAN EFFECT**		
,, ,,	15 to 21 . 10^9		$1.775 . 10^7$	—	Mean of 4 observer's values (see below).

A.d.P., *Ann. der Phys.* ; *P.M.*, *Phil. Mag.* ; *P.R.S.*, *Proc. Roy. Soc.* ; *P.Z.*, *Phys. Zeit.* ; *Rap. C.P.*, *Rapports Congrès à Paris* ; *V.D.P.G.*, *Verh. Deutschs. Phys. Gesell.*

ELECTRONIC e/m FROM ZEEMAN EFFECT

For a spectrum line of wave-length λ, which becomes a normal triplet with a separation of δλ in a magnetic field H (in gauss, *i.e.* E.M.U.), Lorentz has shown that $e/m = 2\pi V\delta\lambda/(\lambda^2 H)$, where V is the velocity of light; e/m is in E.M.U. gm.$^{-1}$. The values 1·79, 1·77, 1·767, 1·771, **mean 1·775 . 10^7 E.M.U. gm.$^{-1}$**, agree well with e/m_0 above.

Line.	e/m	Observer.	Line.	e/m	Observer.
Hg 5791, 5770 5461, 4358	×10^7 1·72 to 2·80 [1900	Blythswood & Marchant, *P.M.* 1900	Zn 4810 4722, 4680.	×10^7, 2×1·767	Cotton & Weiss, *C.R.*, 1907
Zn, Cd	1·6	Reese, *As. Jl.*,	He	1·77	Lohmann, *P.Z.*, 1908
„ „	1·59	Kent, *As. Jl.*, 1901	Hg 5791	1·93	Baeyer & Gehrcke, *A.d.P.*, 1909
Cd 4678 Zn 4680	} 1·71	Färber, *A.d.P.*, 1902	„ 5770 „ 4916	2·06 1·81	
Cd 4678 Zn 4680	} 1·79	Stettenheimer, *A.d.P.*, 1907	„ 5790, 5770 „ 4916, 4358	} 1·771	Gmelin, *A.d.P.*, 1909

ELECTRONIC e/m AND VELOCITY

m_0 is the electromagnetic mass of the negative electron for infinitely small velocities, m the transverse mass for a velocity v; $v/V = \beta$, where V is the velocity of light. (See Lorentz, *L'Eclairage Electrique*, July, 1905, and "The Theory of Electrons," 1909.) On the **theory of Abraham** (*Gött. Nachr.*, 1902),

$$\text{transverse mass } m = m_0 3\left(\frac{1+\beta^2}{2\beta} . \log\frac{1+\beta}{1-\beta} - 1\right)\Big/4\beta^2$$

β	Infinitely small.	0·1	0·5	0·9	0·99	0·999	0·9999	0·999999
m/m_0	1·00	1·015	1·12	1·81	3·28	4·96	6·68	10·1

On the **theory of Lorentz** (*Versl. Kon. Ac. Wet. Am.*, 1904) and the **relativity theory of Einstein** (*A.d.P.*, 1905), $m = m_0(1 - \beta^2)^{-1/2}$. This theory has been confirmed by the experiments of Bucherer (*A.d.P.*, 1909) and Wolz (*ibid.*), using β rays from Ra with velocities from (9 to 21) × 10^9 cm. per sec. Thus the mass of the negative electron is wholly electromagnetic.

β	m/m_0	β	m/m_0	β	m/m_0	β	m/m_0	β	m/m_0	β	m/m_0	β	m/m_0
0·01	1·045	0·34	1·063	0·48	1·140	0·62	1·274	0·76	1·538	0·90	2·294	0·97	4·113
0·05	1·001	0·36	1·072	0·50	1·155	0·64	1·301	0·78	1·598	0·91	2·412	0·98	5·025
0·10	1·005	0·38	1·081	0·52	1·171	0·66	1·331	0·80	1·667	0·92	2·552	0·99	7·089
0·20	1·020	0·40	1·091	0·54	1·188	0·68	1·364	0·82	1·747	0·93	2·721	0·999	22·36
0·25	1·033	0·42	1·102	0·56	1·207	0·70	1·400	0·84	1·843	0·94	2·931		
0·30	1·048	0·44	1·114	0·58	1·228	0·72	1·441	0·86	1·960	0·95	3·203		
0·32	1·056	0·46	1·120	0·60	1·250	0·74	1·487	0·88	2·105	0·96	3·571		

RH AND v: MAGNETIC DEFLECTION

When negative rays of velocity v are deflected by a uniform magnetic field H (at right angles to their direction) into a circular path of radius R, then RH = $vm/e = v\phi(\beta)/(e/m_0)$, where $\phi(\beta) = (1 - \beta^2)^{-\frac{1}{2}}$ on Lorentz's theory (see above), and $e/m_0 = 1·772 \times 10^7$ E.M.U. gm.$^{-1}$.

v is in 10^8 cm. sec.$^{-1}$; RH in gauss cm. **Example.**—If RH = 1210 gauss cm.^{2}, then $v = 174 \times 10^8$ cm./sec.

RH

v	0	6	12	18	24	30	36	42	48	54	60	66	72	78	84
0	0	33·9	67·8	102	136	170	204	239	274	310	346	382	419	456	494
90	532	572	612	653	695	739	784	830	877	926	977	1030	1090	1150	1210
180	1270	1340	1410	1490	1570	1660	1760	1860	1980	2110	2260	2420	2620	2850	3130
270	3490	3970	4660	5800	8330										

A.d.P., Ann. der Phys.; As. Jl., Astrophy. Journ.; C.R., Compt. Rend.; P.M., Phil. Mag.; P.Z. Phys. Ze.t.

RANGE AND VELOCITY OF α RAYS

Range in cms. in air at 76 cm. and $t°$ C. (see Bragg and Kleeman, *Phil. Mag.*, 1905). **Initial velocity** (v) in cms./sec. (Rutherford, *Phil. Mag.*, 1906, 1907). Some of the velocities are calculated from the ranges of the α particles ; RaC, ThC, and Polonium were observed. **Energy** of RaC α ray $= mv^2/2 = \frac{1}{2}v^2 . 2e . m/e_a = 2 \cdot 06^2 . 10^{18} e/(5 \cdot 07 . 10^3) = 8 \cdot 37 . 10^{14} e = 1 \cdot 3 . 10^{-5}$ ergs $= 3 \cdot 1 . 10^{-13}$ calories. Loss of energy in air is proportional to path traversed : thus **initial velocity** of α particle = (velocity of RaC α) $\times \cdot 347 \sqrt{r} + 1 \cdot 25$ cm./sec., where r is the range of particle. Also $v = 1 \cdot 077 r^{1/3} . 10^9$ cm./sec. (Geiger, *P.R.S.*, 1910)

α Ray.	Range.	Initial Vel.	Obs.	α Ray.	Range.	Initial Vel.	Obs.
	cms.	cm./sec.			cms.	cm./sec.	
U . . .	$c.3\cdot4$	$1\cdot56 . 10^9$	Mc. & R.	Rad.Ac .	$4\cdot8$	$1\cdot76 . 10^9$	H.
UX . .	$1\cdot07$?	—	Hess.	AcX .	$6\cdot55$	$2\cdot00$,,	H.
Io . .	$2\cdot8$	—	B.	AcEm .	$5\cdot8$	$1\cdot90$,,	H.
Ra . . .	$3\cdot50/20°C.$	$1\cdot56$,,	B. & K.	AcB .	$5\cdot5$	$1\cdot86$,,	H.
RaEm .	$4\cdot23$,,	$1\cdot70$,,	B. & K.	Th . . .	$3\cdot5$	—	
RaA . .	$4\cdot83$,,	$1\cdot76$,,	B. & K.	Rad.Th .	$3\cdot9$	$1\cdot63$,,	H.
RaC. . .	$7\cdot06$,,	$2\cdot06$,,	B. & K.	ThX . .	$5\cdot7$	$1\cdot89$,,	H.
RaF or	$\{3\cdot95$,,	—	K.	ThEm .	$5\cdot5$	$1\cdot86$,,	H.
Polonium	$\{3\cdot95$,,	—	K. & M.	ThB . .	$5\cdot0$	$1\cdot79$,,	H.
,,	$3\cdot86$,,	$1\cdot62$,,	L.	ThC . .	$8\cdot6$	$2\cdot25$,,	H.

B., Boltwood, *A.J.S.*, May, 1908 ; B. & K., Bragg & Kleeman, *P.M.*, 1905 ; H., Hahn, *P.M.*, 1906 ; Hess, *Wien. Ber.*, 1907 ; K., Kleeman, *P.M.*, 1906 ; K. & M., Kučera & Mašek, *P.Z.*, 1906 ; L., Levin, *A.J.S.*, 1906 ; Mc. & R., McCoy & Ross, *J.A.C.S.*, 1907.

NUMBER OF α PARTICLES FROM Ra

Number of α particles from Ra without its radioactive products $= 3 \cdot 4 . 10^{10}$ per gm. per sec. Number of α particles from Ra with its radioactive products $= 1 \cdot 36 . 10^{11}$ per gm. per sec. (Rutherford and Geiger, *Proc. Roy. Soc.*, 1908).

e/m FOR α RAYS

e/m in E.M.U. per gm. $2 e/m$ for helium $= 2NE/\rho = 4 \cdot 78 . 10^3$ E.M.U./gm. **Mean** for Ra, Pol, RaC $= \mathbf{4 \cdot 82 . 10^3}$ **E.M.U. gm**$^{-1}$. Since the α particle is a helium atom with a charge of $2e$, these values should be equal. * Final velocity of rays used.

Subst.	Velocity.*	e/m	Observer.	Subst.	Velocity.*	e/m	Observer.
	cm./sec.	E.M.U.			cm./sec.	E.M.U.	
Ra .	$1\cdot18$ to $1\cdot74 . 10^9$	$4\cdot6 . 10^3$	Mackenzie, *P.M.*, '05	RaA .	$1\cdot22 . 10^9$	$5\cdot6 . 10^3$	Rutherford, *P.M.*, '06
Pol .	$1\cdot41 . 10^9$	$4\cdot8$,,	Huff (cor?.) *P.R.S* '06	AcB .	$1\cdot0$,,	$4\cdot7$,,	
RaC.	$1\cdot57$,,	$5\cdot07$,,	Rutherford, *P.M.*, '06	ThC .	$1\cdot98$,,	$5\cdot6$,,	Rutherford & Hahn, *P.M.*, '06

STOPPING POWERS OF MATERIALS

If a layer of air of density ρ and thickness t decreases the range of an α particle by the same amount as aluminium foil of density ρ_a and thickness t_a, then the **atomic stopping power**, S, of Al relative to air is given by $S = 27 t \rho / 14 \cdot 4 t_a \rho_a)$ = (number of atoms per cm.^{2} in air layer)/(number of atoms per cm.^{2} in Al foil) (Bragg and Kleeman, *Phil. Mag.*, 1905 ; Bragg, *Phil. Mag.*, 1906).

Metal.	S.	Metal.	S.	Metal.	S.	Gas.	S.	Gas.	S.
(Air at 20° C., 76 cm.)	$1\cdot00$	Ag . .	$3\cdot17$	Ni . .	$2\cdot46$	O_2 . .	$1\cdot055$	C_2H_2 .	$1\cdot11$
		Sn . .	$3\cdot37$	Au . .	$4\cdot45$	N_2O . .	$1\cdot46$	Ethylene	$1\cdot35$
Al . . .	$1\cdot45$	Pt . .	$4\cdot16$	Pb . .	$4\cdot27$	CO_2 . .	$1\cdot47$	Benzene	$3\cdot37$
Cu . . .	$2\cdot43$	Fe . .	$2\cdot26$	H_2 . .	$2\cdot43$	CS_2 . .	$2\cdot18$	Methane	$0\cdot86$

A.J.S., Amer. Journ. Sci.; *J.A.C.S.*, Journ. Amer. Chem. Soc.; *P.M.*, Phil. Mag.; *P.R.S.*, Proc. Roy. Soc. ; *P.Z.*, Phys. Zeit.

NUMBER OF IONS MADE BY AN α PARTICLE

Total number of ions produced by the complete absorption of an α particle with various initial velocities. Observer assumed $e = 4.65 \times 10^{-10}$ E.S.U. (Geiger, *Proc. Roy. Soc.*, 1909).

	Ra	RaEm.	RaA	RaC	RaF
Range in air at 20°C., 76 cm. . .	3·5 cm.	4·33	4·83	7·06	3·86
Number of ions	1.53×10^5	1.74×10^5	1.87×10^5	2.37×10^5	1.62×10^5

IONS PRODUCED AT DIFFERENT VELOCITIES BY AN α PARTICLE

Number of ions made per mm. of path in **air** by an α particle from RaC at various distances from its source. Total number $= 2.37 \times 10^5$ (Geiger, see above).

Distance from RaC in cm.	1	2	3	4	5	6	6·5	7
Ions per mm. of path in air at 12°C. and 76 cm.	2250	2300	2400	2800	3600	5500	7600	4000

TOTAL RELATIVE IONIZATION IN GASES BY α RAYS

I_t = total ionization (relative to air) produced by the complete absorption of α particles in various gases. (B. Bragg, *P.M.*, 1907, used RaC α rays ; B. and C., Bragg and Cook, *P.M.*, 1907 ; L., Laby, *P.R.S.*, 1907, used U α rays ; R., Rutherford, *P.M.*, 1899, used U α rays.)

Gas.	I	Gas.	I_t	Gas.	I_t
Air . .	**1·00**	Methane .	1·16, B. and C.	Et. ether .	1·31, B.; 1·29, L.
O_2 . . .	1·09, B. ; 1·06, R.	Acetylene .	1·26, B ; 1·27, L.	Et. iodide .	1·28, B.
N_2 . . .	0·96, B.	Ethylene .	1·28, B.	Acetaldehyde	1·05, L.
N_2O . .	1·05, B.; 0·99, L.	Pentane .	1·35, B.; 1·345, L.	Chloroform .	1·29, B.
NH_3 . .	1·01, R. ; 0·90, L.	Me. alcohol .	1·22, B.	Carb. tetra-	
CO_2 . . .	1·08, B. ; 1·03, L.	Me. iodide .	1·33, B.	chloride .	1·31, B.
Carbon bi-sulphide .	1·37, B.	Et. alcohol .	1·23, B.		
		Et. chloride .	1·30, B.; 1·18, L.		

RELATIVE VOLUME IONIZATIONS FOR β, γ, AND X RAYS

Relative ionization $= I_r = iP/I_p$, where i is the amount of ionization per unit volume for the gas at a press. p, and I that for air at press. P, the other experimental conditions being the same. In the experiments with γ rays (column headed γ), β rays would also be present. Observers : for β and γ rays, Kleeman, *P.R.S.*, 1907 ; X rays, C., Crowther, *P.C.P.S.*, 1909 ; *P.R.S.*, 1909 ; Mc., McClung, *P.M.*, 1904. I_r for secondary γ rays is much the same as for X rays (see Kleeman, *P.R.S.*, 1909).

Gas.	β	γ	Hard X.	Soft X.	Gas.	β	γ	Hard X.	Soft X.
Air	1·00	1·00	1·00	1·00	Me. alcohol .	1·69	1·75	—	—
H_2	0·16	0·16	0·18, C.	0·01, C.	Me. bromide .	3·73	3·81	—	71, C.
O_2	1·17	1·16	1·17, Mc.	1·3, Mc.	Me. iodide .	5·11	5·37	125, C.	145, C.
NH_3 . . .	0·89	0·90	—	—	Chloroform .	4·94	4·93	—	—
N_2O . . .	1·55	1·55	—	—	CCl_4 . . .	6·28	6·33	71, C.	67, C.
CO_2 . . .	1·60	1·58	1·49, C.	1·57, C.	Et. aldehyde	2·12	2·17	—	—
C_2N_2 . . .	1·86	1·71	—	—	Et. bromide .	4·41	4·63	118	72, C.
SO_2 . . .	2·25	2·27	4·79, Mc.	11·0, Mc.	Et. chloride .	3·24	3·19	17·3, C.	18, C.
CS_2 . . .	3·62	3·66	—	—	Et. ether .	4·39	4·29	—	—
Pentane .	4·55	4·53	—	—	Et. iodide .	5·90	6·47	—	—
Benzene .	3·95	3·94	—	—	Ni. carbonyl .	—	5·98	97, C.	89, C.
Me. acetate .	—	—	3·90, C.	4·95, C.	Hg dimethyl .	—	—	—	425, C.

P.C.P.S., Proc. Camb. Phil. Soc. ; *P.M., Phil. Mag.* ; *P.R.S., Proc. Roy. Soc.*

RELATIVE IONIZATION PER UNIT VOLUME BY α RAYS

Relative ionization = (total ionization) × (stopping power), Metcalfe, *P.M.*, 1909.

| Air . | 1·00 | He . | ·211 | CO . | 1·00 | HCl . | 1·4 | Propane . | 3·05 | Pentane | 4·83 |
| H$_2$. | ·233 | Br$_2$. | 3·9 | NO . | 1·28 | Ethane | 2·08 | Butane . | 4·02 | | |

For calculated **total ionization** when **Röntgen rays** are completely absorbed in various gases, see Crowther, *Proc. Roy. Soc.*, 1909.

HEATING EFFECT OF RADIUM

In calories per sec. per gm. of metallic radium with its radioactive products. E. von Schweidler and Hess, using ·795 gm. Ra enclosed in 1 mm. glass + 5 mm. Cu, obtained **·0328 calorie gm.$^{-1}$ sec.$^{-1}$ = 118 cals. gm.$^{-1}$ hr.$^{-1}$** The heating effect of a radioactive substance is proportional to the ionization it produces (Duane, *Le Radium*, 1909). The heat emission continues at temp. of liquid hydrogen (Curie and Dewar, 1903), and is mainly due to the kinetic energy of the α rays (Rutherford, " Radioactivity ").

Temp. and press. have no effect on heat emission (Schuster, Eve, and Adams, *Nature*, 1907; Rutherford and Petavel, *B.A. Rep.*, 1907; Schmidt, *P.Z.*, 1908).

Heat.	Observer.	Heat.	Observer.
·0278	Curie and Laborde, *C.R.*, 1903	25 %	Produced by Ra }R.&B.
·0292	Runge and Precht., *Berl. Ber.*, 1903	44 % 31 %	„ „ Em + RaB }P.M., „ „ RaC }1904
·0306	Rutherford and Barnes, *Nature*, 1903; *P.M.*, 1904	·0325 ·0372 ·0328	Angström, *P.Z.*, 1905 Precht, *A.d.P.*, 1906 Schweidler and Hess, *Wien. Ber.*, 1908

HEAT EMISSION FROM RaEm, AND THORIUM

The 6 × 10^{-4} c.c. of **RaEm** (with its products) in equilibrium with 1 gm. Ra emit ·75 of the ·0328 calories emitted per sec. by the radium. Thus the total quantity of heat given out by 1 c.c. of RaEm during its whole life = ·75 × ·0328/(λ × 6 × 10^{-4}) = 1·9 × 10^7 calories.

For old (mineral) **thorium** metal, the heat emitted is 5 × 10^{-9} calories per sec. per gm. (Pegram and Webb, *Phy. Rev.*, 1908).

RADIUM EMANATION

Γ is the **period of decay** (in days) to half initial activity. Taking Γ = 3·66 days, then the decay coefficient λ = 2·19 × 10^{-6} sec.$^{-1}$ (see p. 107).

Γ in days.	Observer, etc.	Γ in days.	Observer, etc.
3·77	Rutherford and Soddy, *P.M.*, 1903.	3·75 3·58	Rümelin, *P.M.*, 1907. For first 5 days.
3·88	Bumstead and Wheeler, *A.J.S.*, 1904.	3·75 3·85	During period 5 to 20 days. 20 to 40 days' old emanation.
3·8 to 4·1 3·86	Debierne, *C.R.*, 1909. Sackur, *Ber. C.G.*, 1905.	4·4	One sample Rutherford and Tuomikoski, *P.M.*, 1909.

EQUILIBRIUM VOLUME OF RADIUM EMANATION

Final **volume of radium emanation** at 0° C. and 76 cm. Hg in equilibrium with 1 gm. of metallic radium. **Theoretical** volume = (number of radium atoms breaking up per sec.)/λN = 3·4 × 10^{10}/(2·75 × 10^{19} × 2·19 × 10^{-6}) = 5·64 × 10^{-4} c.c. (Rutherford, " Radioactivity "). The volume of the emanation changes anomalously after it is first formed.

Observed vol.	Observer.	Observed vol.	Observer.
·58 cub. mm. ·601 „	Rutherford, *P.M.*, 1908. Gray & Ramsay, *J.C.S.*, 1909.	·58 cub. mm.	Debierne, *C.R.*, 1909.

A.d.P., Ann. der Phys.; A.J.S., Amer. Journ. Sci.; B.A. Rep., Brit. Ass. Rep.; C.R., Compt. Rend.; J.C.S., Journ. Chem. Sci.; P.M., Phil. Mag.; P.Z., Phys. Zeit.

VAPOUR PRESSURE OF RADIUM EMANATION

Vapour pressure of liquid RaEm. in cm. Hg; melting-point, $-71°$ C. (R., Rutherford, *Nature*, February, 1909 ; G. & R., Gray and Ramsay, *J.C.S.*, June, 1909).

Temp. °C.	R.	$-127°$	$-101°$	$-78°$	$-65° =$ B.P.
Vap. press. cm. Hg		·9	5	25	76

Temp. °C.	G. & R.	$-70°·4$	$-62° =$ B.P.	$-60°·6$	$-55°·8$	$-38°·5$	$-17°·7$	$-10°·2$	$+104°·5$ crit. t.
Vap. press. cm. Hg		50	76	80	100	200	400	500	4745 crit. press.

DIFFUSION OF EMANATIONS

D = coefficient of diffusion (in cm.^{2} sec.$^{-1}$) of the emanation into the gas stated at the pressure p cm. Hg and temp. $t°$ C. indicated. According to J. J. Thomson (*Nature*, November 25, 1909) : "D would only vary slowly with atomic weight," and not as the square root of the molecular weight of the emanation, as is assumed in the table below.

Russ finds $pD =$ const. for AcEm. and for ThEm. Bruhat gives pD/T^2 = const. for AcEm. between 0° and 20°. (Molec. wgt. ThEm.)/(molec. wgt. AcEm.) = 1·42 (Russ). Mol. wgt. of RaEm. = 222 (Gray & Ramsay, 1910).

Gas.	p. and t° C.	D.	Molec. wgt.	Obs.	Gas.	p. and t° C.	D.	Molec. wgt.	Obs.
RADIUM EM.					**ACTINIUM EM.** (*contd.*)				
Air . .	76 ?	·07 to ·09	c. 100	R. & B.	Air . . .	—	·112	70	D.
„ . . .	76, 10°	·10	——	C. & D.	„ . . .	1·4	7·81	——	R.
„ . . .	76, 0°	·101	75 to 100	C.	„ . . .	76·4	·125	——	„
CO_2 . .	——	——	180	B. & W.	„ . . .	76	·123	——	B.
Diff. of Em. into air compared with $O_2, CO_2, SO_2,$ into air . .	——	——	86 to 99	M.	„ . . .	{ 76 to ·9 }0°	·10	70	„ „
Em. into H_2 compared with Hg vap. into H_2 . .	250° { 275° {	{ ·034 Em. ·037 Hg ·0376 Em. ·0407 Hg }	235	P.	**THORIUM EM.**				
					Em. into air, compared with $H_2, O_2,$ $SO_2, CO_2,$ into air . .	——	——	c. 90	M.
ACTINIUM EM.					Air . . .	——	·09	——	Ruth. R.
H_2 . . .	76, 15°	·412	——	B.	„ . . .	76	·103	——	„
H_2 . . .		·33	——	R.	„ . . .	{ 8·2 to 76·1 }	{ ·966 to ·103 }	——	„
SO_2 . .	76, 10°	·062	——	„					
Argon . .	to 18°	·106	——	„	„ . . .	76	·103	——	„
CO_2 . .		·073	——	„	Argon . .	76	·084	——	
CO_2 . . .	76, 15°	·077	——	B.					

B., Bruhat, *Le Radium*, 1909 ; B. & W., Bumstead & Wheeler, *A.J.S.*, 1903 ; C., Chaumont, *Le Radium*, 1909 ; C. & D., Curie & Danne, *C.R.*, 1903 ; D., Debierne, *Le Radium*, 1907 ; M., Makower, *P.M.*, 1905 ; P., Perkins, *A.J.S.* ; R., Russ, *P.M.*, 1909, *Le Radium*, 1909 ; Ruth., Rutherford, "Radioactivity" ; R. & B., Rutherford & Miss Brooks, *C.N.*, 1902.

A.J.S., *Amer. Journ. Sci.* ; *C.N.*, *Chem. News* ; *C.R.*, *Compt. Rend.* ; *J.C.S.*, *Journ. Chem. Soc.* ; *P.M.*, *Phil. Mag.*

EQUILIBRIUM ACTIVITIES IN MINERALS

Relative activity of radioactive products in minerals. Boltwood (*A.J.S.*, April, 1908) found U 2·22 times as active as the Ra alone in minerals (see McCoy and Ross, *A.J.S.*).

Product.	U	Io	Ra	RaEm.	RaA	RaB	RaC	RaF	Ac	Total.
Relative activity . .	1	·34	·45	·62	·54	·04?	·91	·46	·28	4·64

$3·4 \times 10^{-7}$ gm. Ra is in equilibrium with 1 gm. U (Rutherford and Boltwood, *A.J.S.*, 1906). $7·3 \times 10^6$ gms. U equal in activity 1 gm. of Ra + its products to RaC. *i.e.* Ra just over 30 days old (corrected by Boltwood, *A.J.S.*, 1908).

RADIUM AND THORIUM IN ROCKS

Rutherford and Soddy (*P.M.*, May, 1903) and W. E. Wilson (*Nature*, July, 1903) suggested that the heat liberated by radioactive changes is one of the sources of the Earth's heat. Thus the distribution of radium and thorium in the Earth's crust is of geophysical importance. Loss of heat from the Earth's surface = temperature gradient × thermal conductivity of crust × area of Earth's surface = $(1/3200) \times ·004 \times 5·1 \times 10^{18} = 6 \times 10^{12}$ calories per sec. Now, elementary radium in radioactive equilibrium (*i.e.* whole U family) gives out 6×10^{-3} cal./sec. gm. (Rutherford§), and therefore $1·1 \times 10^{14}$ grms. of radium, or $10^{14}/10^{27} = 10^{-13}$ gm. per c.c., throughout the Earth's volume would maintain it at a steady temperature. Thorium contributes 5×10^{-9} cal. /sec. gm. The **total heating effect** in calories per gram of rock per hour is for the lava indicated below by *, 30×10^{-10}; and for the rock indicated by †, $2·9 \times 10^{-10}$; for average igneous rock, 11×10^{-10}.

(See Strutt, *Proc. Roy. Soc.*, 1906-7 ; Joly, "Radioactivity and Geology," 1909.)

Rock, etc.	Obs.	Ra	Th
		gm. per gm. of rock.	
		$\times 10^{-12}$	$\times 10^{-5}$
Igneous rocks	St., 1906	1·7	――
Sedimentary rocks	"	1·1	――
Sandstone	E. M., 1907	·16	――
Clays	" "	·79	――
Devonian.	" "	1 to 4	――
Ordovician	" "	·9	――
Lavas ejected since 1631 * . . .	J., 1909	12·3	2·3
Lava, Mount Erebus	F. F., 1909	2·4	――
126 igneous rocks	J., 1909	7·01	――
64 " "	other obs.	1·3	1·3
Italian igneous rocks	B., 1909	mean	――
Campbell and Auckland Islands, N.Z.	F. F., 1909	{ 1·6 / ·5	igneous / sedimentary
St. Gothard Tunnel— granite	J., 1909	7·7	1·9
schists and altered sedimentary rocks	"	3·4 to 4·9	·5 to 1·2
Simplon Tunnel ‖	"	7·6	――
Transandine Tunnel †	Fl., 1910	·8	·56
Calcareous and dolomitic European rocks	J., 1910	mean of 7 27 samples	·16 < ·05
Deep-sea deposits— Globigerina ooze [1]	"	7·2	――
Radiolarian ooze [2]	"	36·7	――
Red clay [3]	"	27	――

Extent :—[1] 50, [2] 2·5, [3] 5 million square miles.   † 1000 feet below the surface.   § Assuming that the heat due to each member of the family is proportional to the ionization it produces. ‖ Preliminary result. B., Blanc, *P.M.* ; E.M., Eve and McIntosh, *P.M.* ; F.F., Farr and Florance, *P.M.* ; Fl., Fletcher ; J., Joly, *P.M.* ; S., Strutt (above). *A.J.S., Amer. Journ. Sci.* ; *P.M., Phil. Mag.*

RADIUM IN SEA-WATER

In grams per gram of sea-water. Deduced from the observed amount of Ra Em.

Amount.	Place.	Observer.	Amount.	Place.	Observer.
$2\cdot3 \times 10^{-15}$	—	Strutt, *P.R.S.*,'06	4×10^{-15}	Nile	Joly, *P.M.*,1908
·3–·6 ,,	Mid. N. Atlantic	Eve, *P.M.*, 1907	14 ,,	Mediterranean	,, ,, 1909
·9 ,,	Atlantic	,, ,, 1909	5 ,,	Indian Ocean	,, ,, ,,
16 ,,	,,	Joly, *P.M.*, 1908			

RADIUM EMANATION IN ATMOSPHERE

RaEm. per cubic metre of air, expressed in terms of the number of grams of radium with which it would be in equilibrium. The observers below absorbed the emanation by charcoal.

RaEm.	Place.	Observer.	RaEm.	Place.	Observer.
24–27×10^{-12}	Montreal	Eve, *P.M.*, 1907	35–350×10^{-12}	Cam-	Satterly, *P.M.*,
60 ,,	,,	,, 1908	Mean 105 ,,	bridge	1908 and 1910
86–200 ,,	Chicago	Ashman,*A.J.S.*,'08			

MOBILITIES OF NATURAL IONS IN AIR

Mobility or speed K is in cm.^{2} sec.$^{-1}$ volt^{-1} at room temperature and 76 cm. (see p. 95). The ions are named from their velocities: the small ions are assumed to have the velocity of X-ray ions. (See Pollock, *Science*, 1909 ; Eve, *Phil. Mag.*, 19, 1910 ; Lusby, *Proc. Camb. Phil. Soc.*, 1910.)

Ion.	Mean K.	Observer.	Ion.	Mean K.	Observer.
Small . . .	$\{K+ = 1\cdot4\}$ $\{K- = 1\cdot7\}$	Langevin, '03	Large . .	·0003	Langevin,*C.R.*,'05 Pollock, 1908
Intermediate	*c.* ·01	Mean	Large . .	·0003 *	
			Large . .	·0008 †	,, ,,

* Humidity, 19 grms. H_2O per cubic metre. † ·5 grm. H_2O per cubic metre of air. Pollock, *Austl. Ass. Adv. Sci.*, 1908.

ELECTRIC ARCS

Mrs. Ayrton's formula for carbon arcs, $E = a + \beta l + \dfrac{\gamma + \delta l}{i}$, has been shown by Guye and Zébrikoff (*Compt. Rend.*, 1907) to hold for short stable arcs between metals. E is the voltage across the arc, i is the current in amperes, and l the length in mms. of the arc in air at atmospheric pressure. Mrs. Ayrton's formula does not hold for very long arcs, nor for cored carbons. For stability, an arc requires an external resistance R which must be less than $\dfrac{\{E_z - (a + \beta l)\}^2}{4(\gamma + \delta l)}$ ohms, where E_z is the total available voltage ; *or* E_z must exceed $a + \beta l + 2\sqrt{R(\gamma + \delta l)}$. If R is too small the arc hisses, in which case the current is independent of the voltage across the terminals. The constants for carbon refer only to the particular sizes and quality used by Mrs. Ayrton. (See J. J. Thomson, " Conduction of Electricity through Gases.")

Metal.	a	β	γ	δ	Metal.	a	β	γ	δ
C. . . .	38·88	2·074	11·66	10·54	Pd . . .	21·64	3·70	0	21·78
Fe . .	15·73	2·52	9·44	15·02	Ag . . .	14·19	3·64	11·36	19·01
Ni . .	17·14	3·89	0	17·48	Pt . . .	24·29	4·80	0	20·23
Co . .	20·71	2·05	2·07	10·12	Au . . .	20·82	4·62	12·17	20·97
Cu . .	21·38	3·03	10·69	15·24					

A.J.S., *Amer. Journ. Sci.* ; *C.R.*, *Compt. Rend.* ; *P.M.*, *Phil. Mag.* ; *P.R.S.*, *Proc. Roy. Soc.*

ATOMIC AND RADIOACTIVITY CONSTANTS

References : J. J. Thomson's "Conduction of Electricity through Gases," Rutherford's "Radioactivity," H. A. Lorentz, *Éclairage Électrique*, **44**, 1905, "Theory of Electrons," 1909, Jeans' "Dynamical Theory of Gases," and Millikan, *P.M.*, 1917.

Symbol.	Definition.	Value.
e . . .	**Ionic charge,** half charge on an α particle	$4 \cdot 77 . 10^{-10}$ E.S.U.; $1 \cdot 59 . 10^{-20}$
NE . .	Total charge carried in electrolysis by the atoms in $\frac{1}{2}$ c.c. of gas—	[E. M. U. ; $1 \cdot 59 . 10^{-19}$ [coulombs
	For **ideal gas** at $0°$ and 76 cm.	$1 \cdot 2913 . 10^{10}$ E.S.U. cm.$^{-3}$; $\cdot 4304$ E.M.U. cm.$^{-3}$
	„ oxygen „ „ „	$1 \cdot 292 . 10^{10}$ E.S.U. cm.$^{-3}$; $\cdot 4308$ E.M.U. cm.$^{-3}$
	„ hydrogen „ „ „	$1 \cdot 290 . 10^{10}$ E.S.U. cm.$^{-3}$; $\cdot 4300$ E.M.U. cm.$^{-3}$
$N_m E$.	Total charge carried by $\frac{1}{2}$ (gm. molecule) of hydrogen ions	$2 \cdot 894 . 10^{14}$ E.S.U. cm.$^{-3}$; $9 \cdot 647 . 10^{3}$ E.M.U. cm.$^{-3}$
N . .	**Number of molecules** per c.c. of a gas at $0°$ C. and 76 cm. $= NE/e = 1 \cdot 29 . 10^{20}/4 \cdot 77$	$2 \cdot 705 . 10^{19}$ cm.$^{-3}$
N_m .	Number of molecules in 1 gm. molecule of gas	$6 \cdot 062 . 10^{23}$ gm.$^{-1}$
e/m_e .	Ratio of charge to electromagnetic mass for the **negative** electron at small velocities	$5 \cdot 31 . 10^{17}$ E.S.U. gm.$^{-1}$; $1 \cdot 77 . 10^{7}$ E.M.U. gm.$^{-1}$
E/m_H .	The same ratio for the **hydrogen ion** in electrolysis $= 10^{7} \cdot 88/(\cdot 00111827 \times 1 \cdot 008)$	9571 E.M.U. gm.$^{-1}$; $95,706$ coulombs gm.$^{-1}$
e/m_α .	The same ratio for the α **particle**	$4 \cdot 8 . 10^{3}$ E.M.U. gm.$^{-1}$
$2e/m_{He}$.	Calculated for **helium** $= 2NE/\rho = 2 \times \cdot 43 . 10^{-6}/(2 \times 8 \cdot 987)$	$4 \cdot 78 . 10^{3}$ E.M.U. gm.$^{-1}$
m_e . .	Electromagnetic **mass of negative electron** for small velocities $= e/(e/m_0)$	$8 \cdot 8 . 10^{-28}$ gm.
m_H . .	**Mass of hydrogen** atom $= \rho/2N$	$1 \cdot 66 . 10^{-24}$ gm.
m_α . .	**Mass of α particle,** *i.e.* of helium atom	$6 \cdot 56 . 10^{-24}$ gm.
m_H/m_e .	Number of electrons equal in mass to hydrogen atom $= (m_{He})/(m_0 E)$	1850
$a\theta$ . .	Energy of a gas molecule at $\theta°$ C. $= 3\rho/2N$	$a = 2 \cdot 02 . 10^{-16}$ ergs/degree
R . .	For 1 gm. of oxygen, $R = \rho v/\theta = 1 \cdot 0132 . 10^{6}/(273 \cdot 09 . 1 \cdot 429 . 10^{-3})$. Press. in dynes/cm.^{2} ; volume in c.c. (see p. 5)	$\{2 \cdot 5963 . 10^{6}$ cm.^{2}/sec.^{2} $\{2 \cdot 5963 . 10^{6}$ ergs/gm. mol.
	For 1 gm. molecule of an ideal gas, $R = 22 \cdot 412/273 \cdot 09$. Press. in atmos. $= 76$ cm. Hg$(g = 980 \cdot 6)$; vol. in litres (D. Berthelot, *Trav. et Mém. Bur. Intl.*)	$\cdot 08207$ litre atm./gm. mol.
a . . .	The **radius** of a negative **electron** $= 2/3 . e . e/m_0$	$1 \cdot 85 . 10^{-13}$ cm.
	The **diameter** of a hydrogen **molecule** (Sutherland (after Jeans), *Phil. Mag.*, 1910)	$2 \cdot 17 . 10^{-8}$ cm. (see p. 33)
Heat given out by 1 gm. of metallic radium with its products		$\cdot 0328$ cal./sec. ; 118 cal./hr.
Number of α particles emitted by 1 gm. radium without products		$3 \cdot 4 . 10^{10}$ gm.$^{-1}$ sec.$^{-1}$
Initial velocity of α particle from RaC		$2 \cdot 06 . 10^{9}$ cm./sec.
Initial **energy of α particle** from RaC $= mv^2/2 = v^2 e/(2e/m_\alpha) = 2 \cdot 06^2 . 10^{18} \times 1 \cdot 57 . 10^{-20}/(2 \times 5 \cdot 07 . 10^3)$		$1 \cdot 3 . 10^{-5}$ ergs ; $3 \cdot 1 . 10^{-13}$ cal.
Total **number of ions** produced in air by an α ray (RaC)		$2 \cdot 37 . 10^5$
Volume of helium at $0°$ and 76 cm. produced by 1 gm. radium		$5 \cdot 17 . 10^{-9}$ c.c. / (sec. gm.), or 163 mm.^{3}/(yr. gm).
Calculated volume $= 4 \times$ number of α rays emitted/N $= 4 . 3 \cdot 4 . 10^{-9}/2 \cdot 75$		$4 \cdot 94 . 10^{-9}$ c.c./(sec. gm.) ; 156 mm.^{3}/ (yr. gm).
Number of β particles emitted per sec. by the RaC in equilibrium with 1 gm. Ra (Makower, *Phil. Mag.*, 1909)		$5 . 10^{10}$ gm.$^{-1}$ sec.$^{-1}$

CONSTANTS OF RADIOACTIVE SUBSTANCES

Atomic weights: $O = 16$, $U = 238.2$, $Ra = 226.0$, $Th = 232.4$.

Rate of decay: If I is the radioactivity of a substance at a time t, then $I = I_0 \epsilon^{-\lambda t}$, where I_0 is the initial activity when $t = 0$. λ is given below in sec.$^{-1}$. If Γ is the period in which the activity decreases to half its initial value (*i.e.* $I/I_0 = \frac{1}{2}$), then $\lambda = .69315/\Gamma$ sec.$^{-1}$. Γ is given below in secs. (s.), mins. (m.), hrs. (h.), days (d.), or years (y.).

Coefficients of absorption Λ are given in cm.$^{-1}$ for β rays in Al foil and for γ rays in lead foil. If J_0 is the intensity of the rays incident on foil of thickness d cm., and J is the intensity of the emergent rays, then $J = J_0 \epsilon^{-d\Lambda}$.

(See Rutherford's "Radioactive Substances," Camb. Univ. Press, and Wendt, *Phy. Rev.*, 1916, for a complete table.)

Substance.	λ in sec.$^{-1}$.	Half-period. Γ	Rays emitted.	Absorptn. Coef. in cm.$^{-1}$.	
				β Rays. Λ_{Al}	γ Rays. Λ_{Pb}
U	$4.3 \cdot 10^{-18}$	$6 \cdot 10^9$ y.	α	—	—
Rad. U	—	sevl. y.			
U.X	$3.7 \cdot 10^{-7}$	21.5 d.	β, γ	14.4 and 510	.72
Io	$2 \cdot 10^{-12}$	c. 10^4 y.	α		
Ra	$1.1 \cdot 10^{-11}$	2000 y.	α, β	312	
RaEm	$2.08 \cdot 10^{-6}$	3.85 d.	α	—	
RaA	$3.85 \cdot 10^{-3}$	3 m.	α	—	
RaB	$4.33 \cdot 10^{-4}$	26.7 m.	β	13 to 890	
RaC$_1$	$5.93 \cdot 10^{-4}$	19.5 m.	α	13 to 53	.46 to .57
RaC$_2$	—	1.38 m.	β, γ		
RaD	$1.8 \cdot 10^{-9}$	17 y.	β		
RaE$_1$	$1.3 \cdot 10^{-6}$	6.2 d.	β		
RaE$_2$	$1.7 \cdot 10^{-6}$	4.8 d.	β	.44	
RaF (Polonium)	$5.73 \cdot 10^{-8}$	140 d.	α		
Ac	—	30 y.	rayless	—	—
Rad. Ac	$4.1 \cdot 10^{-7}$	19.5 d.	α, β	170	
AcX	$7.6 \cdot 10^{-7}$	10–11 d.	α		
AcEm	$1.8 \cdot 10^{-1}$	3.9 s.	α		
AcA	—	0.002 s.	α		
AcB	$3.20 \cdot 10^{-4}$	36.1 m.	β		
AcC	$5.37 \cdot 10^{-3}$	2.15 m.	α		
AcD	$2.26 \cdot 10^{-3}$	4.71 m.	β	29	2.0 to 3.6
Th	$7 \cdot 10^{-19}$	3.10^{10} y.	α		
MesoTh 1	$4.0 \cdot 10^{-9}$	5.5 y.	rayless	—	—
MesoTh 2	$3.1 \cdot 10^{-5}$	6.2 h.	β, γ	20.2 to 38.5	.5
Rad. Th	$1.09 \cdot 10^{-8}$	737 d.	α		
ThX	$2.17 \cdot 10^{-6}$	3.71 d.	α		
ThEm	$1.31 \cdot 10^{-2}$	53 s.	α		
ThA	—	—	α	140	
ThB	$1.81 \cdot 10^{-5}$	10.6 h.	β		
ThC$_1$	$2.10 \cdot 10^{-4}$	55 m.	α		
ThC$_2$	—	some secs.	α		
ThD	$3.7 \cdot 10^{-3}$	3.1 m.	β	15.7	.46 to .57

PROPERTIES OF RADIOACTIVE SUBSTANCES

Substance.	Properties.	Substance.	Properties.
U . .	Sol. in excess of am. carb. Nitrate soluble in ether and acetone.		Carried down by $PbCO_3$, and by $SnCl_2$ with Hg and Te. RaD, E_1, E_2, and F can be separated by electrolysis.
Rad.U .	Carried down by $BaSO_4$ and ferric hydrate. Soluble in HCl.		
U.X . .	Less volatile than U. Volatile in electric arc. Insoluble in excess of am. carb. Soluble in water and ether. Carried down by barium sulphate, by moist ferric hydrate, and by animal charcoal.	Ac . .	Produces helium. Precipitated by oxalic acid in acid solutions. Oxalate insoluble in HF; accompanies thorium and rare earths.
		Rad.Ac .	Slightly volatile at high temps. Insoluble in NH_4OH. Separated from Ac by electrolysis, by fractional precipitation, by ammonia, and by animal charcoal.
Io . .	Soluble in excess of am. oxalate. Carried down by H_2O_2 in presence of U salts.		
Ra . .	Characteristic spectrum. Spontaneously luminous. Analogous to Ba. $RaCl_2$ and $RaBr_2$ are less soluble than $BaCl_2$ and $BaBr_2$.	AcX . .	Deposited by electrolysis in alkaline solution. Not precipitated by NH_4OH.
		AcEm. .	Behaves as inert gas. Coef. of diffusion in air 0·11. Condenses at $-120°$ C.
RaEm. .	One of group of inert gases. Characteristic spectrum. Coef. of diffusion in air = 0·1 (see p. 103). Mol. wt. = 218.	AcA .	Volatile below 400° C. Soluble in NH_4OH and strong acids.
RaA .	Behaves as a solid. Deposited on cathode in an electric field. Volatile at 800-900° C. Soluble in strong acids.	AcB . .	Volatile below 700° C. Soluble in NH_4OH and strong acids. Deposited by electrolysis of active deposit on the cathode in HCl.
RaB .	Like RaA. Volatile at 600-700° C. Precipitated by $BaSO4$.		
RaC . .	Physically like RaA. Volatile at 800-1300° C. Chemically, like RaB. Deposited on Cu and Ni. Carried down with precipitated copper. Perhaps a mixture of 2 or 3 products.	Th . .	Volatile in electric arc. Colourless salts not spontaneously phosphorescent. Salts pptd. by NH_4OH and oxalic acid.
		Rad.Th .	Carried down by hydrates, precipitated by NH_4OH.
RaD . .	Volatile below 1000° C. Soluble in strong acids. Reactions analogous to those of Pb.	ThX . .	Soluble in NH_4OH. Carried down by iron. Deposited by electrolysis in alkaline soln.
RaE₁ .	Volatile at red heat. Soluble in cold acetic acid. Reactions analogous to those of Pb.	ThEm. .	Inert gas. Condenses just above $-120°$ C. Coefficient of diffusion in air = ·10.
RaE₂ . .	Not volatile at red heat. Reactions analogous to those of bismuth.	ThA . .	Volatile under 630° C. Soluble in strong acids.
RaF(Pol.)	Volatile towards 1000° C. Deposited from its solutions on Bi, Cu, Sb, Ag, Pt.	ThB . .	Volatile below 730° C. Like ThA. Deposited on Ni. Separated from ThA by electrolysis.
		ThC . .	Like ThB.

PHYSICAL CONSTANTS OF CHEMICAL COMPOUNDS

For properties of the **elements**, see : density, p. 20 ; melting and boiling points, p. 48 ; solubility in water, p. 124. **Metallo-organic** compounds are given under "Organic Compounds," p. 118.

Formulæ.—Hydrated forms (which are often crystalline) are indicated thus : CaI_2(and $+ 6H_2O$) ; the properties given are for the anhydrous substance.

Formula (Molecular) Weights are calculated with atomic weights for 1911 (p. 1).

Densities.—When no temp is given, grams. per c.c. at 15° may be assumed. When preceded by "A" the density is relative to that of air ('001293 gram per c.c. at 0° and 760 mms.). To convert this into a density relative to $O = 16$, multiply by 14·47. For those gaseous densities known with accuracy, see p. 26. Other densities on pp. 20–26.

Melting and Boiling Points are for anhydrous substances at 760 mms. mercury unless some other conditions are specified. T = temp. of transition or pseudo-"melting" point of hydrated substance. For fats and waxes, see p. 50.

Solubilities are given as grams of substance in 100 grams of water at the temp. stated. "p" indicates grams per 100 grams of solution. "V" means volumes of substance at 0° and 760 mms. per 100 volumes of water at the temp. stated. "Soluble" infers solubility in either hot or cold water ; "insoluble" indicates solubility in neither. (See also pp. 124, 125.)

For more complete tables, see Van Nostrand's "Chemical Annual" and Biedermann's "Chemiker-Kalender" for current year ; Dammer's "Handbuch der Anorganischen Chemie ;" Beilstein's "Handbuch der Organischen Chemie ;" Watts' "Dictionary of Chemistry ;" and F. W. Clarke's "Specific Gravities."

INORGANIC COMPOUNDS

Formula, formula (molecular) weight, density, melting and boiling points, and solubility in water.

Substance and Formula.	Formula weight (0 = 16).	Density, gms./c.c.	Melting Point, °C.	Boiling Point, °C.	Solubility in Water.
Aluminium—		at./temp.	at./mms.	at./mms.	at./temp.
bromide, Al_2Br_6(and $+ 12H_2O$)	533·7	$\left\{ \begin{array}{c} 2·54 ; \\ A. 18·62 \end{array} \right\}$	93°	263°/747	soluble
chloride, Al_2Cl_6(and $+ 12H_2O$)	267·0	A. 9·34/400°	190°/1910	182°/752	41/15°(p)
iodide, Al_2I_6 (and $+ 12H_2O$) .	815·7	$\left\{ \begin{array}{c} 2·63 ; \\ A. 27 \end{array} \right\}$	185°	360°	soluble
nitrate, $Al(NO_3)_3 . 9H_2O$ . .	375·3	—	T = 73°	dec. 134°	v. soluble
oxide, Al_2O_3	102·2	3·7 — 4	2200	—	insoluble
phosphate, $AlPO_4$	122·1	2·59	infusible	—	insoluble
sulphate, $Al_2(SO_4)_3 . 18H_2O$.	666·7	1·62	decomp.	—	36/20°
Potassium alum, $Al_2(SO_4)_3K_2SO_4 . 24H_2O$	949·1	1·757/20°	84°·5	$\left\{ \begin{array}{c} 23H_2O \\ \text{at } 190° \end{array} \right.$	$\begin{array}{c} 9·6/15° \\ 357/100° \end{array}$
Ammonium—					
ammonia, NH_3	17·03	$\left\{ \begin{array}{c} \text{(liq.) } ·623/0° \\ A. ·5896 \end{array} \right\}$	– 75	– 33·5	see p. 124.
acetate, $NH_4C_2H_3O_2$ . . .	77·07	—	89	—	148/4°
arsenate, $(NH_4)_3AsO_4 . 3H_2O$.	247·1	—	—	—	soluble
bromide, NH_4Br	97·96	$\left\{ \begin{array}{c} 2·33/15° \\ A. 1·64/440° \end{array} \right\}$	diss.	—	$\begin{array}{c} 66/10° \\ 128/100° \end{array}$
carbonate, $(NH_4)_2CO_3 . H_2O$	114·1	—	diss. 85°	—	100/15°
chloride, NH_4Cl	53·50	$\left\{ \begin{array}{c} 1·52/17° \\ A. ·89 \end{array} \right\}$	diss. 350°	—	$\begin{array}{c} 35/15° ; \\ \text{(see p.125.} \end{array}$
chloroplatinate, $(NH_4)_2PtCl_6$.	444·0	3·06	decomp.	—	·67/20°
chromate, $(NH_4)_2CrO_4$ . .	152·2	1·88/11°	decomp.	—	decomp.
iodide, NH_4I	145·0	2·5	sublimes	—	v. soluble
molybdate, $(NH_4)_2MoO_4$ . .	196·1	2·4 — 2·9	decomp.	—	decomp.
nitrate, NH_4NO_3	80·05	1·72/15°	152°	dec. 210°	200/18°

dec. or decomp. = decomposes ; diss. = dissociates ; v. = very ; wh. = white.

INORGANIC COMPOUNDS (contd.)

For general heading, see p. 109.

Substance and Formula.	Formula weight (0 = 16).	Density, gms./c.c.	Melting Point, ° C.	Boiling Point, ° C.	Solubility in Water.
Ammonium (contd.) —		at./temp.	at./mms.	at./mms.	at./temp.
nitrite, NH_4NO_2	64·05	1·7	decomp.	—	soluble
oxalate, $(NH_4)_2C_2O_4 . H_2O$.	142·1	1·5	—	—	4/15°
persulphate, $(NH_4)_2S_2O_8$ . .	228·2	—	decomp.	—	58/0°
phosphomolybdate, $(NH_4)_3PO_4 . 12MoO_3 . 3H_2O$	1931	—	—	—	·03/15°
sulphate, $(NH_4)_2SO_4$.	132·2	1·77/20°	140°	dec. 280°	76/20°
sulphocyanate, NH_4CNS . .	76·12	1·31/13°	159	dec. 170°	162/20°
Antimony—					
bromide, $SbBr_3$	360·0	4·15/23°	93	280°	decomp.
chloride, tri-, $SbCl_3$ . .	226·6	{3 06/26° A. 8·1}	73·2	223	{816/15° ∞ /72°}
„ penta-, $SbCl_5$ . .	297·5	2·35/20°	− 6	102°/68	decomp.
hydride, SbH_3 . . .	123·2	A. 4·3/15°	− 91·5	− 18	20 V.
iodide, tri-, SbI_3 . . .	501·0	{4·85/26° A. 17·6}	167 subl. 114°}	401	decomp.
oxide, tri-, Sb_2O_3 . . .	288·4	5·2–5·7	red heat	1550	·002/15°
„ tetr-, Sb_2O_4 . . .	304·4	4·07	O/800°	—	insoluble
„ pent-, Sb_2O_5 . .	320·4	3·8	O/300°	O_2/800°	insoluble
potassium tartrate, $K(SbO)C_4H_4O_6 . \frac{1}{2}H_2O$	332·3	2·6	$\frac{1}{2}H_2O$/100°	decomp.	{5/9° 36/100°}
sulphide, tri-, Sb_2S_3 . . .	336·6	4·65	fusible	volatilizes	insoluble
„ penta-, Sb_2S_5 .	400·7	4·12/0°	fusible	—	insoluble
Arsenic—					
bromide, $AsBr_3$	314·7	{3·7/15° A. 10·91}	31°	221°	decomp.
chloride, $AsCl_3$	181·3	2·2/0° ; A. 6·3	− 18	130·2	decomp.
fluoride, tri-, AsF_3 . . .	132·0	2·7 ; A. 4·57	− 8·5	63	decomp.
„ penta-, AsF_5 . .	170·0	A. ·415	− 80	− 53	soluble
hydride, AsH_3	77·98	A. 2·7	− 113	− 54·8	slgtly sol.
iodide, di-, AsI_2 . . .	328·8	—	—	—	—
„ tri-, AsI_3 . . .	455·7	4·4/13°	146	{394–414 V.D. 16·1}	30/100°
„ pent-, AsI_5 . . .	709·6	3·93	70	—	decomp.
oxide, tri-, As_2O_3 . . .	197·9	3·6–4·1	subl. 218°	V.D. 13·8	1·7/16°
„ pent-, As_2O_5 . . .	229·9	3·9–4·2	red heat	decomp.	245/12°
Barium—					
bromide, $BaBr_2 . 2H_2O$ . .	333·2	3·85/24°	anhy. 880°	$2H_2O$/100°	103/15°
carbonate, $BaCO_3$ . . .	197·4	4·3	795°	dec. 1450°	·0022/18°
chloride, $BaCl_2 . 2H_2O$ . .	244·3	3·1/24°	anhy. 960°	$2H_2O$/113°	see p.125
hydride, BaH_2	139·4	4·2/0°	volatile	1400°	decomp.
iodide, BaI_2	391·2	4·92	740°	—	170/0°
nitrate, $Ba(NO_3)_2$ . . .	261·4	3·24/23°	575	—	5/0°
oxide, BaO	153·4	4·7 − 5·5	BaO_2/450°	—	1·5/0°
„ per-, BaO_2 . . .	169·4	4·96	BaO/450°	—	insoluble
sulphate, $BaSO_4$	233·4	c. 4·5	infusible	—	·0523/18°
Beryllium—					
bromide, $BeBr_2$	168·9	—	601°	—	soluble
chloride, $BeCl_2$	80·02	—	c. 600	—	v. soluble
sulphate, $BeSO_4 . 4H_2O$ . .	177·2	1·7/10°	dec. r. ht.	$2H_2O$/100°	44/30°

anhy. = anhydrous ; dec. or decomp. = decomposes ; r. ht. = red heat ; subl. = sublimes ;
v. = very ; V.D. = vapour density ; ∞ = soluble in all proportions.

INORGANIC COMPOUNDS (*contd.*)
For general heading, see p. 109.

Substance and Formula.	Formula weight (0 = 16).	Density, gms./c.c.	Melting Point, °C.	Boiling Point, °C.	Solubility in Water.
Bismuth—		at./temp.	at./mms.	at./mms.	at./temp.
bromide, $BiBr_3$	447·76	5·6	200°–215°	453°	decomp.
chloride, tri-, $BiCl_3$ . .	314·38	{ 4·6/11° / A. 11·35 }	· 227	429	decomp.
nitrate, $Bi(NO_3)_3.5H_2O$.	484·11	2·8	74	$5H_2O/80°$	decomp.
oxide, Bi_2O_3	464·0	8·8 — 9	820–860	—	insoluble
sulphide, Bi_2S_3	512·21	7 — 7·8	decomp.	—	insoluble
Boron—					
chloride, BCl_3	117·38	1·35/0°; A.4/17°	—	18°·2	decomp.
fluoride, BF_3	68·0	A. 2·3	−127°	−101	decomp.
oxide, B_2O_3	70·0	1·83/4°	577	—	16/102°
Borax. *See* Sodium borate.					
Boric acid, H_3BO_3 . . .	62·0	1·43/15°	184–186	$H_2O/100°$	4/18°
Cadmium—					
bromide, $CdBr_2$	272·24	4·7–4·9/14°	571	806–812	48·9/18° *p.*
chloride, $CdCl_2$	183·32	3·6/15°	590	*c.* 900	140/20°
nitrate, $Cd(NO_3)_24H_2O$.	308·48	2·4	59·5	132	127/18°
oxide, CdO	128·4	6·9–8·1	infusible	—	insoluble
sulphate, anhy. $CdSO_4$ . .	208·47	4·7/15°	1000°	—	59/23°
„ hydr. $3CdSO_4.8H_2O$	769·54	3·05	—	—	see p.125.
Cæsium—					
carbonate, Cs_2CO_3 . . .	325·62	—	<red heat	dec. 610°	v. soluble
chloride, $CsCl$	168·27	3·97/20°	631°	sublimes	174/10°
hydride, CsH	133·82	2·7	decomp.	—	decomp.
hydroxide, $CsOH$ . . .	149·82	4·02	red heat	—	soluble
nitrate, $CsNO_3$	194·82	3·69/28°	414°	decomp.	15/10°
Calcium—					
bromide, $CaBr_2$	199·93	3·3/20°	760	*c.* 800°	125/0°
carbonate, $CaCO_3$ . . .	100·09	2·7–2·9	dec. 825°	—	·0018 cold
chloride, anhy. $CaCl_2$ . .	111·0	2·3/20°	780°	{4H_2O/30°	63/10°
„ hydr. $CaCl_2.6H_2O$.	219·1	1·65	29	{6H_2O/200°	96/0°
hydride, CaH_2	42·11	1·7	—	—	decomp.
hydroxide, $Ca(OH)_2$ . .	74·11	2·08	—	—	see p.125.
iodide, CaI_2 (and +$6H_2O$) .	293·1	4·9/20°	740	*c.* 710	192/0°
nitrate, $Ca(NO_3)_24H_2O$.	236·17	1·82	561	dec. 132°	54·8/18°
oxide, CaO	56·09	3·08	abt. 2000	—	·13/0°
phosphate, $Ca_3(PO_4)_2$ . .	310·3	3·2	—	—	·003–·008
sulphate, $CaSO_4$	136·16	2·96	—	—	·18/0°
Carbon—					
chloride, tetra-, CCl_4 . .	153·84	1·582/21°	−23°·8	76°·7	insoluble
oxide, sub- (1906), C_3O_2 .	68·00	—	—	7°/761	*
„ mon-, CO . . .	28·00	A. ·967	−207	−190	see p.124.
„ di-, CO_2 . . .	44·00	liq. ·772/20° †	−65	−78·2	see p.124.
sulphide, mono- CS . . .	44·07	1·6–1·83	—	—	—
„ bi-, CS_2 . . .	76·14	1·292/0°	−110	46·2	·2/0°
Cerium—					
chloride (cerous), $CeCl_3$ . .	246·63	3·88/15°·5	v. fusible	—	soluble
oxide (cerous), Ce_2O_3 . .	328·5	6·9–7	—	—	insoluble
„ (ceric), CeO_2 . .	172·25	6·74	—	—	insoluble
sulphate (cerous), $Ce_2(SO_4)_38H_2O$	712·84	3·22	$8H_2O/630°$	—	16·5/0°
Chlorine—					
oxide, mon-, Cl_2O . . .	86·92	{liq. 3·87 / A. 3·007}	explosive	−19	200V/0°

* Forms malonic acid.   † Behn, *Ann. d. Phys.*, 1900.   anhy. = anhydrous ;
dec. or decomp. = decomposes ; hydr. = hydrated ; liq. = liquid ; v. = very.

PHYSICAL CONSTANTS

INORGANIC COMPONDS (contd.)				
For general heading, see p. 109.				

Substance and Formula.	Formula weight (O = 16).	Density, gms./c.c.	Melting Point, °C.	Boiling Point, °C.	Solubility in Water.
Chlorine (contd.)—		at./temp.	at./mms.	at./mms.	at./temp.
oxide, di-, ClO_2	67·46	1·5 ; A. 2·3	− 76°	9·9°/731	20V/4°
Chromium—					
chloride (chromous), $CrCl_2$.	122·92	2·75/14°	—	—	v. soluble
„ (chromic), $CrCl_3$	158·38	{2·76/15° / A. 11/1200°}	—	c. 1300°	slgtly sol.
oxide, Cr_2O_3	152·0	5·04	white heat	—	insoluble
„ tri-, CrO_3 . . .	100·0	2·74	190	decomp.	62·1/0°(p)
sulphate, $Cr_2(SO_4)_3 15H_2O$.	662·65	1·867/17°	$15H_2O$/100°	—	120/20°
Cobalt—					
cobaltous chloride, $CoCl_2$ (and + $6H_2O$)	129·9	2·94	subl. c. 87°	—	29·5/0°
„ hydrate, $Co(OH)_2$	93·02	3·6/15°	—	—	insoluble
„ oxide, CoO . .	74·98	5·7	dec. 100°	—	insoluble
„ sulphate, $CoSO_4 . 7H_2O$	281·2	1·918/15°	96°·8	—	26/3°
cobaltic chloride, $CoCl_3$ . .	165·35	2·94	sublimes	—	soluble
„ oxide, Co_2O_3 . .	165·95	5·1	dec. r. ht.	—	insoluble
„ sulphate, $Co_2(SO_4)_3$	406·15	—	—	—	soluble
Columbium. See Niobium.					
Copper—					
cuprous chloride, Cu_2Cl_2 . .	198·06	{3·7 / A. 6·6/1600°}	410	c. 1000°	insoluble
„ oxide, Cu_2O . .	143·14	5·8–6·1	red heat	—	insoluble
cupric chloride, $CuCl_2$ . .	134·49	3·05	498	decomp.	75/17°
„ nitrate, $Cu(NO_3)_2 3H_2O$	241·64	2·17	114·5	{170° / dec. r. ht.}	(60/25°(p)
„ oxide, CuO . .	79·57	6·30	—	—	insoluble
„ sulphate, $CuSO_4 5H_2O$	249·65	2·28/15°	{$4H_2O$/100° / $5HO_2$/240°}	dec. r. ht.	see p. 125.
Cyanogen, C_2N_2	52·02	{liq. ·866/17° / A. 1·806}	−35°	− 20·7°	4·5 V/20°
Erbium—					
oxide, Er_2O_3	382·8	8·6	infusible	—	insoluble
sulphate, $Er_2(SO_4)_3 8H_2O$.	767·14	3·18	dec. 950°	—	· 23/20°
Gadolinium—					
sulphate, $Gd_2(SO_4)_3$ . . .	602·81	4·14/15°	—	—	2·3/34°
Gallium—					
chloride, tri-, $GaCl_3$ . . .	176·28	A. 12·2/240°	75°·5	220	decomp.
Germanium—					
chloride, tetra-, $GeCl_4$ . .	214·34	1·89/18°	—	86	decomp.
oxide, di-, GeO_2 . .	104·5	4·70/18°	—	—	·4/20°
Glucinum. See Beryllium.					
Gold—					
chloride, $AuCl_3$	303·5	—	288° *	dec. 180°	· 68
Hydrazine, $NH_2 . NH_2$ . . .	32·05	1·01/15°	1·4	113°	v. soluble
„ hydroxide, $N_2H_4 . H_2O$	50·07	1·030/21°	< − 40	119	v. soluble
Hydrobromic acid, HBr . .	80·93	{1·78 / A. 2·79}	− 86	− 68·7	{221/0° / 130/100°}
Hydrochloric acid, HCl . .	36·47	·929/0° †	− 112·5	−83°·1/755	see p. 124.
Hydrocyanic acid, HCN . .	27·02	·697/18°	− 13·8	26·1	∞

* Under chlorine at 1520 mms.　　† Rupert, 1909.　　dec. or decomp. = decomposes ;
liq. = liquid ; r. ht. = red heat ; subl. = sublimes ; v. = very ; ∞ = soluble in all proportions.

INORGANIC COMPOUNDS (contd.)					
For general heading, see p. 109.					

Substance and Formula.	Formula weight (O = 16).	Density, gms./c.c.	Melting Point, °C.	Boiling Point, °C.	Solubility in Water.
		at./temp.	at./mms.	at./mms.	at./temp.
Hydrofluoric acid, HF . . .	20·01	{·988/15° / A. ·691}	−92°·3	19°·4	111/35°
Hydriodic acid, HI	127·93	A. 4·38	−51·3	−36°·7/752	{42,500 / V/10°}
Hydrogen—					
peroxide, H_2O_2	34·02	1·458/0°	−2	80°·2/47	v. soluble
selenide, H_2Se	81·22	A. 2·805	−64	−42°	331V/13°
sulphide, H_2S	34·08	{liq. ·9 / A. 1·178}	−86	−61·6	{305V/15° / see p.124.}
telluride, H_2Te	129·52	A. 4·39	−48	0	soluble
Hydroxylamine, NH_2OH . .	33·03	1·227/14°	33°	70°/60	soluble
Iodine—					
trichloride, ICl_3	233·3	3·11	101°/16atm.	dec. 25°	soluble
Iodic acid, HIO_3	175·93	4·63/0°	½H_2O/170°	——	75/16° p.
Iron—					
carbonyl, $Fe(CO)_5$	195·85	{1·494/0° / A. 6·5}	−19·7	102°·7/764	——
ferrous chloride, $FeCl_2$ . .	126·8	2·99/18°	——	volatilizes	50/19°
„ oxide, FeO	71·85		——		insoluble
„ sulphate, $FeSO_4.7H_2O$	278·03	1·88	64	6H_2O/100°	20·8/10°
„ amm.sulphate, $FeSO_4.(NH_4)_2SO_4.6H_2O$	392·15	1·81	……	——	{18/0° / 78/75°}
oxide (magnetic), Fe_3O_4 . .	231·55	5·5·4	……	——	insoluble
ferric chloride, $FeCl_3$ . . .	162·23	{2·8/11° / A. 11·2/320°}	301	280°–285°	537/100°
„ nitrate, $Fe(NO_3)_3 9H_2O$	404·02	1·683/20°	47·2	decomp.	v. soluble
„ oxide, Fe_2O_3	159·7	5·2–5·3	……	——	insoluble
„ sulphate, $Fe_2(SO_4)_3$ (and $+9H_2O$)	399·91	3·1/18°	……	——	v. slgt. sol.
Lead—					
acetate, $Pb(C_2H_3O_2)_2.3H_2O$	379·2	2·5	3H_2O/75°	280	46/15°
carbonate, $PbCO_3$	267·1	6·4	——	——	decomp.
chloride, $PbCl_2$	277·8	5·8	447°	c. 900	·7/0°
iodide, PbI_2	460·94	6·12	373	861–954	·04/0°
oxide, mon- (litharge), PbO .	223·1	c. 9·3	red heat	——	·002/20°
„ red lead, Pb_3O_4 . . .	685·3	9·09/15°	dc.500°–530°	——	insoluble
„ per- (brown), PbO_2 . .	239·1	8·91–9·5	decomp.	——	insoluble
sulphate, $PbSO_4$	303·2	6·23	937°	——	·004/18°
Lithium—					
carbonate, Li_2CO_3	73·88	2·11	618–710	——	see p.125.
chloride, LiCl	42·40	2–2·07	491–600	dec. w. ht.	72/0°
nitrate, $LiNO_3$	68·95	2·3–2·4	c. 258	——	35/0°
oxide, Li_2O	29·88	2·10/15°	——	——	5/0°
phosphate, $Li_3PO_4.H_2O$ . .	133·8	2·4/15°	857	——	·04
sulphate, Li_2SO_4	110·0	2·21/15°	818–853	——	26/0°
Magnesium—					
carbonate, $MgCO_3$	84·32	3·0	dec. 350°	——	·01
chloride, $MgCl_2.6H_2O$ . . .	203·34	1·56/17°	2H_2O/100°	decomp.	54/20°
nitrate, $Mg(NO_3)_2 6H_2O$ . .	256·44	1·46	90°	143	42/18° p.
oxide, MgO	40·32	3·2–3·7	>2000	——	·001
phosphate, $Mg_3(PO_4)_2.4H_2O$	335·2	1·64/15°	——	——	·02
sulphate, $MgSO_4.7H_2O$ . .	246·5	1·678/16°	5H_2O/150°	——	27/0°

atm. = atmospheres ; dc., dec., or decomp. = decomposes ; liq. = liquid ; slgt. = slightly ; v. = very ; w. ht. = white heat.

I

INORGANIC COMPOUNDS (contd.)					
For general heading, see p. 109.					
Substance and Formula.	Formula weight (O = 16).	Density, gms./c.c.	Melting Point, °C.	Boiling Point, °C.	Solubility in Water.
Manganese—		at./temp.	at./mms.	at./mms.	at./temp.
carbonate, $MnCO_3$ . . .	114·93	3·1–3·7	decomp.	—	v. slgt. sol
chloride, $MnCl_2 . 4H_2O$.	197·9	1·91	87°·6	—	107/10°
nitrate, $Mn(NO_3)_2 . 6H_2O$	287·05	1·82	87·5	dec. 129°·4	54·5/11°p.
oxide, -ous, MnO	70·93	5·1	white heat	—	insoluble
„ -ic, Mn_2O_3 . . .	157·86	4·3–4·8	—	—	insoluble
„ tetr-, Mn_3O_4 . . .	228·79	4·7–4·9	—	—	insoluble
„ di-, MnO_2 . . .	86·93	4·7–5·0	dec. 390	—	insoluble
sulphate,* $MnSO_4 4H_2O$.	223·06	2·1	18° and 30°†	—	111/54°
Mercury—					
mercurous chloride, HgCl	235·46	$\{6·48$ and $7·2\}$ A. 8·21	400–500	sublimes	·0002/18°
„ nitrate, $HgNO_3 . 2H_2O$	298·04	4·78	decomp.	—	v. soluble
„ sulphate, Hg_2SO_4	496·07	7·56	melts, dec.	decomp.	·2 cold
mercuric bromide, $HgBr_2$.	359·84	5·7	244	subl. c. 322°	1/9°
„ chloride, $HgCl_2$.	270·92	$\{5·3–5·5\}$ A. 9·8	287	303–307	$\{5·4/20°(p)\}$ see p.125.
„ iodide, red, HgI_2 .	453·84	$\{6·2–6·3\}$ A. 15·6	241–257	349	·003/17°
„ „ yellow, HgI_2	453·84	$\{5·9–6·1\}$ A. 15·6	241	349	insoluble
„ oxide, HgO . .	216·0	11·14	dec. r. ht.	—	·005/25°
„ sulphate, $HgSO_4$	296·07	6·47	dec. r. ht.	—	decomp.
Molybdenum—					
chloride, $MoCl_5$	273·3	A. 9·5/350°	194°	268°	decomp.
oxide, di-, MoO_2	128·0	6·4/10°	—	—	insoluble
„ tri-, MoO_3 . . .	144·0	4·4/21°	759	sublimes	·2 cold
Nickel—					
carbonyl, $Ni(CO)_4$	170·7	1·318/17°	−25	43°	insoluble
chloride, $NiCl_2$	129·6	2·56	sublimes	—	35/0°(p)
nitrate, $Ni(NO_3)_2 . 6H_2O$.	290·8	2·06/14°	56°·7	136·7	48·5/18°p.
sulphate, $NiSO_4 . 7H_2O$.	280·86	1·98	98–100	—	31·5/9°
Niobium—					
chloride, penta-, $NbCl_5$ . .	270·8	$\{4·4–4·5\}$ A. 9·6/360°	194	240·5	decomp.
Nitrogen—					
nitric acid, HNO_3	63·02	1·53/15°	−41·3	dec. 86	∞
nitrous oxide, N_2O . . .	44·02	$\{1·226/−89°·4\}$ A. 1·614	−102	−89°·4/741	$\{74V/15°\}$ see p.124.
nitric „ NO . . .	30·01	$\{·0013\}$ A. 1·039	−167	−153	$\{5·1V/15°\}$ see p.124.
nitrogen trioxide, N_2O_3 . .	76·02	1·447/−2°	−111	decomp.	soluble
„ peroxide, NO_2 or N_2O_4	46·01	1·49/0° §	−10·1	26°	soluble
„ pentoxide, N_2O_5 .	108·02	1·64/18°	30	dec. 45–50	soluble
„ oxychloride, NOCl.	65·47	1·416/−12°	−60	−5°·6/751	decomp.
Osmium—					
oxide, tetr-, OsO_4	254·9	A. 8·89	20	100	soluble
Ozone, O_3	48·00	$\{·00214\}$ A. 1·659	dec. 270°	−119	v. slgt. sol.
Palladium—					
chloride, $PdCl_2 . 2H_2O$ . .	213·65	—	dec. r. ht.	—	soluble

* The ordinary salt ; also six other hydrates. † Stable between temps. given.
‡ Also anhy. and $6H_2O$. § Density, p. 26. dec. or decomp. = decomposes ;
r. ht. = red heat ; slgt. = slightly ; subl. = sublimes ; v. = very ; ∞ = soluble in all proportions.

INORGANIC COMPOUNDS (contd.)

For general heading, see p. 100.

Substance and Formula.	Formula weight (O = 16).	Density, gms./c.c.	Melting Point, °C.	Boiling Point, °C.	Solubility in Water.
		at./temp.	at./mms.	at./mms.	at./temp.
Perchloric acid, $HClO_4$ . . .	100·47	1·76/22°	−35	19°/11	soluble
Phosphorus—					
bromide, tri-, PBr_3 . . .	270·8	{2·92/0° A. 9·706}	−41°·5	175	decomp.
chloride, tri-, PCl_3 . . .	137·3	{1·612/0° A. 4·875}	−112	76	„
„ penta-, PCl_5 . . .	208·3	A. 3·6/296°	148	162	„
fluoride, tri-, PF_3	88·04	A. 3·02	−160	−95	„
oxide, tri-, P_4O_6	220·2	1·94/25°	22·5	173	soluble
„ tetr-, P_2O_4	126·1	2·54/23°	>100	c. 180	„
„ pent-, P_2O_5	142·1	2·39	subl. r. ht.	—	v. soluble
Phosphine, PH_3	34·06	A. 1·185	−133°	−85	slgtly sol.
„ liquid, P_2H_4 . .	66·11	1·007–1·016	<−10°	57/735	insoluble
Phosphonium chloride, PH_4Cl	70·53	—	26°	sublimes	decomp.
Platinum—					
chloride, tetra-, $PtCl_4$ . . .	337·0	—	decomp.	—	v. soluble
Potassium—					
bromide, KBr . . .	119·02	2·76/20°	750°	subl. w. ht.	see p. 125.
carbonate, K_2CO_3 . . .	138·2	2·29	c. 880	dec. 810°	89/0°
chlorate, $KClO_3$	122·56	2·34/17°	370	dec. 400°	3/0°
chloride, KCl	74·56	1·99/15°	c. 770	subl. w. ht.	see p. 125.
chromate, bi-, $K_2Cr_2O_7$ . .	294·2	2·69/4°	400	decomp.	5/0°
cyanide, KCN . . .	65·11	1·52/16°	red heat	red heat	122/103°
ferricyanide, $K_3Fe(CN)_6$ . .	329·21	1·82/17°	decomp.	—	33/4°
ferrocyanide, $K_4Fe(CN)_6 . 3H_2O$	422·36	1·85/17°	$3H_2O/60$-80	—	28/12°
hydroxide, KOH	56·11	2·04	red heat	subl. w. ht.	see p. 125.
iodate, KIO_3	214·02	3·97/18°	560	—	8/20°
iodide, KI	166·1	{3·04/24° A. 5·5/1320°}	614–723	—	{127/0° see p.125
nitrate, KNO_3	101·11	2·1/4°	c. 345	decomp.	see p. 125.
permanganate, $KMnO_4$ . .	158·03	2·70/10°	dec. 240°	—	6·4/15
sulphate, K_2SO_4	174·27	2·66/20°	1070	sublimes	9·2/10°
„ acid, $KHSO_4$ . .	136·18	2·24 * ; 2·61 †	200	decomp.	36/0°
sulphocyanate, KCNS . .	97·18	1·91	161	—	217/20°
Radium—					
bromide, $RaBr_2$	386·24	—	728	—	soluble
Rubidium—					
carbonate, Rb_2CO_3 . . .	230·9	—	837	dec. 740°	v. soluble
chloride, RbCl	120·9	2·2	710	—	84/10°
sulphate, Rb_2SO_4 . . .	266·97	3·61	—	—	43/10°
Selenium—					
chloride, Se_2Cl_2	229·32	2·91/17°	—	dec. c. 145	decomp.
oxide, SeO_2	111·2	3·95/15°	sub. c. 260	—	v. soluble
Selenious acid, H_2SeO_3 . .	129·22	3·91/15°·7	decomp.	—	„
Selenic acid, H_2SeO_4 . .	145·22	2·95/15°	58	260	„
Silicon—					
chloride, tetra-, $SiCl_4$ . . .	170·14	{1·520 A. 5·94}	−89	57·5	decomp.
fluoride, SiF_4	104·3	A. 3·57	−102	−107	„

* Monoclinic. † Rhombic.
amorph. = amorphous ; cryst. = crystalline ; dec. or decomp. = decomposes ; r. ht. = red heat ; sub. or subl. = sublimes ; v. = very ; w. ht. = white heat.

INORGANIC COMPOUNDS (*contd.*)					
For general heading, see p. 109.					

Substance and Formula.	Formula weight (0 = 16).	Density, gms./c.c.	Melting Point, °C.	Boiling Point, °C.	Solubility in Water.
Silicon (*contd.*)—		at./temp.	at./mms.	at./mins.	at./temp.
oxide (silica), amorph, SiO_2 .	60·3	2·2/16°	} indefinite	——	*c.* ·001
„ „ cryst., SiO_2 .	60·3	2·66	} 1500–1600°	—	insoluble
Silico chloroform, $SiHCl_3$.	135·69	{1·65 A. 4·6}	− 1·3	34°	decomp.
Silver—					
bromide, AgBr	187·8	6·47/25°	427	dec. 700°	·0₄8/20°
chloride, AgCl	143·34	{5·50 A. 5·7/1735°}	460	——	·0₄15/20°
iodide, AgI	234·8	5·67/25°	*c.* 540	—	·0₆3/21°
nitrate, $AgNO_3$	169·89	4·35/19°	218	dec. r. ht.	see p. 125.
sulphate, Ag_2SO_4	311·83	5·4	654–676	decomp.	·77/17°
Sodium—					
borate (borax), $Na_2B_4O_7 . 10H_2O$	382·16	169/17°	red heat	——	soluble
bromide, NaBr	102·92	3·1	733–765	——	77/0°
carbonate, Na_2CO_3 . . .	106·0	2·4–2·5	849	decomp.	see p. 125
„ bi-, $NaHCO_3$ . .	84·01	2·2	$CO_2/270°$	—	8/10°
chloride, NaCl	58·46	2·17/20°	801*	w. heat	see p. 125.
hydroxide, NaOH . . .	40·01	2·13	318	w. heat	63·5/15°
iodide, NaI	149·92	3·65/18°	603–695	——	178/20°
nitrate, $NaNO_3$	85·01	2·27/20°	*c.* 313	——	73/0°
peroxide, Na_2O_2	78·00	2·8	decomp.	——	sol. ; dec.
phosphate, di-, $Na_2HPO_4 . 12H_2O$	358·2	1·52/16°	38	$3H_2O/c.160°$	3·9/10°
sulphate, anhy., Na_2SO_4 . .	142·07	2·67/20°	884	——	see p. 125.
„ hydr., $Na_2SO_4 . 10H_2O$	322·23	1·492/20°	{880° T.32°·383}	$7H_2O/150°$	{5/0° 50·6/32·7°
sulphite, $Na_2SO_3 . 7H_2O$ . .	252·18	1·56	$7H_2O/150°$	decomp.	25/15°
thiosulphate (hypo'), $Na_2S_2O_3 . 5H_2O$	248·22	1·73/17°	32–48	dec. 220°	60/10°
Strontium—					
bromide, $SrBr_2$	247·5	4·2/24°	498–630	——	93/10°
carbonate, $SrCO_3$ . . .	147·6	3·6	dec. 1160°	dec. r. ht.	·001/24°
chloride, $SrCl_2$ (and + $6H_2O$)	158·5	3·05	796–854	{4H_2O/60° 6H_2O/100°}	{48/10° see p.125
nitrate, $Sr(NO_3)_2$ . . .	211·6	3/17°	dec. 645	——	55/10°
oxide, SrO	103·6	3·6	3000	——	35/0°
„ per-, SrO_2	119·6	·546	decomp.	——	decomp.
sulphate, $SrSO_4$	183·7	3·7–4	dec. w. ht.	——	·011/18°
Sulphur—					
dioxide, SO_2	64·07	{1·434/0° A. 2·23}	−76°	− 10°·1	{4730 V./ 15° ; p. 124.
trioxide, SO_3	80·07	{1·97/20° A. 2·77}	14·8	46	decomp.
Sulphuretted hydrogen. *See* Hydrogen sulphide.					
Sulphuric acid, H_2SO_4 . .	98·09	1·834/18°	10·5	dec. 40°	∞
Tellurium—					
chloride, $TeCl_2$	198·42	——	175	327	decomp.
oxide, di-, TeO_2	159·5	5·9/0°	dull r. ht.	< 700	insoluble
„ tri-, TeO_3	175·5	5·07/15°	decomp.		„

* Practically same for ordinary table salt as for pure salt (Hårker).
anhy. = anhydrous ; dec. or decomp. = decomposes ; hydr. = hydrated ; r. ht. = red heat ; w. ht. = white heat ; ∞ = soluble in all proportions.

INORGANIC COMPOUNDS (*contd.*)

For general heading, see p. 109.

Substance and Formula.	Formula weight (0 = 16)	Density, gms./c.c.	Melting Point, °C.	Boiling Point, °C.	Solubility in Water.
Thallium—		at./temp.	at./mms.	at./mms.	at./temp.
carbonate, Tl_2CO_3	468·0	7·1	272°	decomp.	4/15°
chloride, tri-, $TlCl_3$ . . .	310·38	—	25	„	v. soluble
oxide (thallous), Tl_2O . . .	424·0	—	300	—	v. soluble
sulphate, Tl_2SO_4	504·07	6·77	632	decomp.	4·7/15°
Thorium—					
nitrate, $Th(NO_3)_4 . 12H_2O$.	696·2	—	—	—	v. soluble
oxide, ThO_2	264·0	9·87/15°	infusible	—	insoluble
Tin—					
chloride (stannous), $SnCl_2$.	189·92		249°	620°	270/15°
„ (stannic), $SnCl_4$.	260·84	{2·27/20°} A. 9·2	−33	114·1	soluble
oxide (stannous), SnO . .	135·0	6·3	dec. r. ht.	—	insoluble
„ (stannic), SnO_2 . .	151·0	6·6–6·9	1130	—	„
Titanium—					
chloride, tetra-, $TiCl_4$ . . .	189·94	{1·76/0° A. 6·836}	−25	136	decomp.
oxide, di-, TiO_2	80·1	3·7–4·2	c. 1500	—	insoluble
Tungsten—					
chloride, hexa-, WCl_6 . . .	396·76	A. 13·3/350°	275	347	„
oxide, tri-, WO_3	232·0	7·2	red heat	—	„
Uranium					
oxide, di-, UO_2	270·5	10·9	oxidises	—	„
„ (green), U_3O_8 . . .	843·5	7·3	decomp.	—	
„ (yellow), UO_3 . . .	286·5	5·1	decomp.	—	
„ (black), U_2O_5 . . .	557·0	8·4–9·2		—	
Uranyl chloride, UO_2Cl_2 . .	341·42	—	fusible	decomp.	320/18°
„ nitrate, $UO_2(NO_3)_2 . 6H_2O$	502·62	2·81	59°·5	118°	200
Vanadium—					
chloride, tetra-, VCl_4 . . .	192·9	{1·86 A. 6·69}	−18	154	soluble
oxide, pent-, V_2O_5 . . .	182·1	3·5/20°	658	—	0·8/20°
Zinc—					
carbonate, $ZnCO_3$	125·37	4·4	dec. 300°	—	0·001/15°
chloride, $ZnCl_2$	136·29	2·91/25°	262°?	730	330/10°
sulphate, $ZnSO_4 . 7H_2O$ . .	287·55	{1·96 3·4 anhy.}	6H₂O/100°	{7H₂O at} {red heat.}	42/0° 80·8/100°
sulphide, ZnS	97·44	4·0	1050°	—	insoluble
Zirconium—					
oxide, ZrO_2	122·6	5·1–5·7	infusible	—	„

anhy. = anhydrous; dec. or decomp. = decomposes; r. ht. = red heat; v. = very.

FREEZING MIXTURES

Parts by weight.	Temp.	Parts by weight.	Temp.
1 of NH_4NO_3, 1 of water . .	−15° C.	2 of snow or crushed ice, 1 of NaCl	−18°
8 of Na_2SO_4, 5 of water . .	−17	3 of snow, 4 of cryst. $CaCl_2$.	−48

ORGANIC COMPOUNDS
Formula (Molecular) Weight, Density, Melting and Boiling Points.
For general heading, see p. 109.

Substance and Formula.	Formula weight (O = 16).	Density, gms./c.c.	Melting Point, °C.	Boiling Point, °C.
		at./temp.	at./mms.	at./mms.
Acetaldehyde, $CH_3 . CHO$	44·03	·788/16° C.	− 120°	20°·8
Acetic acid, $CH_3 . COOH$	60·03	1·05/20°	16·7	118·5, Y.
Aceto-acetic ether, $CH_3CO . CH_2CO_2$. C_2H_5	130·1	1·028/20°	< −80	181
Acetone, CH_3COCH_3	58·05	·797/15°	−95	56·5
Acetylene, C_2H_2	26·02	$\{$·46/−7° A. ·91$\}$	−81·5/895*	−85
Acrylic acid, $CH_2 : CHCO_2H$ . . .	72·03	1·062/16°	10	140
Alizarine, $C_6H_4(CO)_2C_6H_2(OH)_2$ . .	240·1	—	290	430
Allyl alcohol, $CH_2 : CH . CH_2OH$	58·05	·858/15°	liquid	96·7
,, chloride, $CH_2 : CHCH_2Cl$ . .	76·46	·937/19°	liquid	46
,, thiocyanate, $CH_2 : CHCH_2CNS$	99·08	1·017/10°	liquid	151
Amyl acetate, $C_5H_{11} . CH_3CO_2$ . .	130·1	·879/20°	liquid	148
,, alcohol (n.), $CH_3(CH_2)_3CH_2OH$	88·10	·812/20°	liquid	137
,, ,, (act.), $CH_3C_2H_5CHCH_2$-OH	88·10	·825/0°	liquid	129
,, ,, (sec.),$C_3H_7CH(OH)CH_3$	88·10	·825/0°	liquid	118·5/753
,, ,, (tert.), $(CH_3)_2C(OH)$-C_2H_5	88·10	·814/15°	−12°	102·5
Aniline, $C_6H_5 . NH_2$	93·07	1·023/15°	−8	183·9
Anisol, $C_6H_5OCH_3$	108·1	·99/25°	−37·8	155
Anthracene, $C_6H_4 : C_2H_2C_6H_4$ . . .	178·1	1·15	216	351
Antimony trimethyl, $Sb(CH_3)_3$ . . .	165·3	1·52/15°	liquid	86
Asparagine(l.)$C_2H_3NH_2CO_2H.CONH_2$	132·1	1·55/4°	decomp.	decomp.
Benzaldehyde, C_6H_5CHO	106·1	1·05/15°	− 13°·5	179·5
Benzene, C_6H_6	78·05	·879/20°	5·4	80·2, Y.
Benzoic acid, $C_6H_5 . COOH$ . . .	122·0	1·20/21°	121·4	249·2
Benzophenone, $(C_6H_5)_2CO$	182·1	1·098/50°	48	306
Benzoyl chloride, C_6H_5COCl . . .	140·5	1·212/20°	− 1	198/749
Benzyl alcohol, $C_6H_5CH_2OH$ . . .	108·1	1·043/20°	liquid	206·5
Beryllium ethyl, $Be(C_2H_5)_2$ . . .	67·18	—	—	187
Bismuth triethyl, $Bi(C_2H_5)_3$ . . .	295·1	2·3/18°	—	107
Borneol (i.), $C_{10}H_{17}OH$	154·1	1·01	210	sublimes
Bromo benzene, C_6H_5Br	157·0	1·49/20°	− 31·1	156, Y.
Butyl alcohol(n.),$CH_3(CH_2)_2CH_2 . OH$	74·08	·81/20°	liquid	117·5
,, ,, (sec.),$CH_3CHOH . C_2H_5$	74·08	·819/22°	liquid	99·8
,, carbinol(tert.), $(CH_3)_3C . CH_2OH$	88·10	·812/20°	52	113
,, chloride, $CH_3(CH_2)_3Cl$ . .	92·53	·887/20°	liquid	78
,, ether, $(C_4H_9)_2O$	130·1	·77/20°	—	141
Butyric acid (n.), $CH_3(CH_2)_2COOH$.	88·06	·96/19°	− 8	162·3
,, ,, (iso), $(CH_3)_2CHCOOH$.	88·06	·950/20°	−79	155
Cacodylic acid, $(CH_3)_2AsO . OH$ . .	138·0	—	200	—
Caffeine, $C_8H_{10}N_4O_2 . H_2O$	212·3	1·23/19°	234	sublimes
Camphor, $C_{10}H_{16}O$	152·1	·992/10°	176·4	205·3
Camphoric acid (d.), $C_8H_{14}(COOH)_2$.	200·1	1·19	178	decomp.
Caproic acid, $CH_3(CH_2)_4COOH$ . .	116·1	·929/20°	8	205
Carbolic acid. *See* Phenol.				
Carbon bisulphide, CS_2	76·14	1·292/0°	−110	46·2
,, oxysulphide, COS . . .	60·07	2·104	—	gas
,, tetrachloride, CCl_4	153·8	1·582/21°	−30	76·7, Y.

* Mackintosh, 1907; decomp. = decomposes ; l., = lævo-rotatory (see p. 78). Y., Young, *Journ. de Phys.*, Jan., 1909.

ORGANIC COMPOUNDS (contd.)
For general heading, see p. 109.

Substance and Formula.	Formula weight (0 = 16).	Density, gms./c.c.	Melting Point, ° C.	Boiling Point, ° C.
		at./temp.	at./mms.	at./mms.
Cellulose, $(C_6H_{10}O_5)_x$	162·1	1·525	——	——
Chlor acetic acid, $CClH_2:COOH$.	94·48	1·39/75°	63°	186°
„ benzene, C_6H_5Cl . .	112·5	1·118/10°	—40	132, Y.
Chloral hydrate, $CCl_3.CH(OH)_2$ . .	165·4	1·9	—57	97·5
Chloroform, $CHCl_3$	119·4	1·526/0°	—70	61·2
Chrysene, $C_{18}H_{12}$	228·1	——	250	sublimes
Cineol, $C_{10}H_{18}O$	154·2	·92	—1	176
Cinnamic acid, $C_6H_5CH:CHCOOH$	148·1	1·247	133	300
„ aldehyde, $C_6H_5CH:CH-CHO$.	132·1	1·05/24°	—7·5	——
Citric acid, $(CO_2HCH_2)_2C(OH)CO_2H$ + H_2O	192·1	1·54	153	decomp.
Collidine, α $CH_3.C_5H_3N.C_2H_5$.	121·1	·953/22°	——	180
Coniine (d.), 1:2, $C_5H_{10}N.C_3H_7$	127·2	·849/25°	—2·5	170
Cresol (o.), $CH_3C_6H_4OH$	108·1	1·005	30	191
Cyanic acid, $HCNO$	43·02	1·14/0°	liquid	dec. o
Cyanogen, C_2N_2	52·02	{liq. ·866/17° A. 1·806 }	—35	—20·7
Cymene (p.), $CH_3.C_6H_4.C_3H_7$.	134·12	·852/25°	liquid	175
Dextrin, $C_{12}H_{20}O_{10}$	324·2	1·04	——	——
Diacetyl, $CH_3CO.COCH_3$ . . .	86·05	·973	——	87·7
Dichlor acetic acid, $CHCl_2.COOH$.	128·9	1·522/15°	—4	190
Diethyl amine, $(C_2H_5)_2NH$ . .	73·13	·706/20°	—40	55·5
„ aniline, $(C_2H_5)NC_6H_5$ . .	149·2	·94/18°	liquid	213·5
„ ketone, $C_2H_5COC_2H_5$ . .	86·08	·83/0°	——	103
Dimethyl amine, $(CH_3)_2HN$ . .	45·07	·686/—6°	liquid	8 to 9
„ tartrate, $(CH_3)_2C_4H_4O_6$.	178·1	1·341/15°	48	280
Dinitrobenzene (m.), $C_6H_4(NO_2)_2$.	168·1	1·37	91	297
Diphenyl, $C_6H_5.C_6H_5$	154·1	1·16	70·5	255
Diphenylamine, $(C_6H_5)_2HN$ . .	169·1	1·159	54	310
Epichlorhydrine, C_3H_5ClO . . .	92·49	1·203/0°	——	116
Erythrite, $(CH_2OH.CHOH)_2$ . .	122·1	1·45/17°	112	330
Ethane, $CH_3.CH_3$	30·05	{liq. ·446/0° A. 1·036 }	—171·4	—85·4/749
Ether, $C_2H_5OC_2H_5$	74·08	·718/17°	—117	34·6, Y.
Ethyl acetate, $CH_3CO_2.C_2H_5$ . . .	88·06	·903/18°·5	—83·8	77·1
„ aceto-acetate, $CH_3COCH_2CO_2.C_2H_5$.	130·1	1·028/20°	< —80	181
„ alcohol, C_2H_5OH	46·05	·7937/15°	—112·3	78·3, Y.
„ amine, $C_2H_5H_2N$	45·07	·699/8°	—85	18·7
„ benzoate, $C_6H_5CO_2.C_2H_5$ . .	150·1	1·05/16°	111—116	211·2
„ bromide, $C_2H_5.Br$	108·96	1·45/15°	—116	38·4
„ butyrate, $C_3H_7.COOC_2H_5$ . .	116·1	·898/18°	——	120
„ chloride, C_2H_5Cl	64·50	{·921/0° A. 2·219 }	liquid	12·5
„ cyanide, $C_2H_5.CN$	55·05	·794/7°	—103	97
„ formate, $HCOOC_2H_5$	74·05	·938/0°	——	54·3, Y.
„ iodide, C_2H_5I	156·0	1·944/14°	liquid	72·3
„ isobutyrate $(CH_3)_2CHCOOC_2H_5$	116·1	·890/0°	——	110·1
„ mercaptan, C_2H_5SH	62·11	·839/20°	—22	36·2
„ nitrate, $C_2H_5NO_3$	91·08	1·116/15°	—112	87

dec. or decomp. = decomposes.   Y., Young, *Journ. de Phys.*, Jan., 1909.

PHYSICAL CONSTANTS

ORGANIC COMPOUNDS (*contd.*)
For general heading, see p. 109.

For general heading, see p. 109.

Substance and Formula.	Formula weight ($O = 16$).	Density, gms./c.c.	Melting Point, $^\circ$ C.	Boiling Point, $^\circ$ C.
		at./temp.	at./mms.	at./mms.
Ethyl propionate, $C_2H_5CO_2C_2H_5$	102·1	·896/16°	——	99°·0
„ salicylate, $C_6H_4(HO)CO_2.C_2H_3$	166·1	1·184/20°	——	231·5
„ sulphide, $(C_2H_5)_2S$	90·15	·837/20°	liquid	92·6
„ tartrate (d.), $C_4H_4O_6(C_2H_5)_2$	206·1	1·206/20°	——	280
„ valerate, $C_4H_9CO_2C_2H_5$	130·1	·876/20°	——	144·5
Ethylene, $CH_2 : CH_2$	28·03	{liq. ·61 } {A. ·9784}	−169	−102·7
„ bromide, di-, $CH_2Br.CH_2Br$	187·9	2·19/11°	9·5	131·6
„ chloride, di-, $CH_2Cl.CH_2Cl$	98·93	1·28/0°	−40	83·7
„ oxide, $<(CH_2)_2O$	44·03	·897/0°	liquid	13·5/746
Ethylidene chloride, $CH_3.CHCl_2$	98·93	1·186/12°	liquid	59·9
Eucalyptol, $C_{10}H_{18}O$	154·1	·927/20°	−1	176
Eugenol, $C_6H_3.(OH).OCH_3.C_2H_5$	164·1	1·0779/0°	liquid	247·5
Fluor benzene, C_6H_5F	96·04	1·024/20°	40°	85·2, Y.
Formic acid, $H.COOH$	46·02	1·22/20°	8·6	100·8
Formaldehyde, $H.COH$	30·02	{·815/−20°} {A. 1·6}	——	−21
Fructose (d.), $CH_2OH[CHOH]_3CO\cdot CH_2OH$	180·1	1·55/0°	95	——
Fumaric acid, $(COOH.CH :)_2$	116·0	1·625	286	——
Furfural, $C_4H_3O.COH$	96·03	1·159/20°	liquid	161
Galactose (d.), $CHO[CHOH]_4CH_2OH$	180·1	——	163	——
Glucose (d.), $CHO(HCOH)_4CH_2OH$	198·1	1·54–1·57	146	——
Glutaric acid, $COOH(CH_2)_3COOH$	132·1	——	91	299
Glycerine, $OHCH_2.CHOH.CH_2OH$	92·06	1·26/20°	17	290
Glycocoll, CH_2NH_2COOH	75·08	1·161	c. 234	——
Glycol, $CH_2OH.CH_2OH$	62·05	1·125/25°	−17·4	197·4
Glycollic acid, $CH_2OH.COOH$	76·03	——	78	decomp.
Glyoxal, $CHO.CHO$	58·02	——	——	dec. 160
Glyoxalic acid, $CHO.COOH + H_2O$	92·03	syrup	——	with steam
Grape sugar. *See* Glucose.				
Heptane (n.), $CH_3(CH_2)_5CH_3$	100·1	·688/15°	——	98·4, Y.
Hexane (n.), $CH_3(CH_2)_4CH_3$	86·12	·658/21°	liquid	69, Y.
„ di-isopropyl, $[(CH_3)_2CH]_2$	86·12	·668/17°	liquid	58·1, Y.
Hydrocyanic acid, HCN	27·05	·697/18°	−14	26·1
Indigo, $C_6H_4{<}^{CO}_{NH}{>}C:C{<}^{CO}_{NH}{>}C_6H_4$	262·2	1·35	——	subl. 156°
Indol, $C_6H_4NHCH : CH$	117·1	——	52	245
Iodoform, CHI_3	393·8	2·25/25°	119	subl. & dec.
Isatine, $C_6H_4{<}^{CO}_{N}{>}COH$	147·1	——	2c1	sublimes
Isoamyl acetate, $CH_3.COOC_5H_{11}$	130·1	·876/15°	——	140
„ alcohol, $(CH_3)_2CH(CH_2)_2OH$	88·10	·81/20°	−134	129·7
Isobutane, $(CH_3)_2CHCH_3$	58·08	——	——	116·3
Isobutyl alcohol, $(CH_3)_2CH.CH_2OH$	74·08	·800/18°	liquid	108·4
„ amine, $(CH_3)_2CHCH_2NH_2$	73·13	·736/15°	——	68
Isobutyric acid, $(CH_3)_2CH.COOH$	88·06	·949/20°	−79	155·5
Isopentane, $(CH_3)_2CHCH_2CH_3$	72·10	·628/14°	——	27·9
Isopropyl acetate, $CH_3COOCH(CH_3)_2$	102·1	·917	——	90–93
„ alcohol, $(CH_3)_2HC(OH)$	60·06	·789/20°	liquid	82·8

d., dextro-rotatory (see p. 78); dec. or decomp. = decomposes; subl. = sublimes; Y., Young, *Journ. de Phys.*, Jan., 1909.

ORGANIC COMPOUNDS (*contd.*)
For general heading, see p. 109.

Substance and Formula.	Formula weight ($0 = 16$).	Density, gms./c.c.	Melting Point, °C.	Boiling Point, °C.
		at./temp.°	at./mms.	at./mms.
Isopropyl amine, $(CH_3)_2CHNH_2$ . .	59·11	·690/18°	liquid	31°·5/743
„ cyanide, $(CH_3)_2CHCN$ . .	69·07	—	liquid	107–108
Isoquinoline, $C_6H_4C_3H_3N$	129·1	1·098/20°	24·6	240
Isovaleric acid, $(CH_3)_2CHCH_2COOH$	102·1	·931/20°	− 51	176·3
Lactic acid (i.), $CH_3CHOH . COOH$	90·05	1·248/15°	—	83/1 mm.
Lactose. *See* Milk sugar.				
Maleic acid, $(COOH . CH :)_2$ . . .	116·0	1·59	100	decomp.
Malic acid (i.), $COOH.CHOH.CH_2$-.COOH	134·0	1·60/20°	130–1	—
Malonic acid, $COOH . CH_2 . COOH$.	104·0	—	132	decomp.
Maltose, $C_{12}H_{22}O_{11} + H_2O$ . . .	360·2	1·54/17°	—	—
Mercury methyl, $(CH_3)_2Hg$ . . .	230·0	3·07	liquid	96
Mesitylene, $1 : 3 : 5$, $C_6H_3(CH_3)_3$ . .	120·1	·869/10°	—	164·5
Methane, CH_4	16·03	liq. ·416/−164°	− 184	− 164
Methyl alcohol, CH_3OH	32·03	·796/15°	− 94·9	64·7, Y.
„ acetate, $CH_3COO . CH_3$. . .	74·05	·941/14°	− 101·2	57·1
„ amine, CH_3H_2N	31·08	{ ·699/−11° } { A 1·08 }	gas	− 6·7/756
„ borate, $(CH_3)_3BO_3$	104·1	·94/0°	—	65
„ chloride, CH_3Cl	50·48	{ ·920/18° } { A 1·73 }	—	− 24·1
„ ether, $(CH_3)_2O$	46·05	A 1·62	gas	− 23·6
„ ethyl ether, $CH_3 . O . C_2H_5$. .	60·06	·725/0°	—	10·8
„ formate, $HCOO . CH_3$ . .	60·03	·986/11°	—	31·9, Y.
„ iodide, CH_3I	142·0	2·285/15°	liquid	42·3
„ isobutyrate,$(CH_3)_2CHCOOCH_3$	102·1	·912/0°	—	92·3
„ mercaptan, $CH_3 . SH$ . . .	48·09	—	—	5·8/752
„ nitrate, $CH_3 . NO_3$ . . .	77·03	1·217/15°	liquid	65 explodes
„ nitrite, $CH_3 . NO_2$ . . .	61·03	·991/15°	—	− 12
„ phosphine, CH_3H_2P . . .	48·04	—	gas	− 14
„ propionate, $C_2H_5COO . CH_3$.	88·06	·937/0°	—	79·7
„ salicylate, $C_6H_4(OH)COOCH_3$	152·1	1·182/15°	− 30	224
„ sulphide, $(CH_3)_2S$ . . .	62·12	·845/21°	liquid	c. 38
Methylene bromide, CH_2Br_2 . . .	173·9	2·493	—	98·5
Milk sugar, $C_{12}H_{22}O_{11} + H_2O$ . .	360·2	1·525/20°	203 dec.	decomp.
Morphine, $C_{17}H_{19}NO_3 + H_2O$ . .	303·2	1·32	—	decomp.
Naphthalene, $C_6H_4 : C_4H_4$	128·1	1·152/15°	80	218·1
Naphthol (α), $C_{10}H_7OH$	144·1	1·224/4°	95	c. 279
Naphthyl amine (α), $C_{10}H_7H_2N$ . .	143·1	—	50	300
Nicotine (l.), $C_{10}H_{14}N_2$	162·2	1·01/20°	dec. 250°	246·7/745
Nitro benzene, $C_6H_5NO_2$	123·1	1·187/14°	3·6	209·4/745
„ ethane, $C_2H_5NO_2$	75·08	1·056	194–196	114·4
„ methane, CH_3NO_2	61·07	1·144/15°	liquid	101·7
Octane (n.), $CH_3(CH_2)_6CH_3$. . . .	114·1	·719/0°	liquid	125·8, Y.
Oleic acid, $CH_3(CH_2)_7CH:CH(CH_2)_7$-.COOH	282·3	·891/12°	14	286/100
Palmitic acid, $CH_3(CH_2)_{14}COOH$.	256·3	·846/7·6°	62·6	278/100
Paraldehyde, $(CH_3 . HCO)_3$ . . .	132·1	·994/20°	10·5	124
Penta methylene, $(CH_2)_5$	70·08	·751/20°	—	50·6
„ diamine (cadaverine), $NH_2(CH_2)_5NH_2$	102·2	·917/0°	—	178

dec. or decomp. = decomposes ; l., lævo-rotatory (see p. 78) ; Y., Young, *Journ. de Phys.*, Jan., 1909.

ORGANIC COMPOUNDS (contd.)

For general heading, see p. 109.

Substance and Formula.	Formula weight (0 = 16).	Density, gms./c.c.	Melting Point, °C.	Boiling Point, °C.
		at./temp.	at./mms.	at./mms.
Pentane (n.), $CH_3(CH_2)_3CH_3$ . . .	72·10	·634/15°	—200	36°·2, Y.
Phenetol, $C_6H_5OC_2H_5$	122·08	·963/25°	— 34 ..	171
Phenol, $C_6H_5.OH$	94·05	1·06/33°	42·7	181·5
Phenyl acetic acid, $C_6H_5CH_2COOH$.	136·1	1·23	76·5	265
„ cyanide, C_6H_5CN	103·1	1·008/17°	—17	190
„ hydrazine, $C_6H_5HN.NH_2$	108·1	1·1/23°	23	233
Phloroglucin, 1:3:5, $C_6H_3(OH)_32H_2O$	162·1	—	218 anhy.	sublimes
Phthalic acid, o. $C_6H_4(COOH)_2$.	166·1	1·59	180–200	——
„ anhydride, $C_6H_4<(CO)_2>O$	148·0	1·53/4°	128	284
Picoline (α), $CH_3.C_5H_4N$.	93·07	·933/22°	liquid	129
Picric acid, 1:2:4:6, $C_6H_2OH(NO_2)_3$.	229·1	1·813	122·5	explodes
Propane, $CH_3.CH_2.CH_3$.	44·07	·535	—195	—(38–39)
Propionic acid, $CH_3.CH_2.COOH$.	74·05	·995/20°	—22	140
Propyl acetate (n.), $CH_3COO.C_3H_7$.	102·0	·891/18°	liquid	101·6
„ alcohol (n.), $CH_3CH_2CH_2.OH$	60·06	·804/20°	——	97·2
„ chloride (n.), $CH_3CH_2CH_2Cl$.	78·51	·891/18°	——	46·5
„ formate, $H.COO.C_3H_7$.	88·06	·909/17°	——	80·9, Y.
„ iodide, $CH_3.CH_2.CH_2I$.	170·0	1·745/20°	——	102
Propylene, $CH_3.CH:CH_2$ . .	42·05	A. 1·498	gas	—50·2
Pseudo-cumene, 1:2:4, $C_6H_3(CH_3)_3$	120·1	·879/20°	——	169·8
Pyridine, C_5H_5N	79·08	·985/15°	liquid	117
Pyrogallol (—ic acid, or "pyro"), 1:2:3, $C_6H_3(OH)_3$	126·1	1·46/40°	133	293
Pyrrol, $(CH)_4>NH$	67·08	·967/21°	liquid	131
Quinoline, $C_6H_4<\frac{CH.CH}{N.CH}>$ . .	129·1	1·094/20°	19·5	241
Quinine, $C_{20}H_{24}N_2O_2$	324·3	——	174·9	——
„ sulphate, $(C_{20}H_{24}N_2O_2)_2.H_2SO_4+7H_2O$	872·7	——	205, dry	——
Racemic acid, $(COOH.CH(OH))_2+H_2O$	168·1	1·69/7°	205	——
Rochelle salt (d.), $KNaC_4H_4O_6$ . .	——	——	——	——
Rosaniline (p.), $(C_6H_4NH_2)_3COH$ . .	305·2	——	——	——
Saccharin, $C_6H_4<\frac{CO}{SO_2}>NH$	183·1	.-.	220 dec.	——
Salicylic acid, $OH.C_6H_4.COOH$. .	138·0	1·48/4°	158	sublimes
Sodium ethyl, NaC_2H_5	52·04	——	——	——
Stearic acid, $CH_3(CH_2)_{16}COOH$ . .	284·3	·843/80°	69·3	291/100
Stearine, $(C_{18}H_{85}O_2)_3C_3H_5$ . .	890·9	·924/65°	——	——
Succinic acid, $COOH(CH_2)_2COOH$.	118·0	1·55	185	235
Sugar, cane-, $C_{12}H_{22}O_{11}$ . . .	342·2	1·588/20°	——	185
Sulphanilic acid (p.), $NH_2.C_6H_4.SO_3H$.$2H_2O$.	209·2	——	chars	——
Sulphonal, $(CH_3)_2C(SO_2C_2H_5)_2$ . . .	228·2	——	125	300 dec.
Tartaric acid (i. or meso), $COOH$-$[CHOH]_2COOH.H_2O$ . . .	168·1	1·67	142 anhy.	——
„ „ (d.), $COOH(CHOH)_2COOH$ . .	150·0	1·76/7° P.	170	——
„ „ (l.), $COOH(CHOH)_2COOH$ . .	150·0	1·76	170	——
Terephthalic acid (p.), $C_6H_4(COOH)_2$	166·0	——	——	sublimes
Terpenol, $C_{10}H_{18}O$	154·1	——	70	——

anhy. = anhydrous ; d. = dextro-rotatory (see p. 78) ; P., Perkin ; dec. = decomposes ; l, lævo-rotatory (see p. 78) ; Y., Young.

ORGANIC COMPOUNDS (*contd.*)

For general heading, see p. 109.

Substance and Formula.	Formula weight $(O = 16)$	Density, gms./c.c.	Melting Point, ° C.	Boiling Point, ° C.
		at./temp.	at./mms.	at./mms.
Terpineol, $C_{10}H_{17}HO$	154·1	·936/20°	35°	218°
Tetrabromethylene, $CBr_2 . CBr_2$ . .	343·8	—	53	—
Theobromine, $C_7H_8N_4O_2$	180·2	—	330	decomp.
Thiocyanic acid, HCNS	59·09	—	−12·5	200 dec.
Thiourea, $NH_2 . CS . NH_2$	76·12	1·42	180	—
Thymol, 3 : 2 : 1, $(CH_3)_2 : CH . C_6H_3$-$(CH_3)OH$	150·1	·994/0°	50	232
Tin tetramethyl, $Sn(CH_3)_4$ . . .	179·1	1·314	—	78
Toluene, $C_6H_5 . CH_3$	92·06	·866/20°	−97	111
Toluidine (o.), $CH_3C_6H_4 . NH_2$ . .	107·1	·999/20°	liquid	197
,, (p.), $CH_3C_6H_4NH_2$ . . .	107·1	1·046/—	45	198
Trichloracetic acid, $CCl_3 . COOH$.	163·4	1·63/61°	52·3	195
Triethyl amine, $(C_2H_5)_3N$	101·2	·735/15°	liquid	89
,, arsine, $(C_2H_5)_3As$	162·1	1·15/17°	liquid	{140/736 dec.
,, phosphine, $(C_2H_5)_3P$ . .	118·1	·812/15°	liquid	127/744
Trimethyl amine, $(CH_3)_3N$	59·08	·673/0°	—	3·5
,, arsine, $(CH_3)_3As$	120·0	—	—	<100
,, bismuth, $(CH_3)_3Bi$ . . .	253·1	2·30/18°	—	110
,, carbinol, $(CH_3)_3C . OH$ . .	74·08	·786/20°	25	82·9
,, phosphine, $(CH_3)_3P$ . . .	76·07	>1	liquid	41
Trinitro benzene (s.), 1 : 3 : 5, C_6H_3-$(NO_2)_3$	213·1	—	121·2	decomp.
Turpentine (pinene), $C_{10}H_{16}$ . . .	136·1	·865/15°	—	159
Urea, $NH_2 . CO . NH_2$	60·11	*1·32	132	decomp.
Valeric acid (n.), $CH_3(CH_2)_3 . COOH$	102·1	·943/20°	−58·5	186·4
Xylene (o.), $C_6H_4(CH_3)_2$	106·1	·756/14°	−28	142
,, (m), ,, ...	106·1	·878/0°	−54	139·8
,, (p), ,, ...	106·1	·862/20°	15	138
Zinc ethyl, $Zn(C_2H_5)_2$	123·5	1·182/18°	−28	118
,, methyl, $Zn(CH_3)_2$ . . .	95·42	1·386/10°	−40	46

dec. or decomp. = decomposes.

ELECTROCHEMICAL EQUIVALENTS

Faraday's laws of electrolysis are expressed by $m = izt$, where m is the mass in grammes of an ion liberated in t secs. by a current of i amperes ; z is the electrochemical equivalent of the ion, *i.e.* the mass liberated by 1 ampere in 1 second.

The exactness of Faraday's laws is obscured in many cases by secondary chemical reactions, and the values of the different electrochemical equivalents are practically always derived by calculation from that of silver, which has been accurately determined (see p. 8). Electrochemical equivalents are proportional to chemical equivalents.

$$\text{Chemical equivalent} = \frac{\text{atomic weight of element}}{\text{valency of element for electrolyte used}}$$

Element.	Chemical equivalent.	z.
Silver	107·88/1 . . .	0·0011183 gm. sec.$^{-1}$ amp.$^{-1}$
Copper	63·57/2 . . .	0·0003295 ,, ,,
Hydrogen	1·008/1 . . .	0·00001045 ,, ,, (see p. 106)

SOLUBILITIES

SOLUBILITIES OF GASES IN WATER

AIR IN WATER

1000 c.cs. of water saturated with air at a pressure of 760 mms. contain the following volumes of dissolved oxygen, etc., in c.cs. at 0° and 760 mms.

		Temperature of Water.					
	0° C.	5°	10°	15°	20°	25°	30°
	c.cs.						
Oxygen	10·19	8·9	7·9	7·0	6·4	5·8	5·3
Nitrogen, argon, etc.	19·0	16·8	15·0	13·5	12·3	11·3	10·4
Sum of above	29·2	25.7	22·8	20·5	18·7	17·1	15·7
% of oxygen in dissolved air (by vol.)	34·9%	34·7	34·5	34·2	34·0	33·8	33·6

GASES IN WATER

S indicates the number of c.cs. of gas measured at 0° and 760 mms. which dissolve in 1 c.c. of water at the temperature stated, and when the pressure of the gas plus that of the water-vapour is 760 mms.

A indicates the same, except that the gas itself is at the uniform pressure of 760 mms. when in equilibrium with the water. (For other values, see p. 109.)

Gas.	0° C.	10°	15°	20°	30°	40°	50°	60°
	c cs.							
Ammonia, A	1300	910	802	710	595/28°	—	—	—
Argon, A	·058	·045	·040	·037	·030	·027	—	—
Carbon dioxide, A . .	1·713	1·194	1·019	·878	·66	·53	·44	·36
Carbon monoxide, A . .	·035	·028	·025	·023	·020	·018	·016	·015
Chlorine, S	—	3·09	2·63	2·26	1·77	1·41	1·20	1·0
Helium, A	·0150	·0144	·0139	·0138	·0138	·0139	·0140	—
Hydrogen, A . . . , .	·0215	·0198	·0190	·0184	—	—	—	—
Hydrochloric acid, S . .	506	474	458	442	411	386	362	339
Nitrogen, A	·0239	·0196	·0179	·0164	·0138	·0118	·0106	·0100
Nitrous oxide, A . . .	1·05/5°	·88	·74	·63				
Nitric oxide, A	·074	·057	·051	·047	·040	·035	·031	·029
Oxygen, A	·049	·038	·034	·031	·026	·023	·021	·019
Sulphuretted hydrogen, A	4·68	3·52	3·05	2·67				
Sulphur dioxide, S . .	79·8	56·6	47·3	39·4	27·2	18·8	—	—

Ne, ·0147/20° ; Kr, ·0670 — ·0788/20° ; Xe, ·1109/20° — Antropoff, 1910.

MUTUAL SOLUBILITIES OF LIQUIDS

The data for the uppermost layer of the two solutions in equilibrium are given in the first line in each case. The pressure in some cases exceeds one atmosphere. Numbers are grams per 100 grams of solution. (From data in Seidell's "Solubilities.")

Liquids.	0°C.	10°	20°	30°	40°	50°	60°	70°	80°	00
Water in ether; ethereal layer . . .	1·0	1·1	1·2	1·3	1·5	1·7	1·8	2·0	2·2	—
Ether in water; aqueous layer	12	8·7	6·5	5·1	4·5	4·1	3·7	3·2	2·8	—
Aniline(C₆H₅NH₂) in water; aqueous layer	—	—	3·2	—	3·5	—	3·8	—	4·5	6
Aniline in water; aniline layer	—	—	95·5	—	95	—	95	—	93	92
Phenol (C₆H₅OH) in water; aqueous layer	—	7·5	8·3	8·8	9·6	12	17	33·4	at crit.	
Phenol in water; phenol layer	—	75	72	70	67	63	55	33·4	temp. 68°·3	
Triethylamine in water; amine layer .	51·9	at	72	97	96	96	·6			
Triethylamine[N(C₂H₅)₃]in aqueous layer	51·9	18°·6	14·2	5·8	3·6	2·9	2·2			
CS₂ in methyl alcohol; alcoholic layer .	—	·45	51	58	80·5	at crit. temp.				
CS₂ in CH₃OH; carbon bisulphide layer	—	·98	97	96	80·5	40°·5				

SOLUBILITIES OF SOLIDS IN WATER

s = number of grams of **anhydrous** substance which when dissolved in 100 grams of **water** make a saturated solution at the temperature stated.

p = no. of grams of anhydrous substance per 100 grams of saturated **solution**.

The formula given is that of the solid phase which is in equilibrium with the solution. (See Seidell's "Solubilities," New York, 1907, where the most complete and accurate data will be found for solubilities.) For other solutions, see p. 109.

Substance.		0° C.	10°	15°	20°	40°	60°	80°	100°
Am. chloride, NH_4Cl	s	29·4	33·3	35·2	37·2	45·8	55·2	65·6	77·3
Barium chloride, $BaCl_2.2H_2O$ . .	s	31·6	33·3	34·4	35·7	40·7	46·4	52·4	58·8
Barium hydrate, $Ba(OH)_2.8H_2O$.	s	1·67	2·48	3·23	3·89	8·22	20·9	101·4	——
Bromine (*liquid*), Br.	s	4·22	3·4	3·25	3·20	——	——	——	——
Cadmium sulphate, $CdSO_4.8/3H_2O$.	s	76·5	76·0	76·3	76·6	78·5	83·7	69·7 *	60·77*
Ca. hydrate, $Ca(OH)_2$	s	·185	·176	·170	·165	·141	·116	·094	·077
Copper sulphate, $CuSO_4.5H_2O$ . .	s	14·3	17·4	18·8	20·7	28·5	40·0	55·0	75·0
Li. carbonate, Li_2CO_3	s	1·54	1·43	1·38	1·33	1·17	1·01	·850	·720
Merc. chloride, $HgCl_2$	p	3·50	4·50	5·00	5·40	9·30	14·0	23·1	38·0
Potass. chloride, KCl	s	27·6	31·0	32·4	34·0	40·0	45·5	51·1	56·7
Potass. bromide, KBr	s	53·5	59·5	62·5	65·2	75·5	85·5	95·0	104
Potassium iodide, KI	s	127·5	136	140	144	160	176	192	208
Potassium hydrate, $KOH.2H_2O$ . .	s	97·0	103	107	112	138 §	——	——	178 §
Potass. nitrate, KNO_3	s	13·3	20·9	25·8	32	64	110	169	246
Silv. nitrate, $AgNO_3$	s	122	170	196	222	376	525	669	952
Sodium carbonate, $Na_2CO_3.10H_2O$.	s	7·0	12·5	16·4	21·5	46·1	46·0	45·8	45·5
Sod. chloride, NaCl	s	35·7	35·8	35·9	36·0	36·6	37	38	39·0
Sodium sulphate, $Na_2SO_4.10H_2O$.	s	5·0	9·0	13·4	19·4	49 †	45 †	44 †	42 †
Strontium chloride, $SrCl_2.6H_2O$ . .	s	43	48	50	53	65	82	91 ‡	101 ‡
Succinic acid, $(CH_2)_2(COOH)_2$.	s	2·80	4·50	5·7	6·9	16·2	35·8	70·8	125
Sugar (Cane), $C_{12}H_{22}O_{11}$ . . .	s	179	190	197	204	238	287	362	487

* Solid phase becomes $CdSO_4 . H_2O$ at 74°. † Becomes Na_2SO_4 at 32°·38.
‡ Becomes $SrCl_2 . 2H_2O$ at 70°. § Becomes $KOH.\frac{3}{2}H_2O$ at 32°·5 and $KOH . H_2O$ at 50°.
‖ Becomes $Na_2CO_3 . H_2O$ at 35°.

COMPOSITION OF DRY ATMOSPHERIC AIR

(Ramsay, *Proc. Roy. Soc.*, 1908 ; G. Claude, *Compt. Rend.*, 1909.)

	N_2	O_2	A	CO_2	Kr	Xe	Ne	He
By weight .	75·5	23·2	1·3	·046 to ·4	·028	·005	·0₃86	·0₄56
By volume .	78·05	21·0 *	·95	·03 to ·3	——	——	·0₂123	·0₄40

* 20·91 according to Kreusler.

MOHS' SCALE OF MINERAL HARDNESS
The numbers are not quantitative, but merely indicate the sequence of hardness.

Hardness.	Mineral.	Hardness.	Mineral.	Hardness.	Mineral.
1	Talc	**5**	Apatite	**9**	Corundum
2	Rock salt	**6**	Felspar	**10**	Diamond
3	Calcspar	**7**	Quartz	*c.* **2·5**	Finger-nail
4	Fluor spar	**8**	Topaz	*c.* **6·5**	Penknife

COMPOSITION, DENSITY, AND HARDNESS OF SOME MINERALS
See Dana's "System of Mineralogy" and Appendices, 1892, 1899, and 1909.
Radioactive minerals are indicated thus * ; see Szilard, *Le Radium*, August, 1909.

Name and Formula.	Density.	Hardness.	Name and Formula.	Density.	Hardness.
Albite, $Na_2Al_2Si_6O_{16}$	*c.* 2·6	6–7	Mica (common, Muscovite), $K_2O.3Al_2O_3.6SiO_2.2H_2O$	2·7–3·1	2–2·5
Amber (fossil resin)	1·08	2–2·5			
Anhydrite, $CaSO_4$	2·8–2·9	3–3·5			
Anorthite, $Ca_2Al_4Si_4O_{16}$	*c.* 2·7	6–7	Mica (Biotite, Magnesia mica)	2·7–3·1	2·5–3
Apatite, $Ca_5(Cl,F,OH)(PO_4)_3$	2·9–3·2	5	Monazite,* $(CeLaDi)PO_4$ (1–16% Th)	5	5·2
Aragonite, $CaCO_3$	2·93	3·5–4	Nepheline, $Na_6K_6Al_8Si_9O_{36}$	2·5–2·6	5·5–6
Augite, Mg,Fe,Ca,Al silicate	3·2–3·5	5–6	Olivine, $Mg_2Fe_2SiO_4$	3·3–3·5	6–7
Barytes, Heavy spar, $BaSO_4$	4·5	3–3·5	Orthoclase, $K_2Al_2Si_6O_{16}$	2·4–2·6	6
Beryl, $Be_3Al_2Si_6O_{18}$	2·6–2·7	7–8	Pitchblende,* U_3O_8 with oxides of Pb, and Ca, Fe,Bi,Mn,Mg,Cu,Si, Al, etc. (25–80 % U ; 1–6 % Th)	6·4 (massive) 9·7 (cryst.)	5·5
Bröggerite,* a pitchblende which contains thorium	(56–68% U)	(2–8% Th)			
Calcite, Calcspar, Iceland spar, $CaCO_3$	2·6–2·7	*c.* 3	Pyrites (iron), FeS_2	4·8–5·1	6–6·5
Carnallite, $KCl.MgCl_2.6H_2O$	1·6	1	„ (copper), $CuFeS_2$	4·1–4·3	3·5–4
Carnotite,* $K_2O(U_2O_6)_3V_2O_5.3H_2O$	(*c.* 55% U)	(yellow)	Pyrolusite, MnO_2	4·8–5	2–5·5
			Quartz, SiO_2	2·5–2·8	7
Celestine, $SrSO_4$	3·9	3–3·5	Rock salt, $NaCl$	2·1–2·2	2–2·5
Cerussite, $PbCO_3$	6·4	3–3·5	Rutile, TiO_2	4·2–4·3	6–6·5
Chalcolite,* $Cu(UO_2)_2(PO_4)_2.8H_2O$;	3·4–3·6 (48% U)	2–2·5	Selenite—cryst. gypsum	—	
			Serpentine, $H_4Mg_3Si_2O_9$	*c.* 2·6	3–4
Cléveite *—pitchblende which contains Th & Y	(*c.* 60% U)	(*c.* 4% Th)	Spinel, $MgOAl_2O_3$	3·5–3·6	8
			Sylvine, KCl	1·9–2	2
Corundum, Al_2O_3	3·9–4·2	9	Talc, $H_2Mg_3Si_4O_{12}$	2·5–2·8	1
Dolomite, $CaMgC_2O_6$	2·8–2·9	3·5–4	Thorianite,* Th, U oxides, etc. ; (4–10% U ; *c.* 60% Th) contains He	8–9·7	7 (black cubes)
Felspar, $Al_2K_2Si_6O_{16}$	2·4–2·6	6			
Flint ; agate, SiO_2	2·6	*c.* 6	Thorite,* $ThSiO_4$ (1–9% U ; 40–60% Th)	4·6	(tetragonal)
Fluorspar, Fluorite, CaF_2	3–3·3	4			
Galena, PbS	7·4–7·6	2–3	Tourmaline, hydrated silicate and borate of Al, Na with Li or Fe or Mg	2·9–3·3	7–7·5
Gummite,* Pb,Ca,U, silicate	(50–65% U)				
Gypsum, $CaSO_4.2H_2O$	2·3	1·5–2	Trögerite,* $(UO_2)_3As_2O_8 12H_2O$	(53% U)	(yellow low)
Hæmatite, Fe_2O_3	4·5–5·3	5·5–6·5			
Hornblende, Ca,Mg,Fe,Na,Al, silicate	2·9–3·4	5–6	Uraninite *— crystalline pitchblende (*q.v.*)	(Black	octahedra)
Kainite, $MgSO_4KCl 3H_2O$	2·1	—	Uranite lime,* $CaO(UO_2)_2(PO_4)_2 8H_2O$ (50% U)	3–3·2	2–2·5
Kaolin, $H_4Al_2Si_2O_9$	2·5	1			
Kieserite, $MgSO_4H_2O$	2·55	3	Willemite, Zn_2SiO_4	4	5
Lepidolite (Lithia mica), $(F,OH)_2(Li,K,Na)_2Al_2Si_3O_9$	2·8–3	2·5–4	Wolfram, $(Fe,Mn)WO_4$	7·1–7·9	5–5·5
			Wollastonite, $CaSiO_3$	2·7–2·9	4·5–5
Limestone, $CaCO_3$	2·5–2·8	—	Zeunerite,* Cu,U arsenate	(*c.* 50% U)	(tetragonal)
Magnesite, $MgCO_3$	*c.* 3	3·5–4·5			
Magnetite, Fe_3O_4	4·9–5·2	5·5–6·5	Zircon,* $ZrSiO_4$	4·7	7·5
Meerschaum, $2MgO.3SiO_2.2H_2O$	*c.* 2·6	2–2·5	Zincblende, ZnS	3·9–4·2	3·5–4

FACTORS FOR GRAVIMETRIC ANALYSIS

Calculated with atomic weights for 1911 (p. 1).

Example.—1 gram Al_2O_3 is chemically equivalent to ·5303 gram Al, or 1 gram Al is equivalent to 1/·5303 Al_2O_3. A table of reciprocals is given on p. 136.

(See Van Nostrand's "Chemical Annual," London.)

1 part by weight of	is equivalent (by weight) to	1 part by weight of	is equivalent (by weight) to
Aluminium.		**Calcium** (*contd.*)—	
Al_2O_3	·5303 Al	$Ca_3(PO_4)_2$ . . .	·5422 CaO
"	3·350 $Al_2(SO_4)_3$	$Mg_2P_2O_7$ . . .	1·3935 $Ca_3(PO_4)_2$
Ammonium.		P_2O_5	2·1844 $Ca_3(PO_4)_2$
N	1·216 NH_3	**Carbon.**	
"	1·288 NH_4	CO_2	4·4860 $BaCO_3$
"	3·819 NH_4Cl	"	2·2748 $CaCO_3$
NH_3	2·058 NH_4OH	**Chlorine.**	
Antimony.		AgCl	·2474 Cl
Sb	1·1997 Sb_2O_3	NaCl	·6066 Cl
"	1·3328 Sb_2O_5	**Chromium.**	
Sb_2O_3	1·1109 Sb_2O_5	Cr_2O_3	·6846 Cr
Sb_2O_4	·7897 Sb	"	1·3154 CrO_3
"	·9474 Sb_2O_3	**Cobalt.**	
"	1·0526 Sb_2O_5	Co	1·2713 CoO
Arsenic.		Co_3O_4	·7343 Co
As_2O_3	·7575 As	"	·9336 CoO
"	1·1617 As_2O_5	$Co(NO_2)_3.(KNO_2)_3$	·1306 Co
As_2O_5	·6521 As	"	·1661 CoO
$MgNH_4AsO_4.\frac{1}{2}H_2O$	·3938 As	$(CoSO_4)_2.(K_2SO_4)_3$	·1416 Co
" "	·5199 As_2O_3	**Copper.**	
" "	·6040 As_2O_5	Cu	1·2517 CuO
$Mg_2As_2O_7$ . . .	·4827 As	**Fluorine.**	
"	·6373 As_2O_3	CaF_2	·4866 F
"	·7403 As_2O_5	**Glucinum.** *See*	
Barium.		Beryllium.	
$BaCO_3$	·6960 Ba	**Gold.**	
"	·7771 BaO	Au	1·5395 $AuCl_3$
$BaSO_4$	·5885 Ba	**Hydrogen.**	
"	·6570 BaO	H_2O	·1119 H
"	·7255 BaO_2	**Iodine.**	
Beryllium.		AgI	·5405 I
BeO	·3626 Be	**Iron.**	
Bismuth.		Fe	1·2865 FeO
Bi	1·1154 Bi_2O_3	"	1·4297 Fe_2O_3
Bi_2O_3	·8966 Bi	"	7·0218 $FeSO_4.$
BiOCl	·8017 Bi		$(NH_4)_2SO_4.6H_2O$
"	·8942 Bi_2O_3	FeO	·7773 Fe
Boron.		"	1·1113 Fe_2O_3
B_2O_3	·3143 B	Fe_2O_3	1·4508 $FeCO_3$
"	2·7297 $Na_3B_4O_7.$	"	·9666 Fe_3O_4
	$10H_2O$	CO_2	1·6330 FeO
Bromine.		"	2·6330 $FeCO_3$
AgBr	·4256 Br	**Lead.**	
Cadmium.		Pb	1·0773 PbO
CdO	·8754 Cd	$PbSO_4$	·6831 Pb
Cæsium.		"	·7358 PbO
Cs	1·060 Cs_2O	"	·7887 PbO_2
Cs_2PtCl_6 . . .	·3945 Cs	"	·7536 Pb_3O_4
"	·4184 Cs_2O	**Lithium.**	
Calcium.		Li_2CO_3	·1879 Li
Ca	1·399 CaO	"	·4044 Li_2O
$CaCO_3$	·4005 Ca	Li_3PO_4	·1797 Li
"	·5604 CaO	"	·3868 Li_2O
CO_2	2·275 $CaCO_3$		

FACTORS FOR GRAVIMETRIC ANALYSIS (*contd.*)

1 part by weight of	is equivalent (by weight) to	1 part by weight of	is equivalent (by weight) to
Magnesium.		**Potassium** (*contd.*)	
MgO	·6032 Mg	K_2SO_4	1·1604 KNO_3
$Mg_2P_2O_7$.	·2184 Mg	K_2PtCl_6 . . .	·1609 K
„	·3621 MgO	**Rubidium.**	
Manganese.		Rb_2PtCl_6 . . .	·2953 Rb
MnO	1·1113 Mn_2O_3	**Silicon.**	
Mn_3O_4 . . .	·7203 Mn	SiO_2	·4693 Si
„	·9307 MnO	**Silver.**	
„	1·0350 Mn_2O_3	AgCl	·7526 Ag
„	1·1399 MnO_2	AgBr	·5744 Ag
Mercury.		AgI	·4595 Ag
Hg	1·1603 HgS	**Sodium.**	
HgS	·8963 Hg_2O	AgCl	·4078 NaCl
„	·9308 HgO	$NaHCO_3$ . . .	·3691 Na_2O
Nickel.		Na_2SO_4 . . .	·3238 Na
Ni	1·2727 NiO	„	·4364 Na_2O
Nitrogen.		N_2O_5	1·5740 $NaNO_3$
N	3·8551 N_2O_5	**Strontium.**	
Phosphorus.		$SrCO_3$	·7019 SrO
P_2O_5	·4362 P	$SrSO_4$ . . .	·5641 SrO
$Mg_2P_2O_7$ . . .	·2787 P	**Sulphur.**	
„	·8534 PO_4	$BaSO_4$ . . .	·1460 H_2S
„	·6378 P_2O_5	„	·1374 S
Platinum.		„	·2744 SO_2
K_2PtCl_6 . . .	·4015 Pt	„	·3429 SO_3
„	·6933 $PtCl_4$	„	·4115 SO_4
Potassium.		**Tin.**	
AgCl	·5202 KCl	SnO_2	·7881 Sn
AgBr	·6338 KBr	**Uranium.**	
AgI	·7071 KI	U_3O_8	·8482 U
AgCN	·4863 KCN	„	·9620 UO_2
KCl	·5244 K	UO_2	·8817 U
KBr	·3285 K	**Zinc.**	
KOH	1·2316 K_2CO_3	Zn	1·2448 ZnO
„	·8395 K_2O	ZnO	·8033 Zn
K_2SO_4 . . .	·5403 K_2O		

SOME BOILING-POINT MIXTURES

Boiling-points under 760 mms. of mercury. Percentage compositions by weight. A large number of minimum boiling-point mixtures are known.
(Sidney Young, " Fractional Distillation," 1903.)

	Mixture.		Boiling Points.			% of A in mixt.	Ob-server.
	A.	B.	A.	B.	Mixt.		
Maximum boiling-point mixtures.	Water	Nitric acid	100° C.	86°	125°	32%	Roscoe
	„	Hydrochloric acid	100	*c.* −80	110	80	„
	„	Formic acid	100	100·8	107	23	„
	Me. ether	Hydrochloric acid	−23·6	*c.* −80	−2	61	Friedel
Minimum boiling-point mixtures.	Water	Ethyl alcohol	100	78·3	78·1	4·4	Y. & F.
	Pyridine	Water	117	100	92·5	59	G. & C.
	Benzene	Methyl alcohol	80·2	64·7	58·3	60	Y. & F.
	Me.alcohol	Acetone	64·7	−56·5	55·9	13·5	Pettit

G. & C., Goldschmidt and Constan ; Y. & F., Young and Fortey.

e^{-x}

THE EXPONENTIAL e^{-x}

$e = 2·71828$.   To derive e^x use reciprocals on p. 136. $e^{-·69315} = ·5$.

(Based on Newman, *Trans. Camb. Phil. Soc.*, **13**, 1883.)

For values of x from ·0000 to ·0999.

x	0	·001	·002	·003	·004	·005	·006	·007	·008	·009	·0001	2	3	4	5	6	7	8	9
·00	1·000	9990	9980	9970	9960	9950	9940	9930	9920	9910	1	2	3	4	5	6	7	8	9
·01	9900	9891	9881	9871	9861	9851	9841	9831	9822	9812	1	2	3	4	5	6	7	8	9
·02	9802	9792	9782	9773	9763	9753	9743	9734	9724	9714	1	2	3	4	5	6	7	8	9
·03	9704	9695	9685	9675	9666	9656	9646	9637	9627	9618	1	2	3	4	5	6	7	8	9
·04	9608	9598	9589	9579	9570	9560	9550	9541	9531	9522	1	2	3	4	5	6	7	8—9	
·05	9512	9502	9493	9484	9474	9465	9455	9446	9436	9427	1	2	3	4	5	6	7	8	9
·06	9418	9408	9399	9389	9380	9371	9361	9352	9343	9333	1	2	·3	4	5	6	7	8	9
·07	9324	9315	9305	9296	9287	9277	9268	9259	9250	9240	1	2	3	4	5	6	7	8	8
·08	9231	9222	9213	9204	9194	9185	9176	9167	9158	9148	1	2	3	4	5	6	7	7·8	
·09	9139	9130	9121	9112	9103	9094	9085	9076	9066	9057	1	2	3	4	5	6	6	7	8

For values of x from ·100 to 2·999.

x	0	·01	·02	·03	·04	·05	·06	·07	·08	·09	·001	2	3	4	5	6	7	8	9
·1	9048	8958	8869	8781	8694	8607	8521	8437	8353	8270	9	17	26	34	43	52	60	69	77
·2	8187	8106	8025	7945	7866	7788	7711	7634	7558	7483	8	16	23	31	39	47	55	62	70
·3	7408	7334	7261	7189	7118	7047	6977	6907	6839	6771	7	14	21	28	35	42	49	56	63
·4	6703	6637	6570	6505	6440	6376	6313	6250	6188	6126	6	13	19	26	32	38	45	51	57
·5	6065	6005	5945	5886	5827	5769	5712	5655	5599	5543	6	12	17	23	29	35	40	46	52
·6	5488	5434	5379	5326	5273	5220	5169	5117	5066	5016	5	10	16	21	26	31	37	42	47
·7	4966	4916	4868	4819	4771	4724	4677	4630	4584	4538	5	9	14	19	24	28	33	38	43
·8	4493	4449	4404	4360	4317	4274	4232	4190	4148	4107	4	9	13	17	21	26	30	34	38
·9	4066	4025	3985	3946	3906	3867	3829	3791	3753	3716	4	8	12	15	19	23	27	31	35
1·0	3679	3642	3606	3570	3535	3499	3465	3430	3396	3362	4	7	11	14	18	21	25	28	32
1·1	3329	3296	3263	3230	3198	3166	3135	3104	3073	3042	3	6	9	13	16	19	22	25	29
1·2	3012	2982	2952	2923	2894	2865	2837	2808	2780	2753	3	6	9	11	14	17	20	23	26
1·3	2725	2698	2671	2645	2618	2592	2567	2541	2516	2491	3	5	8	10	13	16	18	21	23
1·4	2466	2441	2417	2393	2369	2346	2322	2299	2276	2254	2	5	7	9	12	14	16	19	21
1·5	2231	2209	2187	2165	2144	2122	2101	2080	2060	2039	2	4	6	8	11	13	15	17	19
1·6	2019	1999	1979	1959	1940	1920	1901	1882	1864	1845	2	4	6	8	10	12	13	15	17
1·7	1827	1809	1791	1773	1755	1738	1720	1703	1686	1670	2	3	5	7	9	10	12	14	16
1·8	1653	1637	1620	1604	1588	1572	1557	1541	1526	1511	2	3	5	6	8	9	11	13	14
1·9	1496	1481	1466	1451	1437	1423	1409	1395	1381	1367	1	3	4	6	7	9	10	11	13
2·0	1353	1340	1327	1313	1300	1287	1275	1262	1249	1237	1	3	4	5	6	8	9	10	12
2·1	1225	1212	1200	1188	1177	1165	1153	1142	1130	1119	1	2	4	5	6	7	8	9	11
2·2	1108	1097	1086	1075	1065	1054	1044	1033	1023	1013	1	2	3	4	5	6	7	8	9
2·3	1003	0993	0983	0973	0963	0954	0944	0935	0926	0916	1	2	3	4	5	6	7	8	9
2·4	0907	0898	0889	0880	0872	0863	0854	0846	0837	0829	1	2	3	3	4	5	6	7	8
2·5	0821	0813	0805	0797	0789	0781	0773	0765	0758	0750	1	2	2	3	4	5	5	6	7
2·6	0743	0735	0728	0721	0714	0707	0699	0693	0686	0679	1	1	2	3	4	4	5	6	6
2·7	0672	0665	0659	0652	0646	0639	0633	0627	0620	0614	1	1	2	3	3	4	4	5	6
2·8	0608	0602	0596	0590	0584	0578	0573	0567	0561	0556	1	1	2	2	3	3	4	5	5
2·9	0550	0545	0539	0534	0529	0523	0518	0513	0508	0503	1	1	2	2	3	3	4	4	5

For values of x from 3·0 to 8·9.

x	0	·1	·2	·3	·4	·5	·6	·7	·8	·9
3	0498	0450	0408	0368	0334	0302	0273	0247	0224	0202
4	0183	0166	0150	0136	0123	0111	0101	0091	0082	0074
5	0067	0061	0055	0050	0045	0041	0037	0033	0030	0027
6	0025	0022	0020	0018	0017	0015	0014	0012	0011	0010
7	0009	0008	0007	0007	0006	0006	0005	0005	0004	0004
8	0003	0003	0003	0002	0002	0002	0002	0002	0002	0001

Subtract Differences.

Mean differences no longer sufficiently accurate.

K

FOUR-FIGURE LOGARITHMS

	0	1	2	3	4	5	6	7	8	9	1	2	3	4	5	6	7	8	9
10	0000	0043	0086	0128	0170	0212	0253	0294	0334	0374	4	9	13	17	21	25	30	34	38
											4	8	12	16	20	24	28	32	36
11	0414	0453	0492	0531	0569	0607	0645	0682	0719	0755	4	8	12	15	19	23	27	31	35
											4	7	11	15	18	22	26	30	33
12	0792	0828	0864	0899	0934	0969	1004	1038	1072	1106	4	7	11	14	18	21	25	28	32
											3	7	10	14	17	20	24	27	31
13	1139	1173	1206	1239	1271	1303	1335	1367	1399	1430	3	7	10	13	16	20	23	26	30
											3	6	9	13	16	19	22	25	28
14	1461	1492	1523	1553	1584	1614	1644	1673	1703	1732	3	6	9	12	15	18	21	24	27
											3	6	9	12	15	18	21	24	27
15	1761	1790	1818	1847	1875	1903	1931	1959	1987	2014	3	6	9	11	14	17	20	23	26
											3	6	8	11	14	17	19	22	25
16	2041	2068	2095	2122	2148	2175	2201	2227	2253	2279	3	5	8	11	13	16	19	21	24
											3	5	8	10	13	16	18	21	23
17	2304	2330	2355	2380	2405	2430	2455	2480	2504	2529	3	5	8	10	13	15	18	20	23
											2	5	7	10	12	15	17	20	22
18	2553	2577	2601	2625	2648	2672	2695	2718	2742	2765	2	5	7	10	12	14	17	19	21
											2	5	7	9	12	14	16	19	21
19	2788	2810	2833	2856	2878	2900	2923	2945	2967	2989	2	5	7	9	11	14	16	18	20
											2	4	7	9	11	13	15	18	20
20	3010	3032	3054	3075	3096	3118	3139	3160	3181	3201	2	4	6	8	11	13	15	17	19
21	3222	3243	3263	3284	3304	3324	3345	3365	3385	3404	2	4	6	8	10	12	14	16	18
22	3424	3444	3464	3483	3502	3522	3541	3560	3579	3598	2	4	6	8	10	12	14	15	17
23	3617	3636	3655	3674	3692	3711	3729	3747	3766	3784	2	4	6	7	9	11	13	15	17
24	3802	3820	3838	3856	3874	3892	3909	3927	3945	3962	2	4	5	7	9	11	12	14	16
25	3979	3997	4014	4031	4048	4065	4082	4099	4116	4133	2	3	5	7	9	10	12	14	15
26	4150	4166	4183	4200	4216	4232	4249	4265	4281	4298	2	3	5	7	8	10	11	13	15
27	4314	4330	4346	4362	4378	4393	4409	4425	4440	4456	2	3	5	6	8	9	11	13	14
28	4472	4487	4502	4518	4533	4548	4564	4579	4594	4609	2	3	5	6	8	9	11	12	14
29	4624	4639	4654	4669	4683	4698	4713	4728	4742	4757	1	3	4	6	7	9	10	12	13
30	4771	4786	4800	4814	4829	4843	4857	4871	4886	4900	1	3	4	6	7	9	10	11	13
31	4914	4928	4942	4955	4969	4983	4997	5011	5024	5038	1	3	4	6	7	8	10	11	12
32	5051	5065	5079	5092	5105	5119	5132	5145	5159	5172	1	3	4	5	7	8	9	11	12
33	5185	5198	5211	5224	5237	5250	5263	5276	5289	5302	1	3	4	5	6	8	9	10	12
34	5315	5328	5340	5353	5366	5378	5391	5403	5416	5428	1	3	4	5	6	8	9	10	11
35	5441	5453	5465	5478	5490	5502	5514	5527	5539	5551	1	2	4	5	6	7	9	10	11
36	5563	5575	5587	5599	5611	5623	5635	5647	5658	5670	1	2	4	5	6	7	8	10	11
37	5682	5694	5705	5717	5729	5740	5752	5763	5775	5786	1	2	3	5	6	7	8	9	10
38	5798	5809	5821	5832	5843	5855	5866	5877	5888	5899	1	2	3	5	6	7	8	9	10
39	5911	5922	5933	5944	5955	5966	5977	5988	5999	6010	1	2	3	4	5	7	8	9	10
40	6021	6031	6042	6053	6064	6075	6085	6096	6107	6117	1	2	3	4	5	6	8	9	10
41	6128	6138	6149	6160	6170	6180	6191	6201	6212	6222	1	2	3	4	5	6	7	8	9
42	6232	6243	6253	6263	6274	6284	6294	6304	6314	6325	1	2	3	4	5	6	7	8	9
43	6335	6345	6355	6365	6375	6385	6395	6405	6415	6425	1	2	3	4	5	6	7	8	9
44	6435	6444	6454	6464	6474	6484	6493	6503	6513	6522	1	2	3	4	5	6	7	8	9
45	6532	6542	6551	6561	6571	6580	6590	6599	6609	6618	1	2	3	4	5	6	7	8	9
46	6628	6637	6646	6656	6665	6675	6684	6693	6702	6712	1	2	3	4	5	6	7	7	8
47	6721	6730	6739	6749	6758	6767	6776	6785	6794	6803	1	2	3	4	5	5	6	7	8
48	6814	6821	6830	6839	6848	6857	6866	6875	6884	6893	1	2	3	4	4	5	6	7	8
49	6902	6911	6920	6928	6937	6946	6955	6964	6972	6981	1	2	3	4	4	5	6	7	8
	0	1	2	3	4	5	6	7	8	9	1	2	3	4	5	6	7	8	9

FOUR-FIGURE LOGARITHMS

	0	1	2	3	4	5	6	7	8	9	1	2	3	4	5	6	7	8	9
50	6990	6998	7007	7016	7024	7033	7042	7050	7059	7067	1	2	3	3	4	5	6	7	8
51	7076	7084	7093	7101	7110	7118	7126	7135	7143	7152	1	2	3	3	4	5	6	7	8
52	7160	7168	7177	7185	7193	7202	7210	7218	7226	7235	1	2	3	3	4	5	6	7	7
53	7243	7251	7259	7267	7275	7284	7292	7300	7308	7316	1	2	2	3	4	5	6	6	7
54	7324	7332	7340	7348	7356	7364	7372	7380	7388	7396	1	2	2	3	4	5	6	6	7
55	7404	7412	7419	7427	7435	7443	7451	7459	7466	7474	1	2	2	3	4	5	5	6	7
56	7482	7490	7497	7505	7513	7520	7528	7536	7543	7551	1	2	2	3	4	5	5	6	7
57	7559	7566	7574	7582	7589	7597	7604	7612	7619	7627	1	2	2	3	4	5	5	6	7
58	7634	7642	7649	7657	7664	7672	7679	7686	7694	7701	1	1	2	3	4	4	5	6	7
59	7709	7716	7723	7731	7738	7745	7752	7760	7767	7774	1	1	2	3	4	4	5	6	7
60	7782	7789	7796	7803	7810	7818	7825	7832	7839	7846	1	1	2	3	4	4	5	6	6
61	7853	7860	7868	7875	7882	7889	7896	7903	7910	7917	1	1	2	3	4	4	5	6	6
62	7924	7931	7938	7945	7952	7959	7966	7973	7980	7987	1	1	2	3	3	4	5	6	6
63	7993	8000	8007	8014	8021	8028	8035	8041	8048	8055	1	1	2	3	3	4	5	5	6
64	8062	8069	8075	8082	8089	8096	8102	8109	8116	8122	1	1	2	3	3	4	5	5	6
65	8129	8136	8142	8149	8156	8162	8169	8176	8182	8189	1	1	2	3	3	4	5	5	6
66	8195	8202	8209	8215	8222	8228	8235	8241	8248	8254	1	1	2	3	3	4	5	5	6
67	8261	8267	8274	8280	8287	8293	8299	8306	8312	8319	1	1	2	3	3	4	5	5	6
68	8325	8331	8338	8344	8351	8357	8363	8370	8376	8382	1	1	2	3	3	4	4	5	6
69	8388	8395	8401	8407	8414	8420	8426	8432	8439	8445	1	1	2	2	3	4	4	5	6
70	8451	8457	8463	8470	8476	8482	8488	8494	8500	8506	1	1	2	2	3	4	4	5	6
71	8513	8519	8525	8531	8537	8543	8549	8555	8561	8567	1	1	2	2	3	4	4	5	5
72	8573	8579	8585	8591	8597	8603	8609	8615	8621	8627	1	1	2	2	3	4	4	5	5
73	8633	8639	8645	8651	8657	8663	8669	8675	8681	8686	1	1	2	2	3	4	4	5	5
74	8692	8698	8704	8710	8716	8722	8727	8733	8739	8745	1	1	2	2	3	4	4	5	5
75	8751	8756	8762	8768	8774	8779	8785	8791	8797	8802	1	1	2	2	3	3	4	5	5
76	8808	8814	8820	8825	8831	8837	8842	8848	8854	8859	1	1	2	2	3	3	4	5	5
77	8865	8871	8876	8882	8887	8893	8899	8904	8910	8915	1	1	2	2	3	3	4	4	5
78	8921	8927	8932	8938	8943	8949	8954	8960	8965	8971	1	1	2	2	3	3	4	4	5
79	8976	8982	8987	8993	8998	9004	9009	9015	9020	9025	1	1	2	2	3	3	4	4	5
80	9031	9036	9042	9047	9053	9058	9063	9069	9074	9079	1	1	2	2	3	3	4	4	5
81	9085	9090	9096	9101	9106	9112	9117	9122	9128	9133	1	1	2	2	3	3	4	4	5
82	9138	9143	9149	9154	9159	9165	9170	9175	9180	9186	1	1	2	2	3	3	4	4	5
83	9191	9196	9201	9206	9212	9217	9222	9227	9232	9238	1	1	2	2	3	3	4	4	5
84	9243	9248	9253	9258	9263	9269	9274	9279	9284	9289	1	1	2	2	3	3	4	4	5
85	9294	9299	9304	9309	9315	9320	9325	9330	9335	9340	1	1	2	2	3	3	4	4	5
86	9345	9350	9355	9360	9365	9370	9375	9380	9385	9390	1	1	2	2	3	3	4	4	5
87	9395	9400	9405	9410	9415	9420	9425	9430	9435	9440	0	1	1	2	2	3	3	4	4
88	9445	9450	9455	9460	9465	9469	9474	9479	9484	9489	0	1	1	2	2	3	3	4	4
89	9494	9499	9504	9509	9513	9518	9523	9528	9533	9538	0	1	1	2	2	3	3	4	4
90	9542	9547	9552	9557	9562	9566	9571	9576	9581	9586	0	1	1	2	2	3	3	4	4
91	9590	9595	9600	9605	9609	9614	9619	9624	9628	9633	0	1	1	2	2	3	3	4	4
92	9638	9643	9647	9652	9657	9661	9666	9671	9675	9680	0	1	1	2	2	3	3	4	4
93	9685	9689	9694	9699	9703	9708	9713	9717	9722	9727	0	1	1	2	2	3	3	4	4
94	9731	9736	9741	9745	9750	9754	9759	9763	9768	9773	0	1	1	2	2	3	3	4	4
95	9777	9782	9786	9791	9795	9800	9805	9809	9814	9818	0	1	1	2	2	3	3	4	4
96	9823	9827	9832	9836	9841	9845	9850	9854	9859	9863	0	1	1	2	2	3	3	4	4
97	9868	9872	9877	9881	9886	9890	9894	9899	9903	9908	0	1	1	2	2	3	3	4	4
98	9912	9917	9921	9926	9930	9934	9939	9943	9948	9952	0	1	1	2	2	3	3	4	4
99	9956	9961	9965	9969	9974	9978	9983	9987	9991	9996	0	1	1	2	2	3	3	3	4
	0	1	2	3	4	5	6	7	8	9	1	2	3	4	5	6	7	8	9

ANTILOGARITHMS

	0	1	2	3	4	5	6	7	8	9	1	2	3	4	5	6	7	8	9
·00	1000	1002	1005	1007	1009	1012	1014	1016	1019	1021	0	0	1	1	1	1	2	2	2
·01	1023	1026	1028	1030	1033	1035	1038	1040	1042	1045	0	0	1	1	1	1	2	2	2
·02	1047	1050	1052	1054	1057	1059	1062	1064	1067	1069	0	0	1	1	1	1	2	2	2
·03	1072	1074	1076	1079	1081	1084	1086	1089	1091	1094	0	0	1	1	1	1	2	2	2
·04	1096	1099	1102	1104	1107	1109	1112	1114	1117	1119	0	1	1	1	1	2	2	2	2
·05	1122	1125	1127	1130	1132	1135	1138	1140	1143	1146	0	1	1	1	1	2	2	2	2
·06	1148	1151	1153	1156	1159	1161	1164	1167	1169	1172	0	1	1	1	1	2	2	2	2
·07	1175	1178	1180	1183	1186	1189	1191	1194	1197	1199	0	1	1	1	1	2	2	2	2
·08	1202	1205	1208	1211	1213	1216	1219	1222	1225	1227	0	1	1	1	1	2	2	2	3
·09	1230	1233	1236	1239	1242	1245	1247	1250	1253	1256	0	1	1	1	1	2	2	2	3
·10	1259	1262	1265	1268	1271	1274	1276	1279	1282	1285	0	1	1	1	1	2	2	2	3
·11	1288	1291	1294	1297	1300	1303	1306	1309	1312	1315	0	1	1	1	2	2	2	2	3
·12	1318	1321	1324	1327	1330	1334	1337	1340	1343	1346	0	1	1	1	2	2	2	2	3
·13	1349	1352	1355	1358	1361	1365	1368	1371	1374	1377	0	1	1	1	2	2	2	3	3
·14	1380	1384	1387	1390	1393	1396	1400	1403	1406	1409	0	1	1	1	2	2	2	3	3
·15	1413	1416	1419	1422	1426	1429	1432	1435	1439	1442	0	1	1	1	2	2	2	3	3
·16	1445	1449	1452	1455	1459	1462	1466	1469	1472	1476	0	1	1	1	2	2	2	3	3
·17	1479	1483	1486	1489	1493	1496	1500	1503	1507	1510	0	1	1	1	2	2	2	3	3
·18	1514	1517	1521	1524	1528	1531	1535	1538	1542	1545	0	1	1	1	2	2	2	3	3
·19	1549	1552	1556	1560	1563	1567	1570	1574	1578	1581	0	1	1	1	2	2	3	3	3
·20	1585	1589	1592	1596	1600	1603	1607	1611	1614	1618	0	1	1	1	2	2	3	3	3
·21	1622	1626	1629	1633	1637	1641	1644	1648	1652	1656	0	1	1	2	2	2	3	3	3
·22	1660	1663	1667	1671	1675	1679	1683	1687	1690	1694	0	1	1	2	2	2	3	3	3
·23	1698	1702	1706	1710	1714	1718	1722	1726	1730	1734	0	1	1	2	2	2	3	3	4
·24	1738	1742	1746	1750	1754	1758	1762	1766	1770	1774	0	1	1	2	2	2	3	3	4
·25	1778	1782	1786	1791	1795	1799	1803	1807	1811	1816	0	1	1	2	2	2	3	3	4
·26	1820	1824	1828	1832	1837	1841	1845	1849	1854	1858	0	1	1	2	2	3	3	3	4
·27	1862	1866	1871	1875	1879	1884	1888	1892	1897	1901	0	1	1	2	2	3	3	3	4
·28	1905	1910	1914	1919	1923	1928	1932	1936	1941	1945	0	1	1	2	2	3	3	4	4
·29	1950	1954	1959	1963	1968	1972	1977	1982	1986	1991	0	1	1	2	2	3	3	4	4
·30	1995	2000	2004	2009	2014	2018	2023	2028	2032	2037	0	1	1	2	2	3	3	4	4
·31	2042	2046	2051	2056	2061	2065	2070	2075	2080	2084	0	1	1	2	2	3	3	4	4
·32	2089	2094	2099	2104	2109	2113	2118	2123	2128	2133	0	1	1	2	2	3	3	4	4
·33	2138	2143	2148	2153	2158	2163	2168	2173	2178	2183	0	1	1	2	2	3	3	4	4
·34	2188	2193	2198	2203	2208	2213	2218	2223	2228	2234	1	1	2	2	3	3	4	4	5
·35	2239	2244	2249	2254	2259	2265	2270	2275	2280	2286	1	1	2	2	3	3	4	4	5
·36	2291	2296	2301	2307	2312	2317	2323	2328	2333	2339	1	1	2	2	3	3	4	4	5
·37	2344	2350	2355	2360	2366	2371	2377	2382	2388	2393	1	1	2	2	3	3	4	4	5
·38	2399	2404	2410	2415	2421	2427	2432	2438	2443	2449	1	1	2	2	3	3	4	4	5
·39	2455	2460	2466	2472	2477	2483	2489	2495	2500	2506	1	1	2	2	3	3	4	5	5
·40	2512	2518	2523	2529	2535	2541	2547	2553	2559	2564	1	1	2	2	3	4	4	5	5
·41	2570	2576	2582	2588	2594	2600	2606	2612	2618	2624	1	1	2	2	3	4	4	5	5
·42	2630	2636	2642	2649	2655	2661	2667	2673	2679	2685	1	1	2	2	3	4	4	5	6
·43	2692	2698	2704	2710	2716	2723	2729	2735	2742	2748	1	1	2	3	3	4	4	5	6
·44	2754	2761	2767	2773	2780	2786	2793	2799	2805	2812	1	1	2	3	3	4	4	5	6
·45	2818	2825	2831	2838	2844	2851	2858	2864	2871	2877	1	1	2	3	3	4	5	5	6
·46	2884	2891	2897	2904	2911	2917	2924	2931	2938	2944	1	1	2	3	3	4	5	5	6
·47	2951	2958	2965	2972	2979	2985	2992	2999	3006	3013	1	1	2	3	3	4	5	5	6
·48	3020	3027	3034	3041	3048	3055	3062	3069	3076	3083	1	1	2	3	4	4	5	6	6
·49	3090	3097	3105	3112	3119	3126	3133	3141	3148	3155	1	1	2	3	4	4	5	6	6
	0	1	2	3	4	5	6	7	8	9	1	2	3	4	5	6	7	8	9

ANTILOGARITHMS

	0	1	2	3	4	5	6	7	8	9	1	2	3	4	5	6	7	8	9
·50	3162	3170	3177	3184	3192	3199	3206	3214	3221	3228	1	1	2	3	4	4	5	6	7
·51	3236	3243	3251	3258	3266	3273	3281	3289	3296	3304	1	2	2	3	4	5	5	6	7
·52	3311	3319	3327	3334	3342	3350	3357	3365	3373	3381	1	2	2	3	4	5	5	6	7
·53	3388	3396	3404	3412	3420	3428	3436	3443	3451	3459	1	2	2	3	4	5	6	6	7
·54	3467	3475	3483	3491	3499	3508	3516	3524	3532	3540	1	2	2	3	4	5	6	6	7
·55	3548	3556	3565	3573	3581	3589	3597	3606	3614	3622	1	2	2	3	4	5	6	7	7
·56	3631	3639	3648	3656	3664	3673	3681	3690	3698	3707	1	2	3	3	4	5	6	7	8
·57	3715	3724	3733	3741	3750	3758	3767	3776	3784	3793	1	2	3	3	4	5	6	7	8
·58	3802	3811	3819	3828	3837	3846	3855	3864	3873	3882	1	2	3	4	4	5	6	7	8
·59	3890	3899	3908	3917	3926	3936	3945	3954	3963	3972	1	2	3	4	5	5	6	7	8
·60	3981	3990	3999	4009	4018	4027	4036	4046	4055	4064	1	2	3	4	5	6	6	7	8
·61	4074	4083	4093	4102	4111	4121	4130	4140	4150	4159	1	2	3	4	5	6	7	8	9
·62	4169	4178	4188	4198	4207	4217	4227	4236	4246	4256	1	2	3	4	5	6	7	8	9
·63	4266	4276	4285	4295	4305	4315	4325	4335	4345	4355	1	2	3	4	5	6	7	8	9
·64	4365	4375	4385	4395	4406	4416	4426	4436	4446	4457	1	2	3	4	5	6	7	8	9
·65	4467	4477	4487	4498	4508	4519	4529	4539	4550	4560	1	2	3	4	5	6	7	8	9
·66	4571	4581	4592	4603	4613	4624	4634	4645	4656	4667	1	2	3	4	5	6	7	9	10
·67	4677	4688	4699	4710	4721	4732	4742	4753	4764	4775	1	2	3	4	5	7	8	9	10
·68	4786	4797	4808	4819	4831	4842	4853	4864	4875	4887	1	2	3	4	6	7	8	9	10
·69	4898	4909	4920	4932	4943	4955	4966	4977	4989	5000	1	2	3	5	6	7	8	9	10
·70	5012	5023	5035	5047	5058	5070	5082	5093	5105	5117	1	2	4	5	6	7	8	9	11
·71	5129	5140	5152	5164	5176	5188	5200	5212	5224	5236	1	2	4	5	6	7	8	10	11
·72	5248	5260	5272	5284	5297	5309	5321	5333	5346	5358	1	2	4	5	6	7	9	10	11
·73	5370	5383	5395	5408	5420	5433	5445	5458	5470	5483	1	3	4	5	6	8	9	10	11
·74	5495	5508	5521	5534	5546	5559	5572	5585	5598	5610	1	3	4	5	6	8	9	10	12
·75	5623	5636	5649	5662	5675	5689	5702	5715	5728	5741	1	3	4	5	7	8	9	10	12
·76	5754	5768	5781	5794	5808	5821	5834	5848	5861	5875	1	3	4	5	7	8	9	11	12
·77	5888	5902	5916	5929	5943	5957	5970	5984	5998	6012	1	3	4	5	7	8	10	11	12
·78	6026	6039	6053	6067	6081	6095	6109	6124	6138	6152	1	3	4	6	7	8	10	11	13
·79	6166	6180	6194	6209	6223	6237	6252	6266	6281	6295	1	3	4	6	7	9	10	11	13
·80	6310	6324	6339	6353	6368	6383	6397	6412	6427	6442	1	3	4	6	7	9	10	12	13
·81	6457	6471	6486	6501	6516	6531	6546	6561	6577	6592	2	3	5	6	8	9	11	12	14
·82	6607	6622	6637	6653	6668	6683	6699	6714	6730	6745	2	3	5	6	8	9	11	12	14
·83	6761	6776	6792	6808	6823	6839	6855	6871	6887	6902	2	3	5	6	8	9	11	13	14
·84	6918	6934	6950	6966	6982	6998	7015	7031	7047	7063	2	3	5	6	8	10	11	13	15
·85	7079	7096	7112	7129	7145	7161	7178	7194	7211	7228	2	3	5	7	8	10	12	13	15
·86	7244	7261	7278	7295	7311	7328	7345	7362	7379	7396	2	3	5	7	8	10	12	13	15
·87	7413	7430	7447	7464	7482	7499	7516	7534	7551	7568	2	3	5	7	9	10	12	14	16
·88	7586	7603	7621	7638	7656	7674	7691	7709	7727	7745	2	4	5	7	9	11	12	14	16
·89	7762	7780	7798	7816	7834	7852	7870	7889	7907	7925	2	4	5	7	9	11	13	14	16
·90	7943	7962	7980	7998	8017	8035	8054	8072	8091	8110	2	4	6	7	9	11	13	15	17
·91	8128	8147	8166	8185	8204	8222	8241	8260	8279	8299	2	4	6	8	9	11	13	15	17
·92	8318	8337	8356	8375	8395	8414	8433	8453	8472	8492	2	4	6	8	10	12	14	15	17
·93	8511	8531	8551	8570	8590	8610	8630	8650	8670	8690	2	4	6	8	10	12	14	16	18
·94	8710	8730	8750	8770	8790	8810	8831	8851	8872	8892	2	4	6	8	10	12	14	16	18
·95	8913	8933	8954	8974	8995	9016	9036	9057	9078	9099	2	4	6	8	10	12	15	17	19
·96	9120	9141	9162	9183	9204	9226	9247	9268	9290	9311	2	4	6	8	11	13	15	17	19
·97	9333	9354	9376	9397	9419	9441	9462	9484	9506	9528	2	4	7	9	11	13	15	17	20
·98	9550	9572	9594	9616	9638	9661	9683	9705	9727	9750	2	4	7	9	11	13	16	18	20
·99	9772	9795	9817	9840	9863	9886	9908	9931	9954	9977	2	5	7	9	11	14	16	18	20
	0	1	2	3	4	5	6	7	8	9	1	2	3	4	5	6	7	8	9

FIVE-FIGURE LOGARITHMS

	0	1	2	3	4	5	6	7	8	9	1	2	3	4	5	6	7	8	9
10	00000	00432	00860	01284	01703						43	85	127	170	212	255	297	340	382
						02119	02531	02938	03342	03743	41	81	121	162	202	243	283	323	364
11	04139	04532	04922	05308	05690						39	77	116	155	193	232	270	309	348
						06070	06446	06819	07188	07555	37	74	111	148	185	222	259	296	333
12	07918	08279	08636	08991	09342						36	71	106	142	177	213	248	284	319
						09691	10037	10380	10721	11059	34	68	102	136	170	205	239	273	307
13	11394	11727	12057	12385	12710						33	66	98	131	164	197	230	262	295
						13033	13354	13672	13988	14301	32	63	95	126	158	190	221	253	284
14	14613	14922	15229	15534	15836						31	61	91	122	152	183	213	244	274
						16137	16435	16732	17026	17319	30	59	88	118	147	177	206	236	265
15	17609	17898	18184	18469	18752						29	57	85	114	142	171	199	228	256
						19033	19312	19590	19866	20140	28	55	83	110	138	166	193	221	248
16	20412	20683	20951	21219	21484						27	53	80	107	134	160	187	214	241
						21748	22011	22272	22531	22789	26	52	78	104	130	156	182	208	233
17	23045	23300	23553	23805	24055						25	50	76	101	126	151	176	201	227
						24304	24551	24797	25042	25285	24	49	73	98	122	147	171	196	220
18	25527	25768	26007	26245	26482						24	48	71	95	119	143	167	190	214
						26717	26951	27184	27416	27646	23	46	70	93	116	139	162	185	209
19	27875	28103	28330	28556	28780						23	45	68	90	113	135	158	181	203
						29003	29226	29447	29667	29885	22	44	66	88	110	132	154	176	198
20	30103	30320	30535	30750	30963	31175	31387	31597	31806	32015	21	42	64	85	106	127	148	170	191
21	32222	32428	32634	32838	33041	33244	33445	33646	33846	34044	20	40	61	81	101	121	141	162	182
22	34242	34439	34635	34830	35025	35218	35411	35603	35793	35984	19	39	58	77	97	116	135	155	174
23	36173	36361	36549	36736	36922	37107	37291	37475	37658	37840	18	37	56	74	92	111	130	148	166
24	38021	38202	38382	38561	38739	38917	39094	39270	39445	39620	18	35	53	71	89	106	124	142	160
25	39794	39967	40140	40312	40483	40654	40824	40993	41162	41330	17	34	51	68	85	102	119	136	153
26	41497	41664	41830	41996	42160	42325	42488	42651	42813	42975	16	33	49	66	82	98	115	131	148
27	43136	43297	43457	43616	43775	43933	44091	44248	44404	44560	16	32	47	63	79	95	111	126	142
28	44716	44871	45025	45179	45332	45484	45637	45788	45939	46090	15	30	46	61	76	91	107	122	137
29	46240	46389	46538	46687	46835	46982	47129	47276	47422	47567	15	29	44	59	74	88	103	118	133
30	47712	47857	48001	48144	48287	48430	48572	48714	48855	48996	14	28	43	57	71	85	100	114	128
31	49136	49276	49415	49554	49693	49831	49969	50106	50243	50379	14	28	41	55	69	83	97	110	124
32	50515	50650	50786	50920	51054	51188	51322	51455	51587	51720	13	27	40	53	67	80	94	107	120
33	51851	51983	52114	52244	52375	52504	52634	52763	52892	53020	13	26	39	52	65	78	91	104	117
34	53148	53275	53403	53529	53656	53782	53908	54033	54158	54283	13	25	38	50	63	76	88	101	113
35	54407	54531	54654	54777	54900	55023	55145	55267	55388	55509	12	24	37	49	61	73	86	98	110
36	55630	55751	55871	55991	56110	56229	56348	56467	56585	56703	12	24	36	48	60	71	83	95	107
37	56820	56937	57054	57171	57287	57403	57519	57634	57749	57864	12	23	35	46	58	70	81	93	104
38	57978	58092	58206	58320	58433	58546	58659	58771	58883	58995	11	23	34	45	56	68	79	90	102
39	59106	59218	59329	59439	59550	59660	59770	59879	59988	60097	11	22	33	44	55	66	77	88	99
40	60206	60314	60423	60531	60638	60745	60853	60959	61066	61172	11	21	32	43	54	64	75	86	97
41	61278	61384	61490	61595	61700	61805	61909	62014	62118	62221	10	21	31	42	52	63	73	84	94
42	62325	62428	62531	62634	62737	62839	62941	63043	63144	63246	10	20	31	41	51	61	72	82	92
43	63347	63448	63548	63649	63749	63849	63949	64048	64147	64246	10	20	30	40	50	60	70	80	90
44	64345	64444	64542	64640	64738	64836	64933	65031	65128	65225	10	20	29	39	49	59	68	78	88
45	65321	65418	65514	65610	65706	65801	65896	65992	66087	66181	10	19	29	38	48	57	67	76	86
46	66276	66370	66464	66558	66652	66745	66839	66932	67025	67117	9	19	28	37	47	56	65	75	84
47	67210	67302	67394	67486	67578	67669	67761	67852	67943	68034	9	18	27	37	46	55	64	73	82
48	68124	68215	68305	68395	68485	68574	68664	68753	68842	68931	9	18	27	36	45	54	63	72	81
49	69020	69108	69197	69285	69373	69461	69548	69636	69723	69810	9	18	26	35	44	53	61	70	79
	0	1	2	3	4	5	6	7	8	9	1	2	3	4	5	6	7	8	9

FIVE-FIGURE LOGARITHMS

	.0	1	2	3	4	5	6	7	8	9	1	2	3	4	5	6	7	8	9
50	69897	69984	70070	70157	70243	70329	70415	70501	70586	70672	9	17	26	34	43	52	60	69	77
51	70757	70842	70927	71012	71096	71181	71265	71349	71433	71517	8	17	25	34	42	51	59	67	76
52	71600	71684	71767	71850	71933	72016	72099	72181	72263	72346	8	17	25	33	41	50	58	65	74
53	72428	72509	72591	72673	72754	72835	72916	72997	73078	73159	8	16	24	32	41	49	57	65	73
54	73239	73320	73400	73480	73560	73640	73719	73799	73878	73957	8	16	24	32	40	48	56	64	72
55	74036	74115	74194	74273	74351	74429	74507	74586	74663	74741	8	16	23	31	39	47	55	63	70
56	74819	74896	74974	75051	75128	75205	75282	75358	75435	75511	8	15	23	31	39	46	54	62	69
57	75587	75664	75740	75815	75891	75967	76042	76118	76193	76268	8	15	23	30	38	45	53	60	68
58	76343	76418	76492	76567	76641	76716	76790	76864	76938	77012	7	15	22	30	37	44	52	59	67
59	77085	77159	77232	77305	77379	77452	77525	77597	77670	77743	7	15	22	29	37	44	51	58	66
60	77815	77887	77960	78032	78104	78176	78247	78319	78390	78462	7	14	22	29	36	43	50	58	65
61	78533	78604	78675	78746	78817	78888	78958	79029	79099	79169	7	14	21	28	36	43	50	57	64
62	79239	79309	79379	79449	79518	79588	79657	79727	79796	79865	7	14	21	28	35	42	49	56	63
63	79934	80003	80072	80140	80209	80277	80346	80414	80482	80550	7	14	21	27	34	41	48	55	62
64	80618	80686	80754	80821	80889	80956	81023	81090	81158	81224	7	13	20	27	34	40	47	54	61
65	81291	81358	81425	81491	81558	81624	81690	81757	81823	81889	7	13	20	27	33	40	46	53	60
66	81954	82020	82086	82151	82217	82282	82347	82413	82478	82543	7	13	20	26	33	39	46	52	59
67	82607	82672	82737	82802	82866	82930	82995	83059	83123	83187	6	13	19	26	32	39	45	51	58
68	83251	83315	83378	83442	83506	83569	83632	83696	83759	83822	6	13	19	25	32	38	44	51	57
69	83885	83948	84011	84073	84136	84198	84261	84323	84386	84448	6	12	19	25	31	37	44	50	56
70	84510	84572	84634	84696	84757	84819	84880	84942	85003	85065	6	12	18	25	31	37	43	49	55
71	85126	85187	85248	85309	85370	85431	85491	85552	85612	85673	6	12	18	24	31	37	43	49	55
72	85733	85794	85854	85914	85974	86034	86094	86153	86213	86273	6	12	18	24	30	36	42	48	54
73	86332	86392	86451	86510	86570	86629	86688	86747	86806	86864	6	12	18	24	30	35	41	47	53
74	86923	86982	87040	87099	87157	87216	87274	87332	87390	87448	6	12	17	23	29	35	41	47	52
75	87506	87564	87622	87679	87737	87795	87852	87910	87967	88024	6	12	17	23	29	35	40	46	52
76	88081	88138	88195	88252	88309	88366	88423	88480	88536	88593	6	11	17	23	29	34	40	46	51
77	88649	88705	88762	88818	88874	88930	88986	89042	89098	89154	6	11	17	22	28	34	39	45	50
78	89209	89265	89321	89376	89432	89487	89542	89597	89653	89708	6	11	17	22	28	33	39	44	50
79	89763	89818	89873	89927	89982	90037	90091	90146	90200	90255	6	11	17	22	28	33	39	44	50
80	90309	90363	90417	90472	90526	90580	90633	90687	90741	90795	5	11	16	22	27	32	38	43	49
81	90848	90902	90956	91009	91062	91116	91169	91222	91275	91328	5	11	16	21	27	32	37	43	48
82	91381	91434	91487	91540	91593	91645	91698	91751	91803	91855	5	11	16	21	26	32	37	42	47
83	91908	91960	92012	92064	92117	92169	92221	92273	92324	92376	5	10	16	21	26	31	36	42	47
84	92428	92480	92531	92583	92634	92686	92737	92788	92840	92891	5	10	15	21	26	31	36	41	46
85	92942	92993	93044	93095	93146	93197	93247	93298	93349	93399	5	10	15	20	26	31	36	41	46
86	93450	93500	93551	93601	93651	93702	93752	93802	93852	93902	5	10	15	20	25	30	35	40	45
87	93952	94002	94052	94101	94151	94201	94250	94300	94349	94399	5	10	15	20	25	30	35	40	45
88	94448	94498	94547	94596	94645	94694	94743	94792	94841	94890	5	10	15	20	25	29	34	39	44
89	94939	94988	95036	95085	95134	95182	95231	95279	95328	95376	5	10	15	19	24	29	34	39	44
90	95424	95472	95521	95569	95617	95665	95713	95761	95809	95856	5	10	14	19	24	29	34	38	43
91	95904	95952	95999	96047	96095	96142	96190	96237	96284	96332	5	9	14	19	24	28	33	38	43
92	96379	96426	96473	96520	96567	96614	96661	96708	96755	96802	5	9	14	19	24	28	33	38	42
93	96848	96895	96942	96988	97035	97081	97128	97174	97220	97267	5	9	14	19	23	28	33	37	42
94	97313	97359	97405	97451	97497	97543	97589	97635	97681	97727	5	9	14	18	23	28	32	37	42
95	97772	97818	97864	97909	97955	98000	98046	98091	98137	98182	5	9	14	18	23	27	32	36	41
96	98227	98272	98318	98363	98408	98453	98498	98543	98588	98632	5	9	14	18	23	27	32	36	41
97	98677	98722	98767	98811	98856	98900	98945	98989	99034	99078	4	9	13	18	22	27	31	36	40
98	99123	99167	99211	99255	99300	99344	99388	99432	99476	99520	4	9	13	18	22	26	31	35	40
99	99564	99607	99651	99695	99739	99782	99826	99870	99913	99957	4	9	13	17	22	26	31	35	39
	0	1	2	3	4	5	6	7	8	9	1	2	3	4	5	6	7	8	9

RECIPROCALS

	0	1	2	3	4	5	6	7	8	9	Subtract Differences.								
											1	2	3	4	5	6	7	8	9
10	1000	9901	9804	9709	9615	9524	9434	9346	9259	9174									
11	9091	9009	8929	8850	8772	8696	8621	8547	8475	8403									
12	8333	8264	8197	8130	8065	8000	7937	7874	7813	7752		Mean differences							
13	7692	7634	7576	7519	7463	7407	7353	7299	7246	7194		not sufficiently							
14	7143	7092	7042	6993	6944	6897	6849	6803	6757	6711		accurate.							
15	6667	6623	6579	6536	6494	6452	6410	6369	6329	6289	4	8	13	17	21	25	29	33	38
16	6250	6211	6173	6135	6098	6061	6024	5988	5952	5917	4	7	11	15	18	22	26	29	33
17	5882	5848	5814	5780	5747	5714	5682	5650	5618	5587	3	6	10	13	16	20	23	26	29
18	5556	5525	5495	5464	5435	5405	5376	5348	5319	5291	3	6	9	12	15	17	20	23	26
19	5263	5236	5208	5181	5155	5128	5102	5076	5051	5025	3	5	8	11	13	16	18	21	24
20	5000	4975	4950	4926	4902	4878	4854	4831	4808	4785	2	5	7	10	12	14	17	19	21
21	4762	4739	4717	4695	4673	4651	4630	4608	4587	4566	2	4	7	9	11	13	15	17	19
22	4545	4525	4505	4484	4464	4444	4425	4405	4386	4367	2	4	6	8	10	12	14	16	18
23	4348	4329	4310	4292	4274	4255	4237	4219	4202	4184	2	4	5	7	9	11	13	14	16
24	4167	4149	4132	4115	4098	4082	4065	4049	4032	4016	2	3	5	7	8	10	12	13	15
25	4000	3984	3968	3953	3937	3922	3906	3891	3876	3861	2	3	5	6	8	9	11	12	14
26	3846	3831	3817	3802	3788	3774	3759	3745	3731	3717	1	3	4	6	7	8	10	11	13
27	3704	3690	3676	3663	3650	3636	3623	3610	3597	3584	1	3	4	5	7	8	9	11	12
28	3571	3559	3546	3534	3521	3509	3497	3484	3472	3460	1	2	4	5	6	7	9	10	11
29	3448	3436	3425	3413	3401	3390	3378	3367	3356	3344	1	2	3	5	6	7	8	9	10
30	3333	3322	3311	3300	3289	3279	3268	3257	3247	3236	1	2	3	4	5	6	7	9	10
31	3226	3215	3205	3195	3185	3175	3165	3155	3145	3135	1	2	3	4	5	6	7	8	9
32	3125	3115	3106	3096	3086	3077	3067	3058	3049	3040	1	2	3	4	5	6	7	8	9
33	3030	3021	3012	3003	2994	2985	2976	2967	2959	2950	1	2	3	4	4	5	6	7	8
34	2941	2933	2924	2915	2907	2899	2890	2882	2874	2865	1	2	3	3	4	5	6	7	8
35	2857	2849	2841	2833	2825	2817	2809	2801	2793	2786	1	2	2	3	4	5	6	6	7
36	2778	2770	2762	2755	2747	2740	2732	2725	2717	2710	1	2	2	3	4	5	5	6	7
37	2703	2695	2688	2681	2674	2667	2660	2653	2646	2639	1	1	2	3	4	4	5	6	6
38	2632	2625	2618	2611	2604	2597	2591	2584	2577	2571	1	1	2	3	3	4	5	5	6
39	2564	2558	2551	2545	2538	2532	2525	2519	2513	2506	1	1	2	3	3	4	4	5	6
40	2500	2494	2488	2481	2475	2469	2463	2457	2451	2445	1	1	2	2	3	4	4	5	5
41	2439	2433	2427	2421	2415	2410	2404	2398	2392	2387	1	1	2	2	3	3	4	5	5
42	2381	2375	2370	2364	2358	2353	2347	2342	2336	2331	1	1	2	2	3	3	4	4	5
43	2326	2320	2315	2309	2304	2299	2294	2288	2283	2278	1	1	2	2	3	3	4	4	5
44	2273	2268	2262	2257	2252	2247	2242	2237	2232	2227	1	1	2	2	3	3	4	4	5
45	2222	2217	2212	2208	2203	2198	2193	2188	2183	2179	0	1	1	2	2	3	3	4	4
46	2174	2169	2165	2160	2155	2151	2146	2141	2137	2132	0	1	1	2	2	3	3	4	4
47	2128	2123	2119	2114	2110	2105	2101	2096	2092	2088	0	1	1	2	2	3	3	4	4
48	2083	2079	2075	2070	2066	2062	2058	2053	2049	2045	0	1	1	2	2	3	3	3	4
49	2041	2037	2033	2028	2024	2020	2016	2012	2008	2004	0	1	1	2	2	2	3	3	4
50	2000	1996	1992	1988	1984	1980	1976	1972	1969	1965	0	1	1	2	2	2	3	3	4
51	1961	1957	1953	1949	1946	1942	1938	1934	1931	1927	0	1	1	2	2	2	3	3	3
52	1923	1919	1916	1912	1908	1905	1901	1898	1894	1890	0	1	1	2	2	2	3	3	3
53	1887	1883	1880	1876	1873	1869	1866	1862	1859	1855	0	1	1	1	2	2	2	3	3
54	1852	1848	1845	1842	1838	1835	1832	1828	1825	1821	0	1	1	1	2	2	2	3	3
	0	1	2	3	4	5	6	7	8	9	1	2	3	4	5	6	7	8	9

Subtract Differences.

	0	1	2	3	4	5	6	7	8	9	1	2	3	4	5	6	7	8	9
											Subtract Differences.								
55	1818	1815	1812	1808	1805	1802	1799	1795	1792	1789	0	1	1	1	2	2	2	3	3
56	1786	1783	1779	1776	1773	1770	1767	1764	1761	1757	0	1	1	1	2	2	2	3	3
57	1754	1751	1748	1745	1742	1739	1736	1733	1730	1727	0	1	1	1	2	2	2	2	3
58	1724	1721	1718	1715	1712	1709	1706	1704	1701	1698	0	1	1	1	1	2	2	2	3
59	1695	1692	1689	1686	1684	1681	1678	1675	1672	1669	0	1	1	1	1	2	2	2	3
60	1667	1664	1661	1658	1656	1653	1650	1647	1645	1642	0	1	1	1	1	2	2	2	3
61	1639	1637	1634	1631	1629	1626	1623	1621	1618	1616	0	1	1	1	1	2	2	2	2
62	1613	1610	1608	1605	1603	1600	1597	1595	1592	1590	0	1	1	1	1	2	2	2	2
63	1587	1585	1582	1580	1577	1575	1572	1570	1567	1565	0	0	1	1	1	1	2	2	2
64	1563	1560	1558	1555	1553	1550	1548	1546	1543	1541	0	0	1	1	1	1	2	2	2
65	1538	1536	1534	1531	1529	1527	1524	1522	1520	1517	0	0	1	1	1	1	2	2	2
66	1515	1513	1511	1508	1506	1504	1502	1499	1497	1495	0	0	1	1	1	1	2	2	2
67	1493	1490	1488	1486	1484	1481	1479	1477	1475	1473	0	0	1	1	1	1	2	2	2
68	1471	1468	1466	1464	1462	1460	1458	1456	1453	1451	0	0	1	1	1	1	2	2	2
69	1449	1447	1445	1443	1441	1439	1437	1435	1433	1431	0	0	1	1	1	1	1	2	2
70	1429	1427	1425	1422	1420	1418	1416	1414	1412	1410	0	0	1	1	1	1	1	2	2
71	1408	1406	1404	1403	1401	1399	1397	1395	1393	1391	0	0	1	1	1	1	1	2	2
72	1389	1387	1385	1383	1381	1379	1377	1376	1374	1372	0	0	1	1	1	1	1	2	2
73	1370	1368	1366	1364	1362	1361	1359	1357	1355	1353	0	0	1	1	1	1	1	2	2
74	1351	1350	1348	1346	1344	1342	1340	1339	1337	1335	0	0	1	1	1	1	1	1	2
75	1333	1332	1330	1328	1326	1325	1323	1321	1319	1318	0	0	1	1	1	1	1	1	2
76	1316	1314	1312	1311	1309	1307	1305	1304	1302	1300	0	0	1	1	1	1	1	1	2
77	1299	1297	1295	1294	1292	1290	1289	1287	1285	1284	0	0	0	1	1	1	1	1	1
78	1282	1280	1279	1277	1276	1274	1272	1271	1269	1267	0	0	0	1	1	1	1	1	1
79	1266	1264	1263	1261	1259	1258	1256	1255	1253	1252	0	0	0	1	1	1	1	1	1
80	1250	1248	1247	1245	1244	1242	1241	1239	1238	1236	0	0	0	1	1	1	1	1	1
81	1235	1233	1232	1230	1229	1227	1225	1224	1222	1221	0	0	0	1	1	1	1	1	1
82	1220	1218	1217	1215	1214	1212	1211	1209	1208	1206	0	0	0	1	1	1	1	1	1
83	1205	1203	1202	1200	1199	1198	1196	1195	1193	1192	0	0	0	1	1	1	1	1	1
84	1190	1189	1188	1186	1185	1183	1182	1181	1179	1178	0	0	0	1	1	1	1	1	1
85	1176	1175	1174	1172	1171	1170	1168	1167	1166	1164	0	0	0	1	1	1	1	1	1
86	1163	1161	1160	1159	1157	1156	1155	1153	1152	1151	0	0	0	1	1	1	1	1	1
87	1149	1148	1147	1145	1144	1143	1142	1140	1139	1138	0	0	0	1	1	1	1	1	1
88	1136	1135	1134	1133	1131	1130	1129	1127	1126	1125	0	0	0	1	1	1	1	1	1
89	1124	1122	1121	1120	1119	1117	1116	1115	1114	1112	0	0	0	1	1	1	1	1	1
90	1111	1110	1109	1107	1106	1105	1104	1103	1101	1100	0	0	0	1	1	1	1	1	1
91	1099	1098	1096	1095	1094	1093	1092	1091	1089	1088	0	0	0	0	1	1	1	1	1
92	1087	1086	1085	1083	1082	1081	1080	1079	1078	1076	0	0	0	0	1	1	1	1	1
93	1075	1074	1073	1072	1071	1070	1068	1067	1066	1065	0	0	0	0	1	1	1	1	1
94	1064	1063	1062	1060	1059	1058	1057	1056	1055	1054	0	0	0	0	1	1	1	1	1
95	1053	1052	1050	1049	1048	1047	1046	1045	1044	1043	0	0	0	0	1	1	1	1	1
96	1042	1041	1040	1038	1037	1036	1035	1034	1033	1032	0	0	0	0	1	1	1	1	1
97	1031	1030	1029	1028	1027	1026	1025	1024	1022	1021	0	0	0	0	1	1	1	1	1
98	1020	1019	1018	1017	1016	1015	1014	1013	1012	1011	0	0	0	0	1	1	1	1	1
99	1010	1009	1008	1007	1006	1005	1004	1003	1002	1001	0	0	0	0	0	1	1	1	1
	0	1	2	3	4	5	6	7	8	9	1	2	3	4	5	6	7	8	9

Subtract Differences.

SQUARES

	0	1	2	3	4	5	6	7	8	9	1	2	3	4	5	6	7	8	9
1·0	1·000	1·020	1·040	1·061	1·082	1·103	1·124	1·145	1·166	1·188	2	4	6	8	10	13	15	17	19
1·1	1·210	1·232	1·254	1·277	1·300	1·323	1·346	1·369	1·392	1·416	2	5	7	9	11	14	16	18	21
1·2	1·440	1·464	1·488	1·513	1·538	1·563	1·588	1·613	1·638	1·664	2	5	7	10	12	15	17	20	22
1·3	1·690	1·716	1·742	1·769	1·796	1·823	1·850	1·877	1·904	1·932	3	5	8	11	13	16	19	22	24
1·4	1·960	1·988	2·016	2·045	2·074	2·103	2·132	2·161	2·190	2·220	3	6	9	12	14	17	20	23	26
1·5	2·250	2·280	2·310	2·341	2·372	2·403	2·434	2·465	2·496	2·528	3	6	9	12	15	19	22	25	28
1·6	2·560	2·592	2·624	2·657	2·690	2·723	2·756	2·789	2·822	2·856	3	7	10	13	16	20	23	26	30
1·7	2·890	2·924	2·958	2·993	3·028	3·063	3·098	3·133	3·168	3·204	3	7	10	14	17	21	24	28	31
1·8	3·240	3·276	3·312	3·349	3·386	3·423	3·460	3·497	3·534	3·572	4	7	11	15	18	22	26	30	33
1·9	3·610	3·648	3·686	3·725	3·764	3·803	3·842	3·881	3·920	3·960	4	8	12	16	19	23	27	31	35
2·0	4·000	4·040	4·080	4·121	4·162	4·203	4·244	4·285	4·326	4·368	4	8	12	16	20	25	29	33	37
2·1	4·410	4·452	4·494	4·537	4·580	4·623	4·666	4·709	4·752	4·796	4	9	13	17	21	26	30	34	39
2·2	4·840	4·884	4·928	4·973	5·018	5·063	5·108	5·153	5·198	5·244	4	9	13	18	22	27	31	36	40
2·3	5·290	5·336	5·382	5·429	5·476	5·523	5·570	5·617	5·664	5·712	5	9	14	19	23	28	33	38	42
2·4	5·760	5·808	5·856	5·905	5·954	6·003	6·052	6·101	6·150	6·200	5	10	15	20	24	29	34	39	44
2·5	6·250	6·300	6·350	6·401	6·452	6·503	6·554	6·605	6·656	6·708	5	10	15	20	25	31	36	41	46
2·6	6·760	6·812	6·864	6·917	6·970	7·023	7·076	7·129	7·182	7·236	5	11	16	21	26	32	37	42	48
2·7	7·290	7·344	7·398	7·453	7·508	7·563	7·618	7·673	7·728	7·784	5	11	16	22	27	33	38	44	49
2·8	7·840	7·896	7·952	8·009	8·066	8·123	8·180	8·237	8·294	8·352	6	11	17	23	28	34	40	46	51
2·9	8·410	8·468	8·526	8·585	8·644	8·703	8·762	8·821	8·880	8·940	6	12	18	24	29	35	41	47	53
3·0	9·000	9·060	9·120	9·181	9·242	9·303	9·364	9·425	9·486	9·548	6	12	18	24	30	37	43	49	55
3·1 {	9·610	9·672	9·734	9·797	9·860	9·923	9·986				6	13	19	25	31	38	44	50	57
								10·05	10·11	10·18	1	1	2	3	3	4	5	5	6
3·2	10·24	10·30	10·37	10·43	10·50	10·56	10·63	10·69	10·76	10·82	1	1	2	3	3	4	5	5	6
3·3	10·89	10·96	11·02	11·09	11·16	11·22	11·29	11·36	11·42	11·49	1	1	2	3	3	4	5	5	6
3·4	11·56	11·63	11·70	11·76	11·83	11·90	11·97	12·04	12·11	12·18	1	1	2	3	3	4	5	6	6
3·5	12·25	12·32	12·39	12·46	12·53	12·60	12·67	12·74	12·82	12·89	1	1	2	3	4	4	5	6	6
3·6	12·96	13·03	13·10	13·18	13·25	13·32	13·40	13·47	13·54	13·62	1	1	2	3	4	4	5	6	7
3·7	13·69	13·76	13·84	13·91	13·99	14·06	14·14	14·21	14·29	14·36	1	2	2	3	4	4	5	6	7
3·8	14·44	14·52	14·59	14·67	14·75	14·82	14·90	14·98	15·05	15·13	1	2	2	3	4	5	5	6	7
3·9	15·21	15·29	15·37	15·44	15·52	15·60	15·68	15·76	15·84	15·92	1	2	2	3	4	5	6	6	7
4·0	16·00	16·08	16·16	16·24	16·32	16·40	16·48	16·56	16·65	16·73	1	2	2	3	4	5	6	6	7
4·1	16·81	16·89	16·97	17·06	17·14	17·22	17·31	17·39	17·47	17·56	1	2	2	3	4	5	6	7	7
4·2	17·64	17·72	17·81	17·89	17·98	18·06	18·15	18·23	18·32	18·40	1	2	3	3	4	5	6	7	8
4·3	18·49	18·58	18·66	18·75	18·84	18·92	19·01	19·10	19·18	19·27	1	2	3	3	4	5	6	7	8
4·4	19·36	19·45	19·54	19·62	19·71	19·80	19·89	19·98	20·07	20·16	1	2	3	4	4	5	6	7	8
4·5	20·25	20·34	20·43	20·52	20·61	20·70	20·79	20·88	20·98	21·07	1	2	3	4	5	5	6	7	8
4·6	21·16	21·25	21·34	21·44	21·53	21·62	21·72	21·81	21·90	22·00	1	2	3	4	5	6	7	7	8
4·7	22·09	22·18	22·28	22·37	22·47	22·56	22·66	22·75	22·85	22·94	1	2	3	4	5	6	7	8	9
4·8	23·04	23·14	23·23	23·33	23·43	23·52	23·62	23·72	23·81	23·91	1	2	3	4	5	6	7	8	9
4·9	24·01	24·11	24·21	24·30	24·40	24·50	24·60	24·70	24·80	24·90	1	2	3	4	5	6	7	8	9
5·0	25·00	25·10	25·20	25·30	25·40	25·50	25·60	25·70	25·81	25·91	1	2	3	4	5	6	7	8	9
5·1	26·01	26·11	26·21	26·32	26·42	26·52	26·63	26·73	26·83	26·94	1	2	3	4	5	6	7	8	9
5·2	27·04	27·14	27·25	27·35	27·46	27·56	27·67	27·77	27·88	27·98	1	2	3	4	5	6	7	8	9
5·3	28·09	28·20	28·30	28·41	28·52	28·62	28·73	28·84	28·94	29·05	1	2	3	4	5	6	7	9	10
5·4	29·16	29·27	29·38	29·48	29·59	29·70	29·81	29·92	30·03	30·14	1	2	3	4	5	7	8	9	10
	0	1	2	3	4	5	6	7	8	9	1	2	3	4	5	6	7	8	9

SQUARES

	0	1	2	3	4	5	6	7	8	9	1	2	3	4	5	6	7	8	9
5·5	30·25	30·36	30·47	30·58	30·69	30·80	30·91	31·02	31·14	31·25	1	2	3	4	6	7	8	9	10
5·6	31·36	31·47	31·58	31·70	31·81	31·92	32·04	32·15	32·26	32·38	1	2	3	5	6	7	8	9	10
5·7	32·49	32·60	32·72	32·83	32·95	33·06	33·18	33·29	33·41	33·52	1	2	3	5	6	7	8	9	10
5·8	33·64	33·76	33·87	33·99	34·11	34·22	34·34	34·46	34·57	34·69	1	2	4	5	6	7	8	9	11
5·9	34·81	34·93	35·05	35·16	35·28	35·40	35·52	35·64	35·76	35·88	1	2	4	5	6	7	8	10	11
6·0	36·00	36·12	36·24	36·36	36·48	36·60	36·72	36·84	36·97	37·09	1	2	4	5	6	7	8	10	11
6·1	37·21	37·33	37·45	37·58	37·70	37·82	37·95	38·07	38·19	38·32	1	2	4	5	6	7	9	10	11
6·2	38·44	38·56	38·69	38·81	38·94	39·06	39·19	39·31	39·44	39·56	1	3	4	5	6	8	9	10	11
6·3	39·69	39·82	39·94	40·07	40·20	40·32	40·45	40·58	40·70	40·83	1	3	4	5	6	8	9	10	11
6·4	40·96	41·09	41·22	41·34	41·47	41·60	41·73	41·86	41·99	42·12	1	3	4	5	6	8	9	10	12
6·5	42·25	42·38	42·51	42·64	42·77	42·90	43·03	43·16	43·30	43·43	1	3	4	5	7	8	9	10	12
6·6	43·56	43·69	43·82	43·96	44·09	44·22	44·36	44·49	44·62	44·76	1	3	4	5	7	8	9	11	12
6·7	44·89	45·02	45·16	45·29	45·43	45·56	45·70	45·83	45·97	46·10	1	3	4	5	7	8	9	11	12
6·8	46·24	46·38	46·51	46·65	46·79	46·92	47·06	47·20	47·33	47·47	1	3	4	5	7	8	10	11	12
6·9	47·61	47·75	47·89	48·02	48·16	48·30	48·44	48·58	48·72	48·86	1	3	4	6	7	8	10	11	13
7·0	49·00	49·14	49·28	49·42	49·56	49·70	49·84	49·98	50·13	50·27	1	3	4	6	7	8	10	11	13
7·1	50·41	50·55	50·69	50·84	50·98	51·12	51·27	51·41	51·55	51·70	1	3	4	6	7	9	10	11	13
7·2	51·84	51·98	52·13	52·27	52·42	52·56	52·71	52·85	53·00	53·14	1	3	4	6	7	9	10	12	13
7·3	53·29	53·44	53·58	53·73	53·88	54·02	54·17	54·32	54·46	54·61	1	3	4	6	7	9	10	12	13
7·4	54·76	54·91	55·06	55·20	55·35	55·50	55·65	55·80	55·95	56·10	1	3	4	6	7	9	10	12	13
7·5	56·25	56·40	56·55	56·70	56·85	57·00	57·15	57·30	57·46	57·61	2	3	5	6	8	9	11	12	14
7·6	57·76	57·91	58·06	58·22	58·37	58·52	58·68	58·83	58·98	59·14	2	3	5	6	8	9	11	12	14
7·7	59·29	59·44	59·60	59·75	59·91	60·06	60·22	60·37	60·53	60·68	2	3	5	6	8	9	11	12	14
7·8	60·84	61·00	61·15	61·31	61·47	61·62	61·78	61·94	62·09	62·25	2	3	5	6	8	9	11	13	14
7·9	62·41	62·57	62·73	62·88	63·04	63·20	63·36	63·52	63·68	63·84	2	3	5	6	8	10	11	13	14
8·0	64·00	64·16	64·32	64·48	64·64	64·80	64·96	65·12	65·29	65·45	2	3	5	6	8	10	11	13	14
8·1	65·61	65·77	65·93	66·10	66·26	66·42	66·59	66·75	66·91	67·08	2	3	5	7	8	10	11	13	15
8·2	67·24	67·40	67·57	67·73	67·90	68·06	68·23	68·39	68·56	68·72	2	3	5	7	8	10	12	13	15
8·3	68·89	69·06	69·22	69·39	69·56	69·72	69·89	70·06	70·22	72·39	2	3	5	7	8	10	12	13	15
8·4	70·56	70·73	70·90	71·06	71·23	71·40	71·57	71·74	71·91	72·08	2	3	5	7	8	10	12	14	15
8·5	72·25	72·42	72·59	72·76	72·93	73·10	73·27	73·44	73·62	73·79	2	3	5	7	9	10	12	14	15
8·6	73·96	74·13	74·30	74·48	74·65	74·82	75·00	75·17	75·34	75·52	2	3	5	7	9	10	12	14	16
8·7	75·69	75·86	76·04	76·21	76·39	76·56	76·74	76·91	77·09	77·26	2	4	5	7	9	11	12	14	16
8·8	77·44	77·62	77·79	77·97	78·15	78·32	78·50	78·68	78·85	79·03	2	4	5	7	9	11	12	14	16
8·9	79·21	79·39	79·57	79·74	79·92	80·10	80·28	80·46	80·64	80·82	2	4	5	7	9	11	13	14	16
9·0	81·00	81·18	81·36	81·54	81·72	81·90	82·08	82·26	82·45	82·63	2	4	5	7	9	11	13	14	16
9·1	82·81	82·99	83·17	83·36	83·54	83·72	83·91	84·09	84·27	84·46	2	4	5	7	9	11	13	15	16
9·2	84·64	84·82	85·01	85·19	85·38	85·56	85·75	85·93	86·12	86·30	2	4	6	7	9	11	13	15	17
9·3	86·49	86·68	86·86	87·05	87·24	87·42	87·61	87·80	87·98	88·17	2	4	6	7	9	11	13	15	17
9·4	88·36	88·55	88·74	88·92	89·11	89·30	89·49	89·68	89·87	90·06	2	4	6	8	9	11	13	15	17
9·5	90·25	90·44	90·63	90·82	91·01	91·20	91·39	91·58	91·78	91·97	2	4	6	8	10	11	13	15	17
9·6	92·16	92·35	92·54	92·74	92·93	93·12	93·32	93·51	93·70	93·90	2	4	6	8	10	12	14	15	17
9·7	94·09	94·28	94·48	94·67	94·87	95·06	95·26	95·45	95·65	95·84	2	4	6	8	10	12	14	16	18
9·8	96·04	96·24	96·43	96·63	96·83	97·02	97·22	97·42	97·61	97·81	2	4	6	8	10	12	14	16	18
9·9	98·01	98·21	98·41	98·60	98·80	99·00	99·20	99·40	99·60	99·80	2	4	6	8	10	12	14	16	18
	0	1	2	3	4	5	6	7	8	9	1	2	3	4	5	6	7	8	9

NATURAL SINES

	0′	6′	12′	18′	24′	30′	36′	42′	48′	54′	1′	2′	3′	4′	5′
0°	·0000	·0017	·0035	·0052	·0070	·0087	·0105	·0122	·0140	·0157	3	6	9	12	15
1	·0175	0192	0209	0227	0244	0262	0279	0297	0314	0332	3	6	9	12	15
2	·0349	0366	0384	0401	0419	0436	0454	0471	0488	0506	3	6	9	12	15
3	·0523	0541	0558	0576	0593	0610	0628	0645	0663	0680	3	6	9	12	15
4	·0698	0715	0732	0750	0767	0785	0802	0819	0837	0854	3	6	9	12	14
5	·0872	0889	0906	0924	0941	0958	0976	0993	1011	1028	3	6	9	12	14
6	·1045	1063	1080	1097	1115	1132	1149	1167	1184	1201	3	6	9	12	14
7	·1219	1236	1253	1271	1288	1305	1323	1340	1357	1374	3	6	9	12	14
8	·1392	1409	1426	1444	1461	1478	1495	1513	1530	1547	3	6	9	12	14
9	·1564	1582	1599	1616	1633	1650	1668	1685	1702	1719	3	6	9	11	14
10	·1736	1754	1771	1788	1805	1822	1840	1857	1874	1891	3	6	9	11	14
11	·1908	1925	1942	1959	1977	1994	2011	2028	2045	2062	3	6	9	11	14
12	·2079	2096	2113	2130	2147	2164	2181	2198	2215	2233	3	6	9	11	14
13	·2250	2267	2284	2300	2317	2334	2351	2368	2385	2402	3	6	8	11	14
14	·2419	2436	2453	2470	2487	2504	2521	2538	2554	2571	3	6	8	11	14
15	·2588	2605	2622	2639	2656	2672	2689	2706	2723	2740	3	6	8	11	14
16	·2756	2773	2790	2807	2823	2840	2857	2874	2890	2907	3	6	8	11	14
17	·2924	2940	2957	2974	2990	3007	3024	3040	3057	3074	3	6	8	11	14
18	·3090	3107	3123	3140	3156	3173	3190	3206	3223	3239	3	6	8	11	14
19	·3256	3272	3289	3305	3322	3338	3355	3371	3387	3404	3	5	8	11	14
20	·3420	3437	3453	3469	3486	3502	3518	3535	3551	3567	3	5	8	11	14
21	·3584	3600	3616	3633	3649	3665	3681	3697	3714	3730	3	5	8	11	14
22	·3746	3762	3778	3795	3811	3827	3843	3859	3875	3891	3	5	8	11	13
23	·3907	3923	3939	3955	3971	3987	4003	4019	4035	4051	3	5	8	11	13
24	·4067	4083	4099	4115	4131	4147	4163	4179	4195	4210	3	5	8	11	13
25	·4226	4242	4258	4274	4289	4305	4321	4337	4352	4368	3	5	8	11	13
26	·4384	4399	4415	4431	4446	4462	4478	4493	4509	4524	3	5	8	10	13
27	·4540	4555	4571	4586	4602	4617	4633	4648	4664	4679	3	5	8	10	13
28	·4695	4710	4726	4741	4756	4772	4787	4802	4818	4833	3	5	8	10	13
29	·4848	4863	4879	4894	4909	4924	4939	4955	4970	4985	3	5	8	10	13
30	·5000	5015	5030	5045	5060	5075	5090	5105	5120	5135	3	5	8	10	13
31	·5150	5165	5180	5195	5210	5225	5240	5255	5270	5284	2	5	7	10	12
32	·5299	5314	5329	5344	5358	5373	5388	5402	5417	5432	2	5	7	10	12
33	·5446	5461	5476	5490	5505	5519	5534	5548	5563	5577	2	5	7	10	12
34	·5592	5606	5621	5635	5650	5664	5678	5693	5707	5721	2	5	7	10	12
35	·5736	5750	5764	5779	5793	5807	5821	5835	5850	5864	2	5	7	9	12
36	·5878	5892	5906	5920	5934	5948	5962	5976	5990	6004	2	5	7	9	12
37	·6018	6032	6046	6060	6074	6088	6101	6115	6129	6143	2	5	7	9	12
38	·6157	6170	6184	6198	6211	6225	6239	6252	6266	6280	2	5	7	9	11
39	·6293	6307	6320	6334	6347	6361	6374	6388	6401	6414	2	4	7	9	11
40	·6428	6441	6455	6468	6481	6494	6508	6521	6534	6547	2	4	7	9	11
41	·6561	6574	6587	6600	6613	6626	6639	6652	6665	6678	2	4	7	9	11
42	·6691	6704	6717	6730	6743	6756	6769	6782	6794	6807	2	4	6	9	11
43	·6820	6833	6845	6858	6871	6884	6896	6909	6921	6934	2	4	6	8	11
44	·6947	6959	6972	6984	6997	7009	7022	7034	7046	7059	2	4	6	8	10
	0′	6′	12′	18′	24′	30′	36′	42′	48′	54′	1′	2′	3′	4′	5′

NATURAL SINES

	0'	6'	12'	18'	24'	30'	36'	42'	48'	54'	1'	2'	3'	4'	5'
45°	·7071	7083	7096	7108	7120	7133	7145	7157	7169	7181	2	4	6	8	10
46	·7193	7206	7218	7230	7242	7254	7266	7278	7290	7302	2	4	6	8	10
47	·7314	7325	7337	7349	7361	7373	7385	7396	7408	7420	2	4	6	8	10
48	·7431	7443	7455	7466	7478	7490	7501	7513	7524	7536	2	4	6	8	10
49	·7547	7559	7570	7581	7593	7604	7615	7627	7638	7649	2	4	6	8	9
50	·7660	7672	7683	7694	7705	7716	7727	7738	7749	7760	2	4	6	7	9
51	·7771	7782	7793	7804	7815	7826	7837	7848	7859	7869	2	4	5	7	9
52	·7880	7891	7902	7912	7923	7934	7944	7955	7965	7976	2	4	5	7	9
53	·7986	7997	8007	8018	8028	8039	8049	8059	8070	8080	2	3	5	7	9
54	·8090	8100	8111	8121	8131	8141	8151	8161	8171	8181	2	3	5	7	8
55	·8192	8202	8211	8221	8231	8241	8251	8261	8271	8281	2	3	5	7	8
56	·8290	8300	8310	8320	8329	8339	8348	8358	8368	8377	2	3	5	6	8
57	·8387	8396	8406	8415	8425	8434	8443	8453	8462	8471	2	3	5	6	8
58	·8480	8490	8499	8508	8517	8526	8536	8545	8554	8563	2	3	5	6	8
59	·8572	8581	8590	8599	8607	8616	8625	8634	8643	8652	1	3	4	6	7
60	·8660	8669	8678	8686	8695	8704	8712	8721	8729	8738	1	3	4	6	7
61	·8746	8755	8763	8771	8780	8788	8796	8805	8813	8821	1	3	4	6	7
62	·8829	8838	8846	8854	8862	8870	8878	8886	8894	8902	1	3	4	5	7
63	·8910	8918	8926	8934	8942	8949	8957	8965	8973	8980	1	3	4	5	6
64	·8988	8996	9003	9011	9018	9026	9033	9041	9048	9056	1	3	4	5	6
65	·9063	9070	9078	9085	9092	9100	9107	9114	9121	9128	1	2	4	5	6
66	·9135	9143	9150	9157	9164	9171	9178	9184	9191	9198	1	2	3	5	6
67	·9205	9212	9219	9225	9232	9239	9245	9252	9259	9265	1	2	3	4	6
68	·9272	9278	9285	9291	9298	9304	9311	9317	9323	9330	1	2	3	4	5
69	·9336	9342	9348	9354	9361	9367	9373	9379	9385	9391	1	2	3	4	5
70	·9397	9403	9409	9415	9421	9426	9432	9438	9444	9449	1	2	3	4	5
71	·9455	9461	9466	9472	9478	9483	9489	9494	9500	9505	1	2	3	4	5
72	·9511	9516	9521	9527	9532	9537	9542	9548	9553	9558	1	2	3	3	4
73	·9563	9568	9573	9578	9583	9588	9593	9598	9603	9608	1	2	2	3	4
74	·9613	9617	9622	9627	9632	9636	9641	9646	9650	9655	1	2	2	3	4
75	·9659	9664	9668	9673	9677	9681	9686	9690	9694	9699	1	1	2	3	4
76	·9703	9707	9711	9715	9720	9724	9728	9732	9736	9740	1	1	2	3	3
77	·9744	9748	9751	9755	9759	9763	9767	9770	9774	9778	1	1	2	3	3
78	·9781	9785	9789	9792	9796	9799	9803	9806	9810	9813	1	1	2	2	3
79	·9816	9820	9823	9826	9829	9833	9836	9839	9842	9845	1	1	2	2	3
80	·9848	9851	9854	9857	9860	9863	9866	9869	9871	9874	0	1	1	2	2
81	·9877	9880	9882	9885	9888	9890	9893	9895	9898	9900	0	1	1	2	2
82	·9903	9905	9907	9910	9912	9914	9917	9919	9921	9923	0	1	1	2	2
83	·9925	9928	9930	9932	9934	9936	9938	9940	9942	9943	0	1	1	1	2
84	·9945	9947	9949	9951	9952	9954	9956	9957	9959	9960	0	1	1	1	1
85	·9962	9963	9965	9966	9968	9969	9971	9972	9973	9974	0	0	1	1	1
86	·9976	9977	9978	9979	9980	9981	9982	9983	9984	9985	0	0	1	1	1
87	·9986	9987	9988	9989	9990	9990	9991	9992	9993	9993	0	0	0	1	1
88	·9994	9995	9995	9996	9996	9997	9997	9997	9998	9998	0	0	0	0	0
89	·9998	9999	9999	9999	9999	1·000	1·000	1·000	1·000	1·000	0	0	0	0	0
	0'	6'	12'	18'	24'	30'	36'	42'	48'	54'	1'	2'	3'	4'	5'

NATURAL COSINES

	0′	6′	12′	18′	24′	30′	36′	42′	48′	54′	1′	2′	3′	4′	5′
											Subtract Differences				
0°	1·000	1·000	1·000	1·000	1·000	1·000	·9999	·9999	·9999	·9999	0	0	0	0	0
1	·9998	9998	9998	9997	9997	9997	9996	9996	9995	9995	0	0	0	0	0
2	·9994	9993	9993	9992	9991	9990	9990	9989	9988	9987	0	0	0	1	1
3	·9986	9985	9984	9983	9982	9981	9980	9979	9978	9977	0	0	1	1	1
4	·9976	9974	9973	9972	9971	9969	9968	9966	9965	9963	0	0	1	1	1
5	·9962	9960	9959	9957	9956	9954	9952	9951	9949	·9947	0	1	1	1	1
6	·9945	9943	9942	9940	9938	9936	9934	9932	9930	9928	0	1	1	1	2
7	·9925	9923	9921	9919	9917	9914	9912	9910	9907	9905	0	1	1	2	2
8	·9903	9900	9898	9895	9893	9890	9888	9885	9882	9880	0	1	1	2	2
9	·9877	9874	9871	9869	9866	9863	9860	9857	9854	9851	0	1	1	2	2
10	·9848	9845	9842	9839	9836	9833	9829	9826	9823	9820	1	1	2	2	3
11	·9816	9813	9810	9806	9803	9799	9796	9792	9789	9785	1	1	2	2	3
12	·9781	9778	9774	9770	9767	9763	9759	9755	9751	9748	1	1	2	3	3
13	·9744	9740	9736	9732	9728	9724	9720	9715	9711	9707	1	1	2	3	3
14	·9703	9699	9694	9690	9686	9681	9677	9673	9668	9664	1	1	2	3	4
15	·9659	9655	9650	9646	9641	9636	9632	9627	9622	9617	1	2	2	3	4
16	·9613	9608	9603	9598	9593	9588	9583	9578	9573	9568	1	2	2	3	4
17	·9563	9558	9553	9548	9542	9537	9532	9527	9521	9516	1	2	3	3	4
18	·9511	9505	9500	9494	9489	9483	9478	9472	9466	9461	1	2	3	4	5
19	·9455	9449	9444	9438	9432	9426	9421	9415	9409	9403	1	2	3	4	5
20	·9397	9391	9385	9379	9373	9367	9361	9354	9348	9342	1	2	3	4	5
21	·9336	9330	9323	9317	9311	9304	9298	9291	9285	9278	1	2	3	4	5
22	·9272	9265	9259	9252	9245	9239	9232	9225	9219	9212	1	2	3	4	6
23	·9205	9198	9191	9184	9178	9171	9164	9157	9150	9143	1	2	3	5	6
24	·9135	9128	9121	9114	9107	9100	9092	9085	9078	9070	1	2	4	5	6
25	·9063	9056	9048	9041	9033	9026	9018	9011	9003	8996	1	3	4	5	6
26	·8988	8980	8973	8965	8957	8949	8942	8934	8926	8918	1	3	4	5	6
27	8910	8902	8894	8886	8878	8870	8862	8854	8846	8838	1	3	4	5	7
28	·8829	8821	8813	8805	8796	8788	8780	8771	8763	8755	1	3	4	6	7
29	·8746	8738	8729	8721	8712	8704	8695	8686	8678	8669	1	3	4	6	7
30	·8660	8652	8643	8634	8625	8616	8607	8599	8590	8581	1	3	4	6	7
31	·8572	8563	8554	8545	8536	8526	8517	8508	8499	8490	2	3	5	6	8
32	·8480	8471	8462	8453	8443	8434	8425	8415	8406	8396	2	3	5	6	8
33	·8387	8377	8368	8358	8348	8339	8329	8320	8310	8300	2	3	5	6	8
34	·8290	8281	8271	8261	8251	8241	8231	8221	8211	8202	2	3	5	7	8
35	·8192	8181	8171	8161	8151	8141	8131	8121	8111	8100	2	3	5	7	8
36	·8090	8080	8070	8059	8049	8039	8028	8018	8007	7997	2	3	5	7	9
37	·7986	7976	7965	7955	7944	7934	7923	7912	7902	7891	2	4	5	7	9
38	·7880	7869	7859	7848	7837	7826	7815	7804	7793	7782	2	4	5	7	9
39	·7771	7760	7749	7738	7727	7716	7705	7694	7683	7672	2	4	6	7	9
40	·7660	7649	7638	7627	7615	7604	7593	7581	7570	7559	2	4	6	8	9
41	·7547	7536	7524	7513	7501	7490	7478	7466	7455	7443	2	4	6	8	10
42	·7431	7420	7408	7396	7385	7373	7361	7349	7337	7325	2	4	6	8	10
43	·7314	7302	7290	7278	7266	7254	7242	7230	7218	7206	2	4	6	8	10
44	·7193	7181	7169	7157	7145	7133	7120	7108	7096	7083	2	4	6	8	10
	0′	6′	12′	18′	24′	30′	36′	42′	48′	54′	1′	2′	3′	4′	5′

Subtract Differences.

NATURAL COSINES

	0'	6'	12'	18'	24'	30'	36'	42'	48'	54'	1'	2'	3'	4'	5'
											| Subtract Differences.				
45°	·7071	·7059	·7046	·7034	·7022	·7009	·6997	·6984	·6972	·6959	2	4	6	8	10
46	·6947	6934	6921	6909	6896	6884	6871	6858	6845	6833	2	4	6	8	11
47	·6820	6807	6794	6782	6769	6756	6743	6730	6717	6704	2	4	6	9	11
48	·6691	6678	6665	6652	6639	6626	6613	6600	6587	6574	2	4	7	9	11
49	·6561	6547	6534	6521	6508	6494	6481	6468	6455	6441	2	4	7	9	11
50	·6428	6414	6401	6388	6374	6361	6347	6334	6320	6307	2	4	7	9	11
51	·6293	6280	6266	6252	6239	6225	6211	6198	6184	6170	2	5	7	9	11
52	·6157	6143	6129	6115	6101	6088	6074	6060	6046	6032	2	5	7	9	12
53	·6018	6004	5990	5976	5962	5948	5934	5920	5906	5892	2	5	7	9	12
54	·5878	5864	5850	5835	5821	5807	5793	5779	5764	5750	2	5	7	9	12
55	·5736	5721	5707	5693	5678	5664	5650	5635	5621	5606	2	5	7	10	12
56	·5592	5577	5563	5548	5534	5519	5505	5490	5476	5461	2	5	7	10	12
57	·5446	5432	5417	5402	5388	5373	5358	5344	5329	5314	2	5	7	10	12
58	·5299	5284	5270	5255	5240	5225	5210	5195	5180	5165	2	5	7	10	12
59	·5150	5135	5120	5105	5090	5075	5060	5045	5030	5015	3	5	8	10	13
60	·5000	4985	4970	4955	4939	4924	4909	4894	4879	4863	3	5	8	10	13
61	·4848	4833	4818	4802	4787	4772	4756	4741	4726	4710	3	5	8	10	13
62	·4695	4679	4664	4648	4633	4617	4602	4586	4571	4555	3	5	8	10	13
63	·4540	4524	4509	4493	4478	4462	4446	4431	4415	4399	3	5	8	10	13
64	·4384	4368	4352	4337	4321	4305	4289	4274	4258	4242	3	5	8	11	13
65	·4226	4210	4195	4179	4163	4147	4131	4115	4099	4083	3	5	8	11	13
66	·4067	4051	4035	4019	4003	3987	3971	3955	3939	3923	3	5	8	11	14
67	·3907	3891	3875	3859	3843	3827	3811	3795	3778	3762	3	5	8	11	14
68	·3746	3730	3714	3697	3681	3665	3649	3633	3616	3600	3	5	8	11	14
69	·3584	3567	3551	3535	3518	3502	3486	3469	3453	3437	3	5	8	11	14
70	·3420	3404	3387	3371	3355	3338	3322	3305	3289	3272	3	5	8	11	14
71	·3256	3239	3223	3206	3190	3173	3156	3140	3123	3107	3	6	8	11	14
72	·3090	3074	3057	3040	3024	3007	2990	2974	2957	2940	3	6	8	11	14
73	·2924	2907	2890	2874	2857	2840	2823	2807	2790	2773	3	6	8	11	14
74	·2756	2740	2723	2706	2689	2672	2656	2639	2622	2605	3	6	8	11	14
75	·2588	2571	2554	2538	2521	2504	2487	2470	2453	2436	3	6	8	11	14
76	·2419	2402	2385	2368	2351	2334	2317	2300	2284	2267	3	6	8	11	14
77	·2250	2233	2215	2198	2181	2164	2147	2130	2113	2096	3	6	9	11	14
78	·2079	2062	2045	2028	2011	1994	1977	1959	1942	1925	3	6	9	11	14
79	·1908	1891	1874	1857	1840	1822	1805	1788	1771	1754	3	6	9	11	14
80	·1736	1719	1702	1685	1668	1650	1633	1616	1599	1582	3	6	9	11	14
81	·1564	1547	1530	1513	1495	1478	1461	1444	1426	1409	3	6	9	12	14
82	·1392	1374	1357	1340	1323	1305	1288	1271	1253	1236	3	6	9	12	14
83	·1219	1201	1184	1167	1149	1132	1115	1097	1080	1063	3	6	9	12	14
84	·1045	1028	1011	0993	0976	0958	0941	0924	0906	0889	3	6	9	12	14
85	·0872	0854	0837	0819	0802	0785	0767	0750	0732	0715	3	6	9	12	14
86	·0698	0680	0663	0645	0628	0610	0593	0576	0558	0541	3	6	9	12	15
87	·0523	0506	0488	0471	0454	0436	0419	0401	0384	0366	3	6	9	12	15
88	·0349	0332	0314	0297	0279	0262	0244	0227	0209	0192	3	6	9	12	15
89	·0175	0157	0140	0122	0105	0087	0070	0052	0035	0017	3	6	9	12	15
	0'	6'	12'	18'	24'	30'	36'	42'	48'	54'	1'	2'	3'	4'	5'

Subtract Differences.

NATURAL TANGENTS

	0'	6'	12'	18'	24'	30'	36'	42'	48'	54'	1'	2'	3'	4'	5'
0°	·0000	·0017	·0035	·0052	·0070	·0087	·0105	·0122	·0140	·0157	3	6	9	12	15
1	·0175	0192	0209	0227	0244	0262	0279	0297	0314	0332	3	6	9	12	15
2	·0349	0367	0384	0402	0419	0437	0454	0472	0489	0507	3	6	9	12	15
3	·0524	0542	0559	0577	0594	0612	0629	0647	0664	0682	3	6	9	12	15
4	·0699	0717	0734	0752	0769	0787	0805	0822	0840	0857	3	6	9	12	15
5	·0875	0892	0910	0928	0945	0963	0981	0998	1016	1033	3	6	9	12	15
6	·1051	1069	1086	1104	1122	1139	1157	1175	1192	1210	3	6	9	12	15
7	·1228	1246	1263	1281	1299	1317	1334	1352	1370	1388	3	6	9	12	15
8	·1405	1423	1441	1459	1477	1495	1512	1530	1548	1566	3	6	9	12	15
9	·1584	1602	1620	1638	1655	1673	1691	1709	1727	1745	3	6	9	12	15
10	·1763	1781	1799	1817	1835	1853	1871	1890	1908	1926	3	6	9	12	15
11	·1944	1962	1980	1998	2016	2035	2053	2071	2089	2107	3	6	9	12	15
12	·2126	2144	2162	2180	2199	2217	2235	2254	2272	2290	3	6	9	12	15
13	·2309	2327	2345	2364	2382	2401	2419	2438	2456	2475	3	6	9	12	15
14	·2493	2512	2530	2549	2568	2586	2605	2623	2642	2661	3	6	9	12	16
15	·2679	2698	2717	2736	2754	2773	2792	2811	2830	2849	3	6	9	13	16
16	·2867	2886	2905	2924	2943	2962	2981	3000	3019	3038	3	6	9	13	16
17	·3057	3076	3096	3115	3134	3153	3172	3191	3211	3230	3	6	10	13	16
18	·3249	3269	3288	3307	3327	3346	3365	3385	3404	3424	3	6	10	13	16
19	·3443	3463	3482	3502	3522	3541	3561	3581	3600	3620	3	7	10	13	16
20	·3640	3659	3679	3699	3719	3739	3759	3779	3799	3819	3	7	10	13	17
21	·3839	3859	3879	3899	3919	3939	3959	3979	4000	4020	3	7	10	13	17
22	·4040	4061	4081	4101	4122	4142	4163	4183	4204	4224	3	7	10	14	17
23	·4245	4265	4286	4307	4327	4348	4369	4390	4411	4431	3	7	10	14	17
24	·4452	4473	4494	4515	4536	4557	4578	4599	4621	4642	4	7	11	14	18
25	·4663	4684	4706	4727	4748	4770	4791	4813	4834	4856	4	7	11	14	18
26	·4877	4899	4921	4942	4964	4986	5008	5029	5051	5073	4	7	11	15	18
27	·5095	5117	5139	5161	5184	5206	5228	5250	5272	5295	4	7	11	15	18
28	·5317	5340	5362	5384	5407	5430	5452	5475	5498	5520	4	8	11	15	19
29	·5543	5566	5589	5612	5635	5658	5681	5704	5727	5750	4	8	12	15	19
30	·5774	5797	5820	5844	5867	5890	5914	5938	5961	5985	4	8	12	16	20
31	·6009	6032	6056	6080	6104	6128	6152	6176	6200	6224	4	8	12	16	20
32	·6249	6273	6297	6322	6346	6371	6395	6420	6445	6469	4	8	12	16	20
33	·6494	6519	6544	6569	6594	6619	6644	6669	6694	6720	4	8	13	17	21
34	·6745	6771	6796	6822	6847	6873	6899	6924	6950	6976	4	9	13	17	21
35	·7002	7028	7054	7080	7107	7133	7159	7186	7212	7239	4	9	13	18	22
36	·7265	7292	7319	7346	7373	7400	7427	7454	7481	7508	5	9	14	18	23
37	·7536	7563	7590	7618	7646	7673	7701	7729	7757	7785	5	9	14	18	23
38	·7813	7841	7869	7898	7926	7954	7983	8012	8040	8069	5	9	14	19	24
39	·8098	8127	8156	8185	8214	8243	8273	8302	8332	8361	5	10	15	20	24
40	·8391	8421	8451	8481	8511	8541	8571	8601	8632	8662	5	10	15	20	25
41	·8693	8724	8754	8785	8816	8847	8878	8910	8941	8972	5	10	16	21	26
42	·9004	9036	9067	9099	9131	9163	9195	9228	9260	9293	5	11	16	21	27
43	·9325	9358	9391	9424	9457	9490	9523	9556	9590	9623	6	11	17	22	28
44	·9657	9691	9725	9759	9793	9827	9861	9896	9930	9965	6	11	17	23	29
	0'	6'	12'	18'	24'	30'	36'	42'	48'	54'	1'	2'	3'	4'	5'

NATURAL TANGENTS

	0'	6'	12'	18'	24'	30'	36'	42'	48'	54'	1'	2'	3'	4'	5'
45°	1·0000	·0035	·0070	·0105	·0141	·0176	·0212	·0247	·0283	·0319	6	12	18	24	30
46	1·0355	0392	0428	0464	0501	0538	0575	0612	0649	0686	6	12	18	25	31
47	1·0724	0761	0799	0837	0875	0913	0951	0990	1028	1067	6	13	19	25	32
48	1·1106	1145	1184	1224	1263	1303	1343	1383	1423	1463	7	13	20	27	33
49	1·1504	1544	1585	1626	1667	1708	1750	1792	1833	1875	7	14	21	28	34
50	1·1918	1960	2002	2045	2088	2131	2174	2218	2261	2305	7	14	22	29	36
51	1·2349	2393	2437	2482	2527	2572	2617	2662	2708	2753	8	15	23	30	38
52	1·2799	2846	2892	2938	2985	3032	3079	3127	3175	3222	8	16	24	31	39
53	1·3270	3319	3367	3416	3465	3514	3564	3613	3663	3713	8	16	25	33	41
54	1·3764	3814	3865	3916	3968	4019	4071	4124	4176	4229	9	17	26	34	43
55	1·4281	4335	4388	4442	4496	4550	4605	4659	4715	4770	9	18	27	36	45
56	1·4826	4882	4938	4994	5051	5108	5166	5224	5282	5340	10	19	29	38	48
57	1·5399	5458	5517	5577	5637	5697	5757	5818	5880	5941	10	20	30	40	50
58	1·6003	6066	6128	6191	6255	6319	6383	6447	6512	6577	11	21	32	43	53
59	1·6643	6709	6775	6842	6909	6977	7045	7113	7182	7251	11	23	34	45	57
60	1·7321	7391	7461	7532	7603	7675	7747	7820	7893	7966	12	24	36	48	60
61	1·8040	8115	8190	8265	8341	8418	8495	8572	8650	8728	13	26	38	51	64
62	1·8807	8887	8967	9047	9128	9210	9292	9375	9458	9542	14	27	41	55	68
63	1·9626	9711	9797	9883	9970	2·0057	2·0145	2·0233	2·0323	2·0413	15	29	44	58	73
64	2·0503	0594	0686	0778	0872	0965	1060	1155	1251	1348	16	31	47	63	79
65	2·1445	1543	1642	1742	1842	1943	2045	2148	2251	2355	17	34	51	68	85
66	2·2460	2566	2673	2781	2889	2998	3109	3220	3332	3445	18	37	55	73	92
67	2·3559	3673	3789	3906	4023	4142	4262	4383	4504	4627	20	40	60	79	99
68	2·4751	4876	5002	5129	5257	5386	5517	5649	5782	5916	22	43	65	87	108
69	2·6051	6187	6325	6464	6605	6746	6889	7034	7179	7326	24	48	71	95	119
70	2·7475	7625	7776	7929	8083	8239	8397	8556	8716	8878	26	52	78	105	131
71	2·9042	9208	9375	9544	9714	9887	3·0061	3·0237	3·0415	3·0595	29	58	87	116	145
72	3·0777	0961	1146	1334	1524	1716	1910	2106	2305	2506	32	64	96	129	161
73	3·2709	2914	3122	3332	3544	3759	3977	4197	4420	4646	36	72	108	144	180
74	3·4874	5105	5339	5576	5816	6059	6305	6554	6806	7062	41	81	122	163	204
75	3·7321	7583	7848	8118	8391	8667	8947	9232	9520	9812	46	93	139	186	232
76	4·0108	0408	0713	1022	1335	1653	1976	2303	2635	2972					
77	4·3315	3662	4015	4374	4737	5107	5483	5864	6252	6646					
78	4·7046	7453	7867	8288	8716	9152	9594	5·0045	5·0504	5·0970					
79	5·1446	1929	2422	2924	3435	3955	4486	5026	5578	6140					
80	5·6713	7297	7894	8502	9124	9758	6·0405	6·1066	6·1742	6·2432					
81	6·3138	3859	4596	5350	6122	6912	7720	8548	9395	7·0264					
82	7·1154	2066	3002	3962	4947	5958	6996	8062	9158	8·0285		Mean differences no longer sufficiently accurate.			
83	8·1443	2636	3863	5126	6427	7769	9152	9·0579	9·2052	9·3572					
84	9·514	9·677	9·845	10·02	10·20	10·39	10·58	10·78	10·99	11·20					
85	11·43	11·66	11·91	12·16	12·43	12·71	13·00	13·30	13·62	13·95					
86	14·30	14·67	15·06	15·46	15·89	16·35	16·83	17·34	17·89	18·46					
87	19·08	19·74	20·45	21·20	22·02	22·90	23·86	24·90	26·03	27·27					
88	28·64	30·14	31·82	33·69	35·80	38·19	40·92	44·07	47·74	52·08					
89	57·29	63·66	71·62	81·85	95·49	114·6	143·2	191·0	286·5	573·0					
	0'	6'	12'	18'	24'	30'	36'	42'	48'	54'					

RADIANS

°	0'	6'	12'	18'	24'	30'	36'	42'	48'	54'	1'	2'	3'	4'	5'
0°	·0000	·0017	·0035	·0052	·0070	·0087	·0105	·0122	·0140	·0157	3	6	9	12	15
1	·0175	0192	0209	0227	0244	0262	0279	0297	0314	0332	3	6	9	12	15
2	·0349	0367	0384	0401	0419	0436	0454	0471	0489	0506	3	6	9	12	15
3	·0524	0541	0559	0576	0593	0611	0628	0646	0663	0681	3	6	9	12	15
4	·0698	0716	0733	0750	0768	0785	0803	0820	0838	0855	3	6	9	12	15
5	·0873	0890	0908	0925	0942	0960	0977	0995	1012	1030	3	6	9	12	15
6	·1047	1065	1082	1100	1117	1134	1152	1169	1187	1204	3	6	9	12	15
7	·1222	1239	1257	1274	1292	1309	1326	1344	1361	1379	3	6	9	12	15
8	·1396	1414	1431	1449	1466	1484	1501	1518	1536	1553	3	6	9	12	15
9	·1571	1588	1606	1623	1641	1658	1676	1693	1710	1728	3	6	9	12	15
10	·1745	1763	1780	1798	1815	1833	1850	1868	1885	1902	3	6	9	12	15
11	·1920	1937	1955	1972	1990	2007	2025	2042	2059	2077	3	6	9	12	15
12	·2094	2112	2129	2147	2164	2182	2199	2217	2234	2251	3	6	9	12	15
13	·2269	2286	2304	2321	2339	2356	2374	2391	2409	2426	3	6	9	12	15
14	·2443	2461	2478	2496	2513	2531	2548	2566	2583	2601	3	6	9	12	15
15	·2618	2635	2653	2670	2688	2705	2723	2740	2758	2775	3	6	9	12	15
16	·2793	2810	2827	2845	2862	2880	2897	2915	2932	2950	3	6	9	12	15
17	·2967	2985	3002	3019	3037	3054	3072	3089	3107	3124	3	6	9	12	15
18	·3142	3159	3176	3194	3211	3229	3246	3264	3281	3299	3	6	9	12	15
19	·3316	3334	3351	3368	3386	3403	3421	3438	3456	3473	3	6	9	12	15
20	·3491	3508	3526	3543	3560	3578	3595	3613	3630	3648	3	6	9	12	15
21	·3665	3683	3700	3718	3735	3752	3770	3787	3805	3822	3	6	9	12	15
22	·3840	3857	3875	3892	3910	3927	3944	3962	3979	3997	3	6	9	12	15
23	·4014	4032	4049	4067	4084	4102	4119	4136	4154	4171	3	6	9	12	15
24	·4189	4206	4224	4241	4259	4276	4294	4311	4328	4346	3	6	9	12	15
25	·4363	4381	4398	4416	4433	4451	4468	4485	4503	4520	3	6	9	12	15
26	·4538	4555	4573	4590	4608	4625	4643	4660	4677	4695	3	6	9	12	15
27	·4712	4730	4747	4765	4782	4800	4817	4835	4852	4869	3	6	9	12	15
28	·4887	4904	4922	4939	4957	4974	4992	5009	5027	5044	3	6	9	12	15
29	·5061	5079	5096	5114	5131	5149	5166	5184	5201	5219	3	6	9	12	15
30	·5236	5253	5271	5288	5306	5323	5341	5358	5376	5393	3	6	9	12	15
31	·5411	5428	5445	5463	5480	5498	5515	5533	5550	5568	3	6	9	12	15
32	·5585	5603	5620	5637	5655	5672	5690	5707	5725	5742	3	6	9	12	15
33	·5760	5777	5794	5812	5829	5847	5864	5882	5899	5917	3	6	9	12	15
34	·5934	5952	5969	5986	6004	6021	6039	6056	6074	6091	3	6	9	12	15
35	·6109	6126	6144	6161	6178	6196	6213	6231	6248	6266	3	6	9	12	15
36	·6283	6301	6318	6336	6353	6370	6388	6405	6423	6440	3	6	9	12	15
37	·6458	6475	6493	6510	6528	6545	6562	6580	6597	6615	3	6	9	12	15
38	·6632	6650	6667	6685	6702	6720	6737	6754	6772	6789	3	6	9	12	15
39	·6807	6824	6842	6859	6877	6894	6912	6929	6946	6964	3	6	9	12	15
40	·6981	6999	7016	7034	7051	7069	7086	7103	7121	7138	3	6	9	12	15
41	·7156	7173	7191	7208	7226	7243	7261	7278	7295	7313	3	6	9	12	15
42	·7330	7348	7365	7383	7400	7418	7435	7453	7470	7487	3	6	9	12	15
43	·7505	7522	7540	7557	7575	7592	7610	7627	7645	7662	3	6	9	12	15
44	·7679	7697	7714	7732	7749	7767	7784	7802	7819	7837	3	6	9	12	15
	0'	6'	12'	18'	24'	30'	36'	42'	48'	54'	1'	2'	3'	4'	5'

147

	0'	6'	12'	18'	24'	30'	36'	42'	48'	54'	1' 2' 3' 4' 5'
45°	·7854	·7871	·7889	·7906	·7924	·7941	·7959	·7976	·7994	·8011	3 6 9 12 15
46	·8029	8046	8063	8081	8098	8116	8133	8151	8168	8186	3 6 9 12 15
47	·8203	8221	8238	8255	8273	8290	8308	8325	8343	8360	3 6 9 12 15
48	·8378	8395	8412	8430	8447	8465	8482	8500	8517	8535	3 6 9 12 15
49	·8552	8570	8587	8604	8622	8639	8657	8674	8692	8709	3 6 9 12 15
50	·8727	8744	8762	8779	8796	8814	8831	8849	8866	8884	3 6 9 12 15
51	·8901	8919	8936	8954	8971	8988	9006	9023	9041	9058	3 6 9 12 15
52	·9076	9093	9111	9128	9146	9163	9180	9198	9215	9233	3 6 9 12 15
53	·9250	9268	9285	9303	9320	9338	9355	9372	9390	9407	3 6 9 12 15
54	·9425	9442	9460	9477	9495	9512	9529	9547	9564	9582	3 6 9 12 15
55	·9599	9617	9634	9652	9669	9687	9704	9721	9739	9756	3 6 9 12 15
56	·9774	9791	9809	9826	9844	9861	9879	9896	9913	9931	3 6 9 12 15
57	·9948	9966	9983	1·0001	1·0018	1·0036	1·0053	1·0071	1·0088	1·0105	3 6 9 12 15
58	1·0123	0140	0158	0175	0193	0210	0228	0245	0263	0280	3 6 9 12 15
59	1·0297	0315	0332	0350	0367	0385	0402	0420	0437	0455	3 6 9 12 15
60	1·0472	0489	0507	0524	0542	0559	0577	0594	0612	0629	3 6 9 12 15
61	1·0647	0664	0681	0699	0716	0734	0751	0769	0786	0804	3 6 9 12 15
62	1·0821	0838	0856	0873	0891	0908	0926	0943	0961	0978	3 6 9 12 15
63	1·0996	1013	1030	1048	1065	1083	1100	1118	1135	1153	3 6 9 12 15
64	1·1170	1188	1205	1222	1240	1257	1275	1292	1310	1327	3 6 9 12 15
65	1·1345	1362	1380	1397	1414	1432	1449	1467	1484	1502	3 6 9 12 15
66	1·1519	1537	1554	1572	1589	1606	1624	1641	1659	1676	3 6 9 12 15
67	1·1694	1711	1729	1746	1764	1781	1798	1816	1833	1851	3 6 9 12 15
68	1·1868	1886	1903	1921	1938	1956	1973	1990	2008	2025	3 6 9 12 15
69	1·2043	2060	2078	2095	2113	2130	2147	2165	2182	2200	3 6 9 12 15
70	1·2217	2235	2252	2270	2287	2305	2322	2339	2357	2374	3 6 9 12 15
71	1·2392	2409	2427	2444	2462	2479	2497	2514	2531	2549	3 6 9 12 15
72	1·2566	2584	2601	2619	2636	2654	2671	2689	2706	2723	3 6 9 12 15
73	1·2741	2758	2776	2793	2811	2828	2846	2863	2881	2898	3 6 9 12 15
74	1·2915	2933	2950	2968	2985	3003	3020	3038	3055	3073	3 6 9 12 15
75	1·3090	3107	3125	3142	3160	3177	3195	3212	3230	3247	3 6 9 12 15
76	1·3265	3282	3299	3317	3334	3352	3369	3387	3404	3422	3 6 9 12 15
77	1·3439	3456	3474	3491	3509	3526	3544	3561	3579	3596	3 6 9 12 15
78	1·3614	3631	3648	3666	3683	3701	3718	3736	3753	3771	3 6 9 12 15
79	1·3788	3806	3823	3840	3858	3875	3893	3910	3928	3945	3 6 9 12 15
80	1·3963	3980	3998	4015	4032	4050	4067	4085	4102	4120	3 6 9 12 15
81	1·4137	4155	4172	4190	4207	4224	4242	4259	4277	4294	3 6 9 12 15
82	1·4312	4329	4347	4364	4382	4399	4416	4434	4451	4469	3 6 9 12 15
83	1·4486	4504	4521	4539	4556	4573	4591	4608	4626	4643	3 6 9 12 15
84	1·4661	4678	4696	4713	4731	4748	4765	4783	4800	4818	3 6 9 12 15
85	1·4835	4853	4870	4888	4905	4923	4940	4957	4975	4992	3 6 9 12 15
86	1·5010	5027	5045	5062	5080	5097	5115	5132	5149	5167	3 6 9 12 15
87	1·5184	5202	5219	5237	5254	5272	5289	5307	5324	5341	3 6 9 12 15
88	1·5359	5376	5394	5411	5429	5446	5464	5481	5499	5516	3 6 9 12 15
89	1·5533	5551	5568	5586	5603	5621	5638	5656	5673	5691	3 6 9 12 15
	0'	6'	12'	18'	24'	30'	36'	42'	48'	54'	1' 2' 3' 4' 5'

INDEX

INDEX

THE END

PRINTED IN GREAT BRITAIN BY WILLIAM CLOWES AND SONS, LIMITED, BECCLES.

Printed in the United States
150660LV00003B/33/A

9 781436 604093